WORKER PROTECTION DURING HAZARDOUS WASTE REMEDIATION

WORKER PROTECTION DURING HAZARDOUS WASTE REMEDIATION

CENTER FOR LABOR EDUCATION AND RESEARCH

Edited by
Lori P. Andrews, P.E.

Contributing Authors

Lori P. Andrews, P.E.
Wifred D. David, M.D.
Jo Carole Dawkins, M.A.
W. Donald Fattig, Ph.D.

Lindsay R. Hill, B.S.
Barbara M. Hilyer, M.S.
Higdon C. Roberts, Jr., Ph.D.
D. Alan Veasey, M.A.

VNR VAN NOSTRAND REINHOLD
_____ New York

Copyright © 1990 by Van Nostrand Reinhold
Library of Congress Catalog Card Number 89-24888
ISBN 0-442-23899-1

All rights reserved. No part of this work covered by the copyright hereon may be reproduced or used in any form or by any means—graphic, electronic, or mechanical, including photocopying, recording, taping, or information storage and retrieval systems—without the written permission of the publisher.

I(T)P Van Nostrand Reinhold is an International Thomson Publishing company.
ITP logo is a trademark under license.

Printed in the United States of America

Van Nostrand Reinhold
115 Fifth Avenue
New York, NY 10003

International Thomson Publishing GmbH
Königswinterer Str. 418
53227 Bonn
Germany

International Thomson Publishing
Berkshire House,168-173
High Holborn, London WC1V 7AA
England

International Thomson Publishing Asia
221 Henderson Bldg. #05-1
Singapore 0315

Thomas Nelson Australia
102 Dodds Street
South Melbourne 3205
Victoria, Australia

International Thomson Publishing Japan
Kyowa Building, 3F
2-2-1 Hirakawacho
Chiyoda-ku, Tokyo 102
Japan

Nelson Canada
1120 Birchmount Road
Scarborough, Ontario
M1K 5G4, Canada

16 15 14 13 12 11 10 9 8 7 6 5 4 3

Library of Congress Cataloging-in-Publication Data
Worker protection during hazardous waste remediation/edited by Lori
  P. Andrews . . .[et al.].
     p. cm.
   ISBN 0-442-23899-1
   1. Hazardous waste sites—Safety measures. 2. Industrial safety.
  I. Andrews, Lori P.
TD1050.S24W67   1990
363.72'88—dc20   89-24888
                                                             CIP

Dedication

To our families who have supported us through this endeavor.

Related Titles from Van Nostrand Reinhold:

Quick Selection Guide to Chemical Protective Clothing, Second Edition
by Krister Forsberg and S.Z. Mansdorf

Hazardous Waste Site Remediation:
The Engineer's Perspective
by O'Brien & Gere Engineers

Emergency Responder Training for the Hazardous Materials Technician
by the Center for Labor Education and Research

A Practical Guide to Chemical Spill Response
by John W. Hosty with Patricia Foster

Hazardous Waste Management Compliance Handbook
by the Environmental Resource Center

The Hazardous Waste Q&A, Revised Edition:
An In-depth Guide to the Resource Conservation and Recovery Act and the
Hazardous Materials Transportation Act
by Travis P. Wagner

The Complete Guide to Hazardous Waste Regulations, Second Edition:
A Comprehensive Step-By-Step Guide to the Regulation of Hazardous Wastes
under RCRA, TSCA, HMTA, OSHA and Superfund
by Travis P. Wagner

Hazardous Waste Identification and Classification Manual
by Travis P. Wagner

A Comprehensive Guide to the Hazardous Properties of Chemical Substances
by Pradyot Patnaik

Sax's Dangerous Properties of Industrial Materials, Eighth Edition
by Richard J. Lewis, Sr.

Hazardous Chemicals Desk Reference, Third Edition
by Richard J. Lewis, Sr.

Hawley's Condensed Chemical Dictionary, Twelfth Edition
by Richard J. Lewis, Sr.

Foreword

To achieve a number of its national goals our society needs a well trained workforce, which is capable of accomplishing its tasks safely, and knowledgeable enough to take those measures required to minimize potentially toxic, health endangering exposures. Healthy, productive, knowledgeable workers are needed, if we are to be productive and competitive as a nation.

There are several situations where our country has moved to meet the need for such a workforce. Worker training programs incorporating health and safety have long been required for underground miners. Likewise, certain standards and programs established under the Occupational Safety and Health Act have provided an impetus for worker training. The newly established Hazardous Waste Worker Training Program supported by the National Institute for Environmental Health Sciences (NIEHS) in the Department of Health and Human Services is playing a major role in stimulating, innovative worker safety and health training programs and extending these programs to the construction trades. Clearly, NIEHS has moved swiftly and deftly in responding to the Congressional mandate for such training.

The University of Alabama at Birmingham (UAB) can be justly proud of the leadership role which its Center for Labor Education and Research and its Deep South Educational Resource Center for Occupational Health are playing in the national Hazardous Waste Worker Training Program. Their training manual plays a key role in that effort and it represents the contribution of many dedicated individuals. Hopefully, the manual will not only prove to be an indispensable guide for workers during training, but also a ready reference source once workers are on the job. The manual can be expected to change as experience is gained in cleanup operations and as new scientific knowledge is developed. Workers completing periodic refresher training will be able to benefit fully from the manual updates that will be necessary from time to time and they will take this new knowledge and new resource back to work with them.

There are other benefits which workers completing the training called for in this manual will accrue. First, they will be better able to assist fellow workers in knowledge sharing or pooling. Having a team of well trained workers at cleanup sites helps protect individual workers, employers, and the neighboring community. Second, when they are not working at hazardous waste cleanup sites, workers completing the training called for in this manual are likely to find themselves among a minority who have good insight and knowledge about a variety of worksite safety and health problems. The carry over of the benefits of the training program to other construction jobs is likely to be considerable. Third, workers completing the training called for by this manual should have improved job opportunities. For example, it is likely that similar training will be called for in the removal of lead based paint from public housing and the cleanup of nuclear facility waste sites.

The Hazardous Waste Worker Training Program and this manual are the result of enlightened Congressional legislation, wise administration, and a responsive university. The resulting benefits to workers, employers and our society should extend over a period of many years.

John F. Finklea, M.D.
Visiting Scientist
Centers for Disease Control

Preface

The Center for Labor Education and Research (CLEAR) has been involved in training workers in Occupational Safety and Health since its inception in 1972. This particular aspect of our labor education effort has remained a constant and growing part of our program up to the present. In the early years we worked exclusively within the State of Alabama. However, in 1979 we were funded under the Department of Labor's New Directions Grant program and this enabled us to increase our staff and expand our efforts throughout the South.

Following our five years with the New Directions program, we were designated the official, exclusive workers' Occupational Safety and Health training program for the state funded asbestos removal projects. We coupled this work with several contracts from the Department of Labor under which we developed a series of films, slide shows and videos to help train Occupational Safety and Health Administration (OSHA) inspectors who would be conducting hazardous and toxic waste site investigations.

In 1987, The Center and the School of Public Health, Deep South Educational Resource Center for Occupational Safety and Health were selected as one of the eleven National Institute of Environmental Health Science grantees in the Superfund cleanup program. We are responsible for training workers in all the Southern States. As this book goes to press, we are in the third year of the projected five-year grant period.

Throughout the education and training efforts outlined above we have had to develop or adapt a wide range of both teaching techniques and student learning materials. The diversity of our student clientele, and the rapidly changing and expanding field of occupational safety and health has required ongoing evaluation, review, and restructuring of our classroom and field work. The Deep South Educational Resource Center has assisted us by providing a significant amount of this effort for the grant. This book is then a product of our general experience and learning in the broad aspects of occupational safety and health training and our particular expertise acquired in the area of hazardous and toxic waste.

Worker to us is a generic term. We interpret it broadly to mean anyone who, in the normal course of their job, may be exposed to hazardous or toxic substances. Whether you are an engineer, technician, heavy equipment operator, inspector, or laborer there are fundamental commonalities to the potential risks of exposure. There are also very specific occupational or job related risks involved. This book is designed to be a comprehensive yet basic guide to working safely on hazardous waste remediation sites.

Higdon C. Roberts, Jr., Ph.D.
Director, Center for Labor Education and Research

Acknowledgments

The efforts of the entire CLEAR staff are greatly appreciated. Special thanks go to Karen Blackwood who prepared the manuscript through all its many revisions and to Sheri Lares for keeping the office organized during this project.

Contents

FOREWORD / VII

PREFACE / IX

ACKNOWLEDGMENTS / X

1. INTRODUCTION / 1

- HISTORICAL PERSPECTIVE / 1
- ABOUT THIS BOOK / 2
- AVAILABLE RESOURCES / 4
- CONCLUSION / 5

2. RIGHTS AND RESPONSIBILITIES OF WORKERS / 6

- INTRODUCTION / 6
- OBJECTIVES / 6
- THE OCCUPATIONAL SAFETY AND HEALTH ACT OF 1970 / 6
  Background and Intent / 6
  Organizations Created by the OSHAct / 7
  OSHA Standards / 7
  Rights and Responsibilities of Employees Under the OSHAct / 8
- OSHA SAFETY STANDARDS APPLICABLE TO WORK ON HAZARDOUS WASTE SITES / 9
- OSHA REGULATIONS UNDER TITLE 29 CFR PART 1910.120: HAZARDOUS WASTE OPERATIONS AND EMERGENCY RESPONSE / 10
  Intent and Applicability / 10
  General Requirements / 11
  Site-Specific Safety and Health Plan / 11
  Site Characterization and Analysis / 13
  Site Control / 14
  Training / 14
  Medical Surveillance / 16
  Engineering Controls, Work Practices, and Personal Protective Equipment / 17
  Site Monitoring / 18
  Informational Programs / 19

Handling Drums and Containers / 19
Decontamination / 19
Emergency Response / 20
Site Illumination / 21
Sanitation at Temporary Workplaces / 21
New Technology Programs / 22
- SUMMARY OF EMPLOYEE RIGHTS AND RESPONSIBILITIES UNDER 29 CFR 1910.120 / 22

3. MEDICAL SURVEILLANCE / 25

- INTRODUCTION / 25
- OBJECTIVES / 25
- REASONS FOR CONDUCTING MEDICAL SURVEILLANCE / 26
- GENERAL CHARACTERISTICS OF A MEDICAL SURVEILLANCE PROGRAM / 26
- PRE-EMPLOYMENT MEDICAL EXAMINATION / 29
    Medical and Occupational History / 30
    Physical Examination / 30
    Laboratory Tests / 31
- EXAMINATION FOLLOW-UP/CONSULTATION / 32
- PERIODIC MEDICAL EXAMINATION / 32
- TERMINATION EXAMINATION / 34
- MEDICAL RECORDS / 34
- PROGRAM REVIEW / 35

Appendix 3-1 Special Blood and Urine Testing / 35

4. HAZARD RECOGNITION / 41

- INTRODUCTION / 41
- OBJECTIVES / 41
- RISK ASSESSMENT / 41
- CHEMICAL HAZARDS / 42
    Objectives / 42
    Basic Chemistry / 42
    Acids and Bases / 48
    Reactions Involving Strong Oxidizing Agents / 51
    Practical Considerations / 51
    Organic Chemicals / 51
- FIRE, EXPLOSION, AND CHEMICAL REACTION CONTROL / 56
    Objectives / 56

Definitions / 58
Basic Concepts / 59
Fire and Explosion / 60
Various Flammables Classifications / 60
NFPA Hazard Identification System / 63
Controlling Chemical Reactions / 65
Categorization of Hazardous Wastes by Compatibility / 66
Compatibility Staging / 67
- OXYGEN DEFICIENCY / 69
    Introduction / 69
    Objectives / 70
    Importance of Oxygen / 70
    Percentage of Oxygen in Air / 71
    Where May Oxygen Deficiency be Likely? / 71
    Measuring Oxygen Concentration / 72
- RADIATION / 72
    Introduction / 72
    Objectives / 72
    Radiation / 73
    Important Concepts Associated with Radioactive Isotopes / 75
    General Cautions Regarding Radioactivity / 76
    Biological Effects of Radiation / 76
    How Radiation Is Measured / 79
    Potential Sources of Radiation at Hazardous Waste Sites / 79
    Regulatory Agencies / 80
- NOISE / 81
    Introduction / 81
    Objectives / 81
    Effects, Measurement, and Protection / 81
- ELECTRICITY / 82
    Introduction / 82
    Objective / 82
    The Nature of Electricity / 83
    Electrical Hazards May Be Divided Into Five Categories / 83
    Resistance / 83
    Prevention of Electrical Accidents / 84
- HEAT AND COLD / 85
    Introduction / 85
    Objectives / 85
    Normal Mechanisms / 86
    Contributing Factors / 86
    Symptoms / 87

Prevention / 88
Monitoring / 89
Treatment / 89
- BIOLOGICAL HAZARDS / 90
Introduction / 90
Objectives / 92
Microorganisms / 92
Plants and Animals / 93

5. TOXICOLOGY / 95

- INTRODUCTION / 95
- OBJECTIVES / 95
- MEASUREMENT OF TOXICITY / 96
- FORMS OF TOXIC SUBSTANCES / 97
- THE LD_{50} CONCEPT / 98
- BIOLOGICAL RESPONSE TO EXPOSURE TO MORE THAN ONE CHEMICAL / 98
- TESTS FOR OTHER TOXIC EFFECTS / 99
- ACUTE VERSUS CHRONIC TOXICITY / 100
- ROUTES OF ENTRY / 100
- TOXIC CHEMICALS AND TARGET TISSUES / 103
- BIOLOGICAL TOXIC EFFECTS / 104
- LIMITING EXPOSURE TO TOXICANTS / 104
- THRESHOLD LIMIT VALUES / 105
- CARCINOGENS / 106
- THE MATERIAL SAFETY DATA SHEET / 108

6. ENGINEERING CONTROLS / 113

- INTRODUCTION / 113
- OBJECTIVES / 114
- SITE CHARACTERIZATION / 114
  Historical Research / 115
  Examples of Databases (Levine, Martin, 1985) / 116
  Site Map / 117
  Perimeter Reconnaissance / 117
  On-Site Survey and Hazard Assessment / 118
- ZONING / 119
- TRENCHING / 122
  Trenching Safety Precautions / 123
  Cave-in Hazards / 124
- DIKING / 126
  Dikes and Diking Systems / 126

Assessing the Stability of Existing Dikes / 128
Dikes and Spill Containment / 129
Construction of Dike Systems / 129

7. SAFE WORK PRACTICES / 131

- INTRODUCTION / 131
- OBJECTIVES / 131
- UNSAFE ACTS / 131
- UNSAFE CONDITIONS / 134
  The Site / 134
  Clothing / 135
  Use of Tools / 136
  Fuel / 136
  Fire Prevention and Response / 136
- HEAVY EQUIPMENT AND DRILL RIG SAFETY (NIOSH, 1982; National Drilling Federation, n.d.) / 136
  Problems with Drilling Operations at Hazardous Waste Sites / 138
- CONFINED SPACES / 139
  Definitions / 139
  Safe Entry Procedures / 140
  Entry Permit / 140
- HANDLING DRUMS AND CONTAINERS / 142
  Handling Drums and Containers—The Law: Requirements under 29 CFR 1910.120 / 143
  Occasions for Container Handling / 145
  Reasons for Concern / 145
  Minimization of Danger / 145
  The Container Handling Process / 146
  Preliminary Classification / 147
  Planning / 147
  Moving of Containers / 147
  Opening Containers / 148
  Staging of Containers / 150
  Bulking of Waste / 151
  Special Case Problems / 152

8. SAFE USE OF FIELD EQUIPMENT / 154

PART ONE. DIRECT-READING AIR MONITORING INSTRUMENTS / 154

- INTRODUCTION / 154
- OBJECTIVES / 155

- PURPOSE OF AIR MONITORING AT HAZARDOUS WASTE SITES / 155
    - Preliminary Site Survey / 155
    - Periodic Monitoring / 156
- IMPORTANT CONSIDERATIONS FOR DIRECT-READING AIR MONITORING / 156
    - Portability / 156
    - Reliable, Useful Results / 156
    - Selectivity/Sensitivity / 157
    - Inherent Safety / 157
- SOME COMMONLY USED DIRECT-READING INSTRUMENTS / 157
    - Radioactive Atmospheres / 157
    - Oxygen Deficient/Combustible Atmospheres / 158
    - Toxic Atmospheres / 161

PART TWO. PERSONAL SAMPLING INSTRUMENTS / 169

- INTRODUCTION / 169
- OBJECTIVES / 169
- CHOOSING SAMPLING METHODS / 169
- ACTIVE SAMPLERS / 170
    - General Considerations / 170
    - Sampling Pumps / 171
    - Sample Collection Devices / 171
- PASSIVE SAMPLERS / 177
- CALIBRATION / 178

9. PERSONAL PROTECTIVE EQUIPMENT / 180

- INTRODUCTION / 180
- OBJECTIVES / 180

PART 1: RESPIRATORS / 180

- RESPIRATORY PROTECTIVE REQUIREMENTS / 180
    - The Need for Respiratory Protection / 180
    - OSHA Requirements / 181
- CLASSIFICATION OF RESPIRATORY PROTECTIVE EQUIPMENT / 181
    - Facepiece Type / 182
    - Method of Protection / 182

- SELECTION OF RESPIRATORY PROTECTIVE EQUIPMENT / 184
  - Selection Considerations for Air-Purifying Respirators / 186
  - Selection Considerations for Supplied-Air (or Airline) Respirators / 190
  - Selection Considerations for Self-Contained Breathing Apparatus / 190
- THE IMPORTANCE OF RESPIRATOR FIT / 191
  - Fit and Fit Testing / 191
  - Assigned Protection Factors / 195
  - Positive-Pressure Versus Negative-Pressure Modes of Respirator Operation / 196
- WHAT CONSTITUTES EFFECTIVE RESPIRATORY PROTECTION? / 197

PART 2. CHEMICAL PROTECTIVE CLOTHING AND ACCESSORIES / 198

- SELECTION OF CHEMICAL PROTECTIVE CLOTHING (CPC) / 198
  - Selection Considerations / 198
  - Attacks on CPC / 199
  - Resistance to Chemical Attacks / 199
  - Availability of Information on Performance Characteristics of CPC / 199
  - Problems with Information Available on CPC / 201
  - Basic Principles of CPC Selection / 202
  - CPC Materials and Technologies / 202
  - Responsibility of the Employer / 203
- TYPES OF CPC AND ACCESSORIES / 203
  - Chemical Protective Clothing / 204
  - Protective Clothing for Unique Hazards / 207
  - Accessories / 207

PART 3. LEVELS OF PROTECTION / 207

- LEVEL A PROTECTION / 207
- LEVEL B PROTECTION / 213
- LEVEL C PROTECTION / 213
- LEVEL D PROTECTION / 214
- MODIFIED LEVELS OF PROTECTION / 214

PART 4. USE OF PPE / 218

- WRITTEN PPE PROGRAM / 219
- TRAINING IN PPE USE / 223
- WORK MISSION DURATION / 224
    Air Supply Consumption / 224
    Permeation and Penetration of Protective Clothing
      or Equipment / 225
    Ambient Temperature Extremes / 226
- HEAT STRESS AND OTHER PHYSIOLOGICAL
  FACTORS / 226
    Monitoring the Effects of Heat Stress / 226
    Heat Injury Prevention / 228
- PERSONAL FACTORS AFFECTING RESPIRATOR USE / 228
- DONNING PPE / 228
- INSPECTION OF PPE / 229
    Inspecting CPC / 229
- IN-USE MONITORING OF PPE / 230
- DOFFING PPE / 231
- STORAGE OF PPE / 231
- REUSE OF CPC / 231
- MAINTENANCE OF PPE / 232

10. SAFE SAMPLING TECHNIQUES / 234

- INTRODUCTION / 234
- OBJECTIVES / 234
- PURPOSE OF SAMPLING / 234
- WHY SAMPLING IS DANGEROUS / 235
- DEVELOPMENT OF A SAMPLING PLAN / 235
    Background Information / 236
    Sampling Location / 236
    How Many Samples Per Sampling Point / 236
    Volume Per Sample / 237
    Sample Containers / 237
    Selection of Sampling Equipment / 238
    Personal Protective Equipment / 244
    Sample Integrity / 244
    Decontamination of Sampling Equipment / 245
    Recordkeeping / 246
- IMPLEMENTATION OF SAMPLING PLAN / 246

- PACKAGING, MARKING, LABELING, AND SHIPPING OF HAZARDOUS MATERIAL SAMPLES / 246
    - Introduction / 246
    - Environmental Samples Versus Hazardous Materials Samples / 248
    - Environmental Samples / 248
    - Rationale: Hazardous Material Samples / 249
    - Procedures: Samples Classified as Flammable Liquid (or Solid) / 251
    - Procedures: Samples Classified as Poison "A" / 254
    - Sample Identification / 255

11. GROUNDWATER PRINCIPLES AND MONITORING CONSIDERATIONS / 256

- INTRODUCTION / 256
- OBJECTIVES / 256
- CHARACTERISTICS OF GROUNDWATER SYSTEMS / 257
    - Groundwater Hydrology / 257
    - The Hydrologic Cycle / 257
    - Water Beneath the Earth's Surface / 258
    - Groundwater Flow / 258
    - Factors Affecting Groundwater Systems / 258
    - Properties of Geologic Units / 258
    - Aquifers / 259
    - Effects of Man's Activities on Hydrologic Systems / 261
- CONTAMINATION OF GROUNDWATER / 262
    - Sources of Contamination / 262
    - Movement of Contaminants in Groundwater / 262
    - Predicting Contaminant Migration / 266
- SPECIFIC CONSIDERATIONS FOR WASTE SITE MONITORING WELLS / 266
- GROUNDWATER SAMPLING PROCEDURES / 267
    - Purging Monitoring Wells Prior to Sampling / 268
    - Equipment Used for Removing Groundwater / 269
    - Collecting Groundwater Samples / 270
    - Specific Purging and Sampling Procedures / 270

12. EMERGENCY PROCEDURES / 276

- INTRODUCTION / 276
- OBJECTIVES / 276

- LAWS / 276
    - Community Emergencies / 276
    - Site Emergencies / 290
- UNDERSTANDING AND RESPONDING TO EMERGENCIES / 290
    - Emergencies: Overview / 290
    - Fire Extinguishing, Suppressants and Protection / 292
    - Essential Prevention Measures / 298
    - Spills and Spill Response / 299
    - Basic Containment Principles / 301
- MEDICAL EMERGENCY RESPONSE AND FIRST AID / 305
    - Primary Survey / 305
    - Secondary Survey / 307

13. TRANSPORTATION OF HAZARDOUS WASTES / 310

- INTRODUCTION / 310
- OBJECTIVES / 310
- DIFFERENTIATION OF TERMS / 310
- GOVERNMENT REGULATIONS / 311
    - Special Situations / 313
- LABELING CONTAINERS / 316
- PLACARDING VEHICLES / 317
- MANIFEST SYSTEM / 317
- TRUCKS ON SITE / 318
- TRANSPORTATION ACCIDENTS / 320
- COOPERATION WITH THE TSD FACILITY / 320

14. DECONTAMINATION / 322

- INTRODUCTION / 322
- OBJECTIVES / 322
- PROGRAM DESIGN / 322
- METHODS / 324
    - Physical Decontamination / 324
- DECONTAMINATION PROCEDURES / 325
    - Decontamination of Protective Clothing / 325
    - Decontamination of Tools / 331
    - Decontamination of Vehicles and Heavy Equipment / 331
- PROTECTING DECONTAMINATION PERSONNEL / 332
- MEASURING EFFECTIVENESS / 332

15. SITE SAFETY PLAN / 334

- INTRODUCTION / 334
- THE ROLE OF OSHA AND EPA / 334
- ANATOMY OF A BASIC PLAN / 334
    Plan Description / 334
    Putting the Site Safety Plan Into Action / 336
    Requirements for Informing Employees of Provisions of the Plan / 337
    Methods of Interpretation / 344
- BENEFITS ANALYSIS / 345
    Hazard Awareness Review / 347
    Checklist for Safe Site Entry / 349

Glossary / 353

Index / 377

1
Introduction

Never before in history has the need for responsible use of natural and synthetic resources, management of technologies, and protection of the environment been so great. This and many future generations yet to come will face a world full of ever growing, ever changing opportunities as the human race wrestles with the challenging problems that come about anytime a balance is sought between environmental concerns and industrial, technological, and economic necessities. One thing is sure: increasing numbers of human beings will be affected one way or another by ecological issues and developments.

One optimistic step leading us toward a harmonious coexistence between humanity and the world in which we live took place in 1976 when the United States Congress enacted the Resource Conservation and Recovery Act (RCRA). Significant progress was further achieved when Congress allocated government resources and operating funds to clean up the thousands of hazardous waste sites that have been silently festering for many years. This action was mandated through the Comprehensive Environmental Response, Compensation and Liability Act of 1980 or, as it is commonly known, the Superfund Act. Both of these acts require that the Environmental Protection Agency (EPA) develop and enforce specific standards to achieve the goals established within them.

A major industry evolved from the enforcement of these EPA standards to provide the site assessments and remediations necessary to remove the hazards from the suspected and the environmentally unsound waste sites. Because of the demand for a large workforce, workers were employed with little or no experience in handling toxic chemicals or hazardous wastes. It became obvious that regulations needed to be developed to provide the necessary protection for worker safety and health at these sites. Congress stepped in and promulgated the modified Superfund Act of 1980 with the Superfund Amendment and Reauthorization Act of 1986 (SARA). This book and CLEAR's training program have been developed to help the hazardous waste remediation industry train their personnel and to comply with the aspects of these standards. The following narrative provides more detail on the three acts previously discussed.

HISTORICAL PERSPECTIVE

As mentioned above, in 1976 the Resource Conservation and Recovery Act mandated the management of hazardous waste. Essentially, this legislation regulates facilities presently involved in the treatment, storage, and disposal (TSD) of hazardous waste. Furthermore, it requires an all encompassing hazardous waste tracking system. This cradle-to-grave monitoring program attributes the generator with the ultimate responsibility for their hazardous waste throughout its journey to disposal. The waste tracking document, referred to as the Hazardous Waste Manifest, is displayed in Chapter 13 of this book.

In section 40CFR261, RCRA defines hazardous wastes in detail. Hazardous waste are grouped into four classifications: (1) ignitable, (2) corrosive, (3) reactive, and (4) extraction procedure toxic. Additional hazardous waste lists are provided for specific waste streams and for waste producing processes.

Several years later, the Superfund Act of 1980 mandated thorough and all encompassing cleanup procedures designed primarily to prevent future long-term environmental damage from unsound hazardous waste dumps. It also provided for emergency removal of hazardous wastes from accidental chemical releases. Finally, the act established Federal funding provisions for cleanup operations at those hazardous waste sites having no identifiable responsible parties.

In 1986 Superfund was reauthorized by Congress and is now called SARA. Congress, with the aid of leading environmental organizations, modified the new act in light of what was learned from experience with Superfund (CERCLA). SARA therefore represents a hybrid that is being adapted to meet the present needs of these clean-ups. SARA sets forth provisions guaranteeing the safety and protecting the health of all workers involved in activities on RCRA and CERCLA sites as well as emergency response incidents involving hazardous substances, wastes, or materials.

Under SARA, the Occupational Safety and Health Administration (OSHA) was directed to develop standard procedures and strategies designed to ensure the health and safety of all workers involved in hazardous waste remediations (see Chapter 2 for more information).

ABOUT THIS BOOK

Hazardous waste sites are dangerous not only for the obvious reasons (such as potential exposure to toxic wastes), but also for some less obvious reasons. For example, these sites are unpredictable—nobody knows for sure what is buried in an abandoned site until the site assessment is complete and surprises can still occur. Also, site characteristics change as materials are removed; specifically, trenches are excavated causing a confined space hazard or bulking of compatible containerized wastes create waste hazards. This book addresses these specific problems and provides solutions with practical procedures and applications for the special work practices needed to maintain a safe work environment and to maintain compliance with SARA and the applicable OSHA standards. In each of the fourteen chapters that follow an introduction is provided as an overview of applicability to hazardous waste sites and objectives designed to help the reader establish individualized learning goals. The following discussion is an overview of each chapter with its significance to you the hazardous waste worker.

Chapter 2 describes all the OSHA regulations governing the employer and employee at a hazardous waste cleanup. It is important to thoroughly understand your rights and responsibilities while working on these sites to protect yourself from needless exposure to hazards.

A major concern of workers and employers is the desire to know what is happening to their bodies because of this work. Medical Surveillance, Chapter 3, discusses the OSHA requirements for determining worker's health through medical tests. This chapter defines these tests and how to interpret the results.

Hazard Recognition, Chapter 4, is crucial in prevention of worker exposure. Understanding the principles, chemical relationships, compatibility, signs, and symptoms of hazards can greatly minimize your exposures. This chapter deals with recognition of hazards and future chapters address the evaluation and control methods.

Understanding how chemical exposures actually occur can also prevent unnecessary worker exposures. Toxicology, Chapter 5, deals with the basics of how chemicals enter the body, types of effects, and terminology. Also, this chapter provides an introduction to Material Safety Data Sheets (MSDS). They are developed for specific chemicals and can be referenced throughout any cleanup for information to assist the worker in determining methods to reduce exposure.

Chapters 6 and 7, Engineering Controls and Safe Work Practices, enlighten the worker on proper procedures to use on sites. By understanding engineering controls the individual can avoid accidental exposure such as entering a hot zone without correct personal protective equipment. By understanding proper procedures for safe work practices and some common problems to avoid, workers can work not only more safely but also more efficiently.

Safe Use of Field Equipment, Safe Sampling, and Groundwater Basics/Sampling (Chapters 8, 10, and 11) deal with specific capabilities of field equipment and sampling equipment on hazardous waste remediation. Hopefully these chapters will assist samplers and surveillance personnel in the use of this equipment.

One of the most important chapters in this book is Chapter 9, Personal Protective Equipment (PPE). This chapter details protective equipment from head to toe and correlates PPE with environments by degree of hazard.

Understanding and having a *second nature* knowledge of emergency procedures specific to your cleanup site *can save your life.* Quick action in cleanup of spills can prevent further hazards, fast response to fires can prevent spreading to the community or to other vulnerable areas (hot zones) of the site, and quick attention to an injured worker can save a life. These procedures are detailed in Chapter 12.

4 WORKER PROTECTION DURING HAZARDOUS WASTE REMEDIATION

Knowledge of the Department of Transportation regulations and specifics of how to read the Hazard Class Table 49 CFR 172.101 can aid the worker in identifying an existing hazard. Also, by providing correctly labeling and placarding containers and vehicles can prevent workers future hazards such as mixing of incompatible wastes. These topics and practical tips concerning working safely around dump trucks and tractor-tailor rigs are discussed in Chapter 13 on Transportation.

Decontamination procedures for personnel and equipment are described in Chapter 14. Equipment needed for decon stations is described and practical flow charts are included to aid the worker in setting up a decon facility applicable to the specific site.

Finally, Chapter 15, Site Safety Plans provides an analysis of a generic site safety plan and discusses the benefits of implementing a good plan. Also, a checklist for safe entry into hazardous waste sites is included. By knowing the specifics of the plan and using the checklist for your site you can maintain a level of control over the hazards you are exposed to during the actual work period.

AVAILABLE RESOURCES

This book is a comprehensive guide to working safely during hazardous waste remediations. It is important to consider including other references to complement this book in an on-site library. They include:

- Department of Transportation. Hazardous Materials Table, 172.101.
- National Fire Protection Agency, 1986. *Fire Protection Guide On Hazardous Materials*, 9th Edition. Quincy, MA: NFPA.
- NIOSH/OSHA/USCG/EPA, 1985. *Occupational Safety and Health Guidance Manual for Hazardous Waste Site Activities*. Washington, DC: U.S. Government Printing Office.
- OSHA. 29 CFR 1910.1200. Material Safety Data Sheets.
- Sax, N. Irving. 1984. *Dangerous Properties of Industrial Materials*, 6th Edition. New York, NY: Van Nostrand Reinhold Company.
- Sax, N. Irving, and Lewis, Richard J., Sr. 1987. *Hawley's Condensed Chemical Dictionary*, 11th Edition. New York, NY: Van Nostrand Reinhold Company.
- Schwope, A. D., Costas, P. P., Jackson, J. O., Stulb, J. O., and Welitzman, D. J., 1987. *Guidelines for the Selection of Chemical Protective Clothing*. Cincinnati, OH: American Conference of Governmental Industrial Hygienists.
- U.S. DHHS, Public Health Service, 1985. *NIOSH Pocket Guide to Chemical Hazards*. Washington, DC: U.S. Government Printing Office.

CONCLUSION

This book is dedicated to a healthy world and healthy people. Since we are convinced that hazardous waste control and elimination is essential, but most certainly not without risk, it has fallen to us to prepare this text as a resource guide to be used often by anyone and everyone whose profession dictates some sort of interaction with these hazardous wastes. We at CLEAR are convinced that through proper and thorough training of American workers the chances increase that the quality of life will be improved and enhanced for all of us.

REFERENCES

Andrews, Richard M. Personal conversation with author, 12 June 1989.

Comprehensive Environmental Response, Compensation and Liability Act of 1980. U.S. Congress, Washington, DC: U.S. Government Printing Office.

EPA 40 CFR 261. 1980. Office of the Federal Register. Washington, DC: U.S. Government Printing Office.

Epstein, Samuel S., M. D., Brown, Lester O., and Pope, Carl. 1982. *Hazardous Waste in America*. San Francisco, CA: Sierra Club Books.

OSHA 29 CFR 1910.120. 1986. Office of the Federal Register. Washington, DC: U.S. Government Printing Office.

Resource Conservation and Recovery Act of 1976. U.S. Congress. Washington, DC: U.S. Government Printing Office.

Superfund Amendments Reauthorization Act 1986. U.S. Congress. Washington, DC: U.S. Government Printing Office.

2

Rights and Responsibilities of Workers

INTRODUCTION

According to federal law, the American worker has certain rights and responsibilities. Under the Occupational Safety and Health Act (OSHAct) of 1970, workers involved in any given work activity are covered by a general set of rights and responsibilities. Also, workers engaged in certain work activities are covered by specific regulatory standards created by the Occupational Safety and Health Administration (OSHA). Many of these standards are applicable to cleanup activities on hazardous wastes sites. In addition, an OSHA standard specifically applicable to hazardous waste clean-up operations has been implemented (29 CFR 1910.120).

This chapter is intended to inform the reader of the rights and responsibilities of workers, with special emphasis on those rights and responsibilities specifically applicable to work on hazardous waste sites.

OBJECTIVES

- Understand the role of OSHA in protecting the health and safety of workers.
- Be aware of OSHA safety standards applicable to work on hazardous waste sites.
- Know the applicable provisions of 29 CFR 1910.120.
- Know and understand the rights and responsibilities of employers and employees involved in clean-up operations on hazardous waste sites.

THE OCCUPATIONAL SAFETY AND HEALTH ACT OF 1970

Background and Intent

The Occupational Safety and Health Act (OSHAct) of 1970 went into effect on April 28, 1971. Prior to that time no uniform federal safety and health regulations existed. State regulations varied widely, and enforcement proceedings against violators of existing regulations were almost nonexistent. As a result of unsafe and unhealthy workplace conditions, unacceptably large numbers of

American workers were experiencing illness, injury, or death. The OSHAct was passed by the United States Congress as a means of addressing this problem (Olishifski, 1979).

The OSHAct was intended to ensure safe and healthful conditions in the American workplace. The act requires that employers take steps to protect employees from recognized workplace hazards that are likely to cause illness or injury. If practical, recognized hazards should be completely eliminated from the workplace, such as through the use of engineering controls. If elimination of a hazard is not practical, the employer must provide other measures, such as personal protective equipment, to protect employees. The act also requires that employees comply with all applicable rules, regulations, and standards pertaining to occupational safety and health (OSHA, 1985; Hammer, 1985).

Organizations Created by the OSHAct

A number of organizations came into existence as a result of the OSHAct. Two of these organizations, the Occupational Safety and Health Administration and the National Institute of Occupational Safety and Health are of particular importance and will be referred to frequently within this text (Hammer, 1985).

The Occupational Safety and Health Administration. The Occupational Safety and Health Administration (OSHA) was created within the Department of Labor to act as the primary guardian of worker safety and health. As such, OSHA was given the authority to develop and implement workplace safety and health standards.

OSHA is also responsible for enforcing compliance with standards. Toward this end OSHA has the authority to conduct inspections, issue citations, and levy fines.

The National Institute for Occupational Safety and Health. The National Institute for Occupational Safety and Health (NIOSH) was created as a research agency within the Department of Health and Human Services. NIOSH conducts research on occupational hazards, and makes recommendations to OSHA regarding the creation or revision of OSHA standards based on research results. NIOSH has the very important responsibility of evaluating items of personal protective equipment and hazard measuring instruments, such as those which are commonly used on hazardous waste sites. NIOSH also provides education and training in occupational safety and health.

OSHA Standards

OSHA Standards are legally enforceable sets of industry-specific regulations intended to address concerns for the safety and health of workers. These standards are developed and revised on a constant basis. The opportunity to comment on proposed new standards or revisions is extended to employers, employees, and all other interested parties. OSHA standards applicable to work on hazardous waste sites are tabulated in the following section of this chapter and referenced throughout the remainder of this text.

Rights and Responsibilities of Employees Under the OSHAct

Employee Rights. The OSHAct gave certain rights to the American worker. Most significantly, workers have the general right to a safe and healthy work environment. The act also provides for specific worker rights (OSHA, 1985). As an employee you have the right to:

- Be informed of your rights and responsibilities, as listed on the OSHA poster (OSHA 2203) which your employer is required to post within the workplace.
- Review copies of appropriate OSHA standards, rules, regulations, and requirements that the employer is required to make available at the workplace.
- Request information from your employer on safety and health hazards in the work area, on precautions that may be taken, and on procedures to be followed if an employee is involved in an accident or is exposed to toxic substances.
- Request the OSHA Area Director to conduct an inspection if you believe hazardous conditions or violations of standards exist in your workplace.
- Have your name withheld from your employer, upon request to OSHA, if you file a written and signed complaint.
- Be advised of OSHA actions regarding your complaint and have an informal review, if requested, of any decision not to inspect or not to issue a citation.
- Have an authorized employee representative accompany the OSHA compliance officer during the inspection tour.
- Respond directly to questions from the OSHA compliance officer during an inspection tour.
- Observe any monitoring or measuring of hazardous materials and examine resulting records, as specified under the act.
- Have an authorized representative, or yourself, review the Log and Summary of Occupational Injuries (OSHA No. 200) at a reasonable time and place.
- Request a closing discussion with the compliance officer following an inspection.

RIGHTS AND RESPONSIBILITIES OF WORKERS

- Submit a written request to NIOSH for information on whether any substance in your workplace has potentially toxic effects in the concentration present, and have your name withheld from your employer, if you so request.
- Object to an abatement period set in a citation issued to your employer by writing to the OSHA Area Director within 15 working days of the issuance of the citation.
- Be notified by your employer if he or she applies for a variance from an OSHA standard, testify at a variance hearing, and appeal the final decision.
- Submit information or comment to OSHA on the issuance, modification, or revocation of OSHA standards and request a public hearing.
- Exercise the rights listed above without being punished or discriminated against by the employer.

Employee Responsibilities. In addition to specific rights, the OSHAct also gave specific responsibilities to the American worker (OSHA, 1985). As an employee, you have the responsibility to:

- Read the OSHA poster at your job site.
- Comply with all applicable OSHA standards.
- Follow all employer safety and health rules and regulations, and wear or use prescribed protective equipment while engaged in work.
- Report hazardous conditions to your supervisor.
- Report any job-related injury or illness to your employer, and seek treatment promptly.
- Cooperate fully with the OSHA compliance officer during an inspection if he or she inquires about safety and health conditions in your workplace.
- Exercise your rights under the Act in a responsible manner.

OSHA SAFETY STANDARDS APPLICABLE TO WORK ON HAZARDOUS WASTE SITES

Numerous safety standards developed by OSHA to protect the safety and health of workers involved in construction activities and general labor are also applicable to activities on hazardous waste sites. Specific safety topics and applicable standards are shown in Table 2-1. Standards are referenced according to their location within the Code of Federal Regulations (CFR). For example, regulations limiting employee exposure to noise are contained in part 1910.95 of Title 29 of the Federal Code. From the standpoint of hazardous waste work, the most significant OSHA standard is 29 CFR 1910.120. This standard will be described at length in the following section of this chapter.

10 WORKER PROTECTION DURING HAZARDOUS WASTE REMEDIATION

Table 2-1. OSHA Safety Standards Applicable to Work on Hazardous Waste Sites.

Safety Topics	Applicable Standards			
	Labor (29 CFR 1910)		Construction (29 CFR 1926)	
Ventilation		⎧ 1910.94		⎧ 1926.57
Noise	Subpart G	⎨ 1910.95	Subpart D	⎨ 1926.52
Ionizing radiation		⎩ 1910.96		⎩ 1926.53
Hazardous materials	Subpart H			
Personal protective equipment				
General		⎧ 1910.132		
Eye/face		1910.133		⎧ 1926.102
Hearing		1910.95		1926.101
Respiratory	Subpart I	⎨ 1910.134	Subpart E	⎨ 1926.103
Head		1910.135		⎩ 1926.100
Foot		⎩ 1910.136		
Fire protection	Subpart L		Subpart F	
Materials handling and storage	Subpart N		Subpart H	
Electrical	Subpart S		Subpart K	
Toxic/hazardous substances	Subpart Z			
Trenching and excavation			Subpart P	

OSHA REGULATIONS UNDER TITLE 29 CFR PART 1910.120: HAZARDOUS WASTE OPERATIONS AND EMERGENCY RESPONSE

Title 29 CFR 1910.120 is an OSHA standard promulgated specifically to address critical concerns for the health and safety of personnel involved in hazardous waste activities. The standard is a comprehensive document, relating virtually all aspects of hazardous waste site remediation. Thus, it is important that all personnel involved in remedial actions on these sites have an understanding of the specific provisions of the standard. The provisions of the final version of the standard (effective March 6, 1990) are presented here in abbreviated form.

Intent and Applicability

OSHA regulations contained in 29 CFR 1910.120 were created to protect the safety and health of workers who have the potential for exposure to hazardous waste materials during hazardous waste site cleanup operations, work activities at TSD facilities, and emergency response activities. With regard to site cleanup operations, the standard is applicable to:

- Cleanup operations on CERCLA (superfund) sites.
- Corrective actions involving cleanup operations on RCRA sites.

- Site cleanup activities mandated by local, State, or Federal governmental bodies.
- Voluntary cleanup operations on sites recognized as uncontrolled hazardous waste disposal sites by local, State, or Federal governmental bodies.

The provisions of 29 CFR 1910.120 detailed in this section are applicable to workers engaged in these types of cleanup activities.

General Requirements

The standard requires that the employer develop and implement a safety and health program for employees involved in hazardous waste operations. The safety and health program is intended to identify, evaluate, and control site hazards and provide for appropriate response to emergency situations that may arise on site. As a minimum, the safety and health program must include the following components:

- Organizational structure chapter.
- Comprehensive work plan chapter.
- Site-specific safety and health plan chapter.
- Safety and health training program.
- Medical surveillance program.
- Standard operating procedures for safety and health.

Figure 2-1 shows the organization and components of the employer's safety and health program as specified in the standard.

Site-Specific Safety and Health Plan

The standard requires that a written, site-specific Safety and Health Plan be developed as a separate chapter of the employer's safety and health program for each site on which the employer's personnel are involved in cleanup operations. In order to comply with the standard, the plan must adequately address the specific safety and health hazards of each phase of operation onsite and state worker-protection provisions required for safe work on the site. Topics which the site safety plan must address are shown in Fig. 2-1 and specific health and safety requirements pertaining to these topics are detailed throughout this section.

All potentially affected employees must be informed of the specific provisions of the site safety and health plan. At a minimum, employee briefings shall be conducted prior to initial site entry and at other times thereafter as often as needed to ensure compliance with the plan. The plan must be available on site for inspection by all parties involved, including employees and their authorized

12 WORKER PROTECTION DURING HAZARDOUS WASTE REMEDIATION

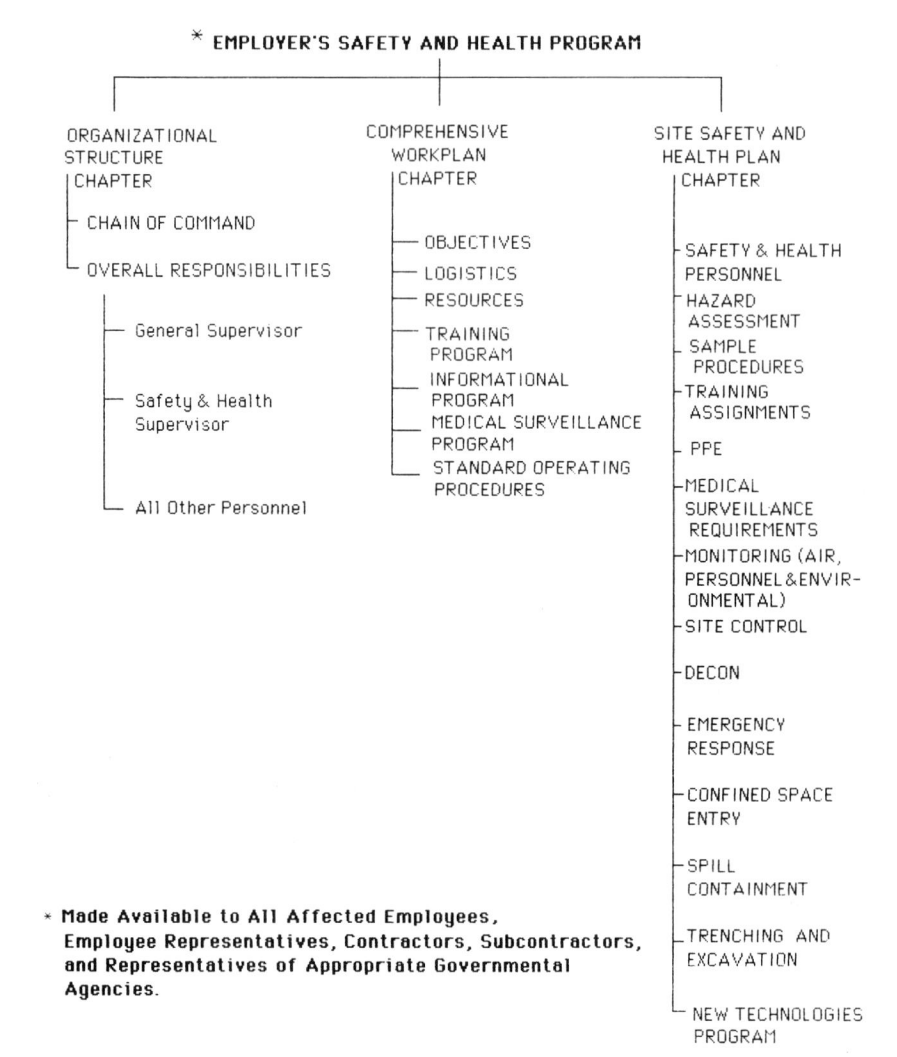

```
                    * EMPLOYER'S SAFETY AND HEALTH PROGRAM
       ┌─────────────────────────┼─────────────────────────┐
ORGANIZATIONAL              COMPREHENSIVE              SITE SAFETY AND
STRUCTURE                   WORKPLAN                   HEALTH PLAN
  CHAPTER                     CHAPTER                    CHAPTER

 ├─ CHAIN OF COMMAND
 │                           ├─ OBJECTIVES              ├─ SAFETY & HEALTH
 └─ OVERALL RESPONSIBILITIES ├─ LOGISTICS                  PERSONNEL
                             ├─ RESOURCES               ├─ HAZARD
    ├─ General Supervisor    ├─ TRAINING                   ASSESSMENT
    │                        │  PROGRAM                 ├─ SAMPLE
    │                        ├─ INFORMATIONAL              PROCEDURES
    ├─ Safety & Health       │  PROGRAM                 ├─ TRAINING
    │  Supervisor            ├─ MEDICAL SURVEILLANCE       ASSIGNMENTS
    │                        │  PROGRAM                 ├─ PPE
    │                        └─ STANDARD OPERATING
    └─ All Other Personnel      PROCEDURES             ├─ MEDICAL
                                                          SURVEILLANCE
                                                          REQUIREMENTS
                                                       ├─ MONITORING (AIR,
                                                          PERSONNEL & ENVIR-
                                                          ONMENTAL)
                                                       ├─ SITE CONTROL
                                                       ├─ DECON
                                                       ├─ EMERGENCY
                                                          RESPONSE
                                                       ├─ CONFINED SPACE
                                                          ENTRY
                                                       ├─ SPILL
                                                          CONTAINMENT
* Made Available to All Affected Employees,
  Employee Representatives, Contractors, Subcontractors, ├─ TRENCHING AND
  and Representatives of Appropriate Governmental           EXCAVATION
  Agencies.
                                                       └─ NEW TECHNOLOGIES
                                                          PROGRAM
```

Fig. 2-1. Organization and components of employer's safety and health program.

representatives, contractors and subcontractors, and representatives of appropriate governmental agencies.

Under 29 CFR 1910.120 the safety and health supervisor is responsible for developing and implementing the safety and health plan, verifying compliance with the plan, and conducting inspections to evaluate the effectiveness of the plan. The safety and health supervisor is also responsible for modifying the plan as work progresses and site conditions change or additional information becomes available about the site.

The site safety plan is a very important tool for protecting the safety and health of workers during clean-up operations on hazardous waste sites. For this reason, the subject of site safety plans will be covered in detail, with reference to the standard, in Chapter 15.

Site Characterization and Analysis

A thorough site characterization and analysis is required before cleanup operations may begin on a hazardous waste site. This allows the identification of specific hazards present so that appropriate protective measures can be taken. The standard requires that site characterization be carried out in stages as described in the following paragraphs.

A preliminary evaluation should be performed *before* the initial site entry to aid in the selection of protective measures required for safe entry. The preliminary evaluation should be designed to provide as much information as possible before site entry. This information should include:

- Hazards involved (especially IDLH conditions).
- Location, size, accessibility, and topography of site.
- Potential pathways of dispersion.
- Emergency response capability.
- Description and expected duration of work activities on the site.

The initial entry to a site must be well planned and carefully executed. Personal protective equipment (PPE) must be used during the initial entry as required to keep employee exposure below applicable exposure limits for hazardous substances known or expected to be on site based on the preliminary evaluation. If the preliminary assessment indicates the need for respirators on site, a five minute escape air supply must be immediately available to all employees involved in the initial entry unless positive-pressure self-contained breathing apparatus (SCBA) is used. If site hazards are not adequately identified by the preliminary evaluation, personnel making the initial entry will use at least Level B PPE and monitor for hazardous conditions during the entry. Air monitoring will be conducted using direct-reading instruments to identify hazardous conditions as listed below in the section on site monitoring.

A detailed evaluation must be conducted immediately after the initial entry to further identify site hazards and allow selection of appropriate engineering controls, work practices, and PPE. Employees shall be fully informed of all risks associated with hazardous substances on the site as soon as the nature of those risks has been established. Also, an air monitoring program for ongoing hazard assessment must be implemented after the site is determined to be safe for the beginning of cleanup operations.

Site Control

OSHA requires that a site control program be developed for each site. This program must be designed to adequately control the exposure of employees to hazardous substances on site, and prevent the migration of contaminants to "clean" areas of the site. The site control program must be developed during the planning stages, implemented before cleanup work begins, and modified as often as required by changing site conditions during cleanup operations.

The standard requires that the site control program include or address, as a minimum, the following items or topics:

- A site map.
- Site work zones.
- Use of the buddy system on site.
- Site communications (including emergency alarm procedures).
- Safe work practices.
- Identification of the nearest source of medical assistance.

Training

General Training Requirements. The standard includes training requirements applicable to all employees who may be exposed to site hazards. Training requirements vary according to job assignment and potential for exposure to hazardous substances.

General site workers, such as laborers and equipment operators, who engage in activities which have a high exposure potential are required to complete:

- 40 hours of off-site instruction.
- 3 days of on-the-job training under the direct supervision of a trained, experienced supervisor.
- 8 hours of annual refresher training.

Employees who work only in areas which have been monitored and fully characterized, indicating that no PPE is required and that emergencies are unlikely (for example, the site support zone) are required to complete:

- 24 hours of off-site instruction.
- 1 day of on-the-job training under the direct supervision of a trained, experienced supervisor.
- 8 hours of annual refresher training.

The same requirements apply to employees who make site visits occasionally to perform specific tasks (for example, groundwater monitoring or land surveying) and are unlikely to experience exposure in excess of applicable limits. If, at some time after initial training, employees such as these are to be transferred into a job involving a higher exposure potential, they must complete an additional 16 hours of off-site training and 2 days of on-site training in order to upgrade to full certification.

Supervisors must complete:

- The same (or equivalent) training as required for the employees they supervise.
- 8 additional hours of specialized offsite supervisory training.
- 8 hours of annual refresher training.

Scope of Training. The scope of training should be such that all employees are well versed in the following topics:

- Names of all site safety and health personnel and alternates.
- Site hazards.
- Use of PPE.
- Safe work practices.
- Safe use of engineering controls and site equipment.
- Medical surveillance requirements.
- Symptoms which may indicate overexposure to site hazards.
- Site control measures.
- Decontamination procedures.
- Provisions of the emergency response plan.
- Confined space entry procedures.
- Spill containment procedures.

The Supervisor's additional training should cover topics such as:

- The employer's safety and health program.
- Employee training program.
- PPE program.
- Health hazard monitoring techniques.
- Spill containment program.

OSHA requires that all employees be adequately trained in order to do their jobs safely despite what those jobs may be. Thus, additional training as required for jobs involving exceptional hazards must be provided (for example, handling shock-sensitive materials or segregating incompatibles contained in lab packs). Also, certified employees who begin work on an unfamiliar site should receive site-specific training sufficient to familiarize them with any unfamiliar hazards. See the checklist in Chapter 15 for safe site entry and on-site training. Furthermore, the standard requires that ongoing, site-specific training assignments be included in the site safety plan.

Medical Surveillance

OSHA standard 29 CFR 1910.120 requires that each employer involved in cleanup activities institute a program of medical surveillance. Coverage under this program must be extended to all employees who are:

- Potentially exposed to hazardous substances at or above applicable exposure limits, without regard to the use respirators, for 30 days or more per year.
- Required to wear a respirator for 30 days or more per year.
- Injured due to overexposure to hazardous substances during an emergency incident.
- Assigned to hazardous materials response (HAZMAT) teams.

The standard outlines the minimum requirements, as presented here, for a medical surveillance program. A much more comprehensive coverage of the topic is offered by Chapter 3, which is devoted entirely to medical surveillance.

The standard includes specific requirements for medical examinations performed under the medical surveillance program. Medical examinations must be conducted:

- Before assignment of new employees.
- At least annually during employment, unless the attending physician believes a longer interval (not to exceed two years) is appropriate.
- At the time of reassignment to an area or job which does not require medical surveillance, if more than six months has passed since the most recent examination.
- As soon as possible after accidental overexposure or the appearance of symptoms which may be exposure-related.
- Whenever deemed necessary by the physician.
- At the time of termination, if more than six months has passed since the most recent examination.

These examinations must be performed at no cost or loss of pay to the employee, at a reasonable time and place, and by, or under the direct supervision of, a licensed physician.

Specific content or focus of the medical examination must be determined by the examining physician based on conditions in the workplace.

Exams shall include a complete or updated medical and work history and focus on any symptoms which may be related to chemical exposure. Fitness for duty under site conditions (such as use of required PPE under expected temperature extremes) should be emphasized.

Under 29 CFR 1910.120, an employee covered by the medical surveillance program is entitled to receive a written physician's opinion. The results of specific exams and tests will be included if requested in writing by the employee. The opinion must state any medical conditions which may require treatment or place the worker at greater risk due to site hazards or work conditions. Any recommended work-assignment limitations will also be included and any resulting work reassignment should entail no loss of pay or seniority.

The standard mandates confidentiality of medical examination results. Therefore, specific findings unrelated to occupational exposure cannot be revealed by the examining physican to the employer.

Engineering Controls, Work Practices, and Personal Protective Equipment

The standard requires that engineering controls, work practices, and personal protective equipment (PPE) be used as required to protect employees from site hazards. In addition to conventional safety hazards, employees must be protected from exposure to hazardous substances in excess of applicable exposure limits. As always, PPE can be used only as a last resort for situations in which engineering controls and work practices are not a feasible option for eliminating hazardous conditions. Employee rotation cannot be used to comply with exposure limits, except in situations in which there is no other feasible means for complying with dose limits for ionizing radiation.

Selection and Use of PPE. Selection of PPE must be based on site-specific conditions and updated as those conditions change or additional information is generated about the site. A written PPE program is required, and must incorporate at a minimum the following topics:

- Selection.
- Use and limitations.
- Work mission duration.
- Maintenance.
- Storage.
- Decontamination and disposal.
- Training and proper fitting.
- Donning and doffing.
- Inspection procedures.
- Limitations during temperature extremes.
- Program evaluation.

Specific Requirements for PPE. Under 29 CFR 1910.120, employees must be provided with one of the following methods of respiratory protection for work in "immediately dangerous to life and health" (IDLH) atmospheres:

- Positive-pressure self-contained breathing apparatus (SCBA) fitted with a full facepiece.
- Positive-pressure air-line respirator fitted with a full facepiece and escape air supply.

For work in areas of skin-absorption hazards which may result in an IDLH situation, totally-encapsulating chemical-protective (TECP) suits must be used. These suits must be able to maintain positive internal pressure and capable of preventing inward gas leakage in excess of 0.5%. Methods to be used in testing TECP suits are described in Appendix A of the standard.

Site Monitoring

The standard requires that atmospheric conditions be monitored on site whenever there is any question as to the degree of employee exposure to hazardous substances. All sites must be adequately monitored so as to ensure adequate protection of site personnel and allow proper utilization of engineering controls, work practices, and PPE.

Initial Air Monitoring. Unless site hazards have been adequately characterized prior to the initial site entry, personnel making the entry must use direct-reading instruments to monitor for the following conditions:

- IDLH conditions.
- Atmospheres containing contaminants in excess of applicable exposure limits.
- Radiation above dose levels.
- Flammable atmospheres.
- Oxygen-deficient atmospheres.

Periodic Air Monitoring. Monitoring to determine the level of contamination of the air onsite should be conducted on a regular basis. As a minimum, periodic air monitoring must be conducted during cleanup operations whenever there is reason to believe that an IDLH condition or flammable atmosphere may have developed, or that exposure levels have increased above applicable exposure limits since prior monitoring.

Personnel Monitoring. Personnel monitoring of workers most likely to be overexposed shall be conducted frequently during cleanup operations. If those employees are found to be exposed in excess of applicable exposure limits, then the monitoring program must be expanded in order to characterize the degree of exposure of the worker population as a whole. A representative sampling procedure may be used for this characterization.

Informational Programs

An information program shall be developed and implemented by the employer to inform employees, contractors, and subcontractors of potential chemical exposure risks associated with site operations. The program could range from something as simple as a directory of "generic" material safety data sheets for chemicals being handled to a computer data base such as those listed in the Engineering Controls chapter. However, personnel working outside of the operations part of a site are not covered by this requirement.

Handling Drums and Containers

OSHA Standard 29 CFR 1910.120 contains a number of specific rules and procedures to be followed on sites in which hazardous waste materials are stored in drums and other containers. As a minimum, site drum and container handling procedures must incorporate the provisions specified in this section of the standard. See the Safe Work Practices chapter for specific regulations on handling drums and containers.

Decontamination

Under 29 CFR 1910.120, minimum provisions for decontamination during site cleanup operations are mandated. These provisions require that all personnel, clothing, and equipment be decontaminated before leaving the contaminated area of a site. Also, standard operating procedures (SOP) must be developed in order to minimize employee contamination. Decon procedures used shall be communicated to all employees before entry. All PPE used in a contaminated area must be properly decontaminated or else disposed of in compliance with hazardous waste management regulations (RCRA). Likewise, all decontamination equipment and solvents shall be decontaminated, or else disposed of as above. Furthermore, all decontamination areas will be located so as to minimize contaminant migration and minimize contact between contaminated and uncontaminated employees and equipment.

The standard requires that any worker wearing permeable clothing which becomes wet with hazardous wastes remove the clothing immediately and shower thoroughly. If commercial laundries and cleaning services are utilized for decontamination of PPE, they shall be warned of the hazards posed by the contaminants involved. If required by decon procedures, regular showers and change rooms must be provided on site. Also, the effectiveness of decon procedures shall be monitored by the site safety and health officer.

Emergency Response

Under 29 CFR 1910.120, the employer is required to address concerns for emergency preparedness during clean-up operations. Toward this end, the standard requires the employer to develop an emergency response plan as a separate section of the site safety and health plan. The emergency response plan must be designed to handle any emergencies which could reasonably be anticipated to occur on the site. This plan must be developed and implemented by the employer *before* cleanup operations begin, and made available in writing to all employees, employee representatives, and OSHA personnel.

Employers can be exempted from the requirement for emergency response plan development if they intend to immediately evacuate all employees in the event of any emergency and will not permit their employees to engage in emergency response operations. However, an emergency evacuation plan is required in this case.

The standard requires that the site emergency response plan address each of the following topics:

- Pre-emergency planning.
- Personnel roles, lines of authority, training, and communications for emergencies.
- Emergency recognition and prevention.
- Safe distances and places of refuge.
- Site security and control.
- Site topography, layout, and prevailing weather conditions.
- Procedures for reporting emergency incidents to local, State, and Federal agencies.
- Evacuation routes and procedures.
- Emergency decontamination procedures.
- Emergency medical treatment and first aid.
- Emergency alerting and response procedures.
- Critique of response and followup.
- Periodic plan review and amendment.
- PPE and emergency equipment for emergency response.
- Comptability with the disaster, fire, and emergency response plans of local, State, and Federal agencies.

The standard also requires that emergency response training be provided to all potential responders. This training must be adequate to allow employees to respond effectively and safely to any anticipated emergencies to which the employer expects the employees to respond. Some employees may be exempted from training requirements based on previous training and work experience. The overall site training program must include regular rehearsals of the emergency response plan.

In the event of an emergency on site, emergency response procedures must be carried out in accordance with the emergency response plan. An alarm system shall be available to notify employees of an emergency on site. Visible or audible alarms may be used as long as the alarm signal is clearly perceptible above background levels of noise, light, or other activity on the site. As soon as the alarm system is activated, employees will cease work, lower background noise levels to enhance communication, and begin emergency response plan procedures. Based on information available at the time of the emergency, the employer will evaluate the incident, evaluate on-site response capabilities, and take appropriate steps to implement the site emergency response plan.

Site Illumination

The standard requires that all site work areas have adequate lighting. As shown in Table 2-2, minimum lighting requirements vary according to work activities performed in a given area.

Sanitation at Temporary Workplaces

Minimum requirements for sanitation at hazardous waste cleanup sites are also mandated by 29 CFR 1910.120. These requirements pertain to water supplies, toilet facilities, temporary sleeping quarters, washing facilities, and related concerns.

Table 2-2. Site Illumination.

Foot candles	Area of operations
5	General site areas
3	Excavation and waste areas, accessways, active storage areas, loading platforms, refueling fields, and maintenance areas
5	Indoors: warehouse, corridors, hallways, and exitways
5	Tunnels, shafts, and general underground work areas. (Exception: Minimum of 10 foot candles is required at tunnel and shaft heading during drilling, mucking, and scaling. Mine Safety and Health Administration approved cap lights shall be acceptable for use in the tunnel heading.)
10	General shops (e.g., mechanical and electrical equipment rooms, active store rooms, barracks or living quarters, locker or dressing rooms, dining areas, and indoor toilets and workrooms).
30	First aid stations, infirmaries, and offices

22 WORKER PROTECTION DURING HAZARDOUS WASTE REMEDIATION

Site Water Supplies. An adequate supply of potable water must be provided for employees, and kept free from contamination. Nonpotable water outlets must be clearly marked, and no potential crossconnections between potable and nonpotable water systems are allowed.

Toilet Facilities. Adequate toilet facilities, as specified in Table 2-3, must be provided for employees engaging in cleanup activities. However, these requirements are not applicable to mobile crews having transportation readily available to nearby toilet facilities.

Temporary Sleeping Quarters. If provided, temporary sleeping quarters must be heated, ventilated, and lighted.

Washing Facilities. The employer must provide adequate washing facilities for all employees involved in operations where hazardous substances may be harmful to the employee. These facilities shall be close to the work site, located in areas where exposures are below applicable exposure limits, and equipped so as to allow employees to remove hazardous substances from themselves. Regular showers and change rooms are required for sites of six months or greater cleanup duration. These facilities should consist of two change rooms, one for doffing work clothes and one for donning street clothes, which are separated by a shower room.

New Technology Programs

Under 29 CFR 1910.120 the employer is required to develop procedures for implementing and evaluating new technologies, equipment, and control measures designed to enhance worker protection as part of the site safety and health plan.

SUMMARY OF EMPLOYEE RIGHTS AND RESPONSIBILITIES UNDER 29 CFR 1910.120

As an employee involved in hazardous waste site cleanup operations, you are afforded certain legal rights by 29 CFR 1910.120. These rights entitle you to:

Table 2-3. Toilet Facilities.

Number of Employees	Minimum Number of Facilities
20 or less	One
More than 20, less than 200	One toilet seat and 1 urinal per 40 employees.
More than 200	One toilet seat and 1 urinal per 50 employees

- Information contained in your employer's safety and health program.
- Information contained in the current safety and health plan of any site on which you work.
- Information regarding all risks associated with site operations.
- Protection from site hazards through engineering controls, work practices, and PPE, including:
  — Positive-pressure SCBA, or air-line respirators with an escape air supply, for IDLH conditions.
  — TECP suits for areas where skin contact may result in an IDLH situation.
  — A five-minute escape air supply to be kept on hand during initial site entry (unless SCBA is used), if preliminary site evaluation indicates that breathing hazards requiring the use of respirators are present.
  — Protection equivalent to Level B for initial site entry, unless site hazards have been positively identified beforehand.
- Direct-reading instruments used to check for IDLH conditions during initial entry, unless hazards have been indentified adequately beforehand.
- Training in areas such as hazard recognition and safe work practices, sufficient to enable you to work safely and effectively.
- Coverage under a program of medical surveillance adequate to monitor your health, your fitness for duty, and the effectiveness of protective measures used on site. You are entitled to:
  — Examinations performed at no cost to you, at a reasonable time and place, and by (or under the direct supervision of) a licensed physician.
  — A written physician's opinion covering your general state of health, any medical treatment needed, your fitness for work on site, and any recommended limitations on work assigned to you.
  — Confidentiality of all findings which are not related to occupational exposure.
- Knowledge of site decontamination procedures.
- Full decontamination upon exiting contaminated areas on site.
- Immediate removal of wet clothing, followed by a shower, if you become wet with hazardous wastes while wearing clothing which is not designed to provide protection from liquids.
- Information contained in the site emergency response plan.
- Specific training in emergency response, if you are expected to respond during a site emergency.
- Work areas which have adequate lighting.
- An adequate supply of water suitable for drinking and washing.
- Adequate facilities for washing.
- Adequate toilet facilities.
- Showers and change rooms for jobs lasting six months or longer.
- Relative comfort and sanitation of all facilities provided.

Note: Workers have certain rights under the hazard communication standard (29 CFR 1910.1200). Provisions of the hazard communication standard will be covered in Chapter 4.

As an employee involved in hazardous waste site cleanup operations, you also have certain responsibilities. These responsibilities are:

- Familiarity with the chain of command and your role in all on-site activities.
- Familiarity and compliance with all provisions of the site safety and health plan.
- Full cooperation with the site safety and health supervisor.
- Utilization of all engineering controls, work practices, and PPE required for protection from site hazards.
- Utilization of safe work practices.
- Utilization of direct-reading instruments, as instructed by superiors, to monitor site conditions.
- Full cooperation with all site monitoring procedures (especially personnel monitoring).
- Complete decontamination upon exiting a contaminated area.
- Familiarity with all aspects of the site emergency response plan (including the emergency chain of command and your emergency response role).
- Full utilization of washing facilities prior to leaving work at the end of your shift.

REFERENCES

Hammer, W. 1985. *Occupational Safety Management and Engineering*, 3rd Ed. Englewood Cliffs, NJ: Prentice-Hall, Inc.

Olishifski, J.B., Editor. 1979. *Fundamentals of Industrial Hygiene*, 2nd Ed. Chicago, IL: National Safety Council.

OSHA, 1985. *All About OSHA*. OSHA Publication No. 2056. Washington, DC: U.S. Government Printing Office.

U.S. Department of Labor. Title 29 Part 1910. Washington, DC: U.S. Government Printing Office.

U.S. Department of Labor Title 29 Part 1926. Washington, DC: U.S. Government Printing Office.

3
Medical Surveillance

INTRODUCTION

Workers involved in remedial activities on hazardous waste sites may be exposed to a number of potentially injurious conditions. These conditions may involve toxic chemicals, biologic hazards, radiation, heavy work loads, work in PPE, and heat or cold stress. Thus it is vital that the physical condition of hazardous waste workers be carefully assessed prior to employment and carefully monitored during employment. This requires a comprehensive program of medical surveillance. For this reason, medical surveillance of workers involved in hazardous waste site cleanup activities is required by OSHA standard 29 CFR 1910.120. This chapter is intended to incorporate requirements for medical surveillance included in the standard while also presenting additional information and considerations. This supplemental information is important for an understanding of the significance of medical surveillance and medical monitoring results (David 1988; NIOSH 1985).

OBJECTIVES

- Understand the importance of medical surveillance in safeguarding the health of hazardous waste workers.
- Realize the benefits to both employees and employers of full cooperation in a program of medical surveillance.
- Understand the practical aspects behind legal requirements for medical surveillance under OSHA standards.
- Be aware of the potential adverse effects of overexposure to chemicals commonly encountered on hazardous waste sites.
- Be able to understand specific medical monitoring procedures and testing protocols used in medical surveillance.
- For managerial personnel: Be able to work effectively with experts in occupational medicine to set up a comprehensive program of medical surveillance designed to address site-specific needs.

REASONS FOR CONDUCTING MEDICAL SURVEILLANCE

While many valid reasons could be given for conducting medical surveillance, the primary purpose is to assess and monitor the health and fitness of workers. In order to be effective, monitoring must be conducted both prior to and during employment. Also, the program must be comprehensive enough to allow the early detection of any occupational health problems so that long-term harmful effects can be minimized.

Medical surveillance also provides accurate information, in the form of medical records. This information may be used to:

- Conduct epidemiological studies.
- Adjudicate claims.
- Serve as evidence in litigation.
- Report the medical conditions of workers to appropriate agencies as required by law.
- Assess the effectiveness of engineering controls, work practices, and personal protective equipment in safeguarding the health of hazardous waste site workers.

GENERAL CHARACTERISTICS OF A MEDICAL SURVEILLANCE PROGRAM

A specific medical surveillance program should be developed for each site, since each site represents a unique set of potential health hazards. Table 3-1 relates chemicals which may be encountered on hazardous waste sites to target organs, potential health effects, and appropriate medical monitoring. Hazards due to factors such as extreme climatic conditions and the amount of personal protective equipment required on site must also be considered.

Due to the complexities involved, medical surveillance should be conducted by, or under the direct supervision of, a qualified physician who has extensive knowledge of occupational medicine or else utilizes the services of an occupational medicine consultant. Medical personnel involved must have extensive expertise in order to tailor the medical surveillance program to the specific hazards and characteristics of a given site. For example, specific medical examinations and laboratory testing may be needed to monitor exposure to specific chemicals on a given site. Also, specific medical conditions may preclude an employees use of personal protective equipment required for a given job on site.

Medical surveillance of each employee should consist of:

Table 3-1. Common Chemical Toxicants Found at Hazardous Waste Sites, Their Health Effects and Medical Monitoring.

HAZARDOUS SUBSTANCE OR CHEMICAL GROUP	COMPOUNDS	USES	TARGET ORGANS	POTENTIAL HEALTH EFFECTS	MEDICAL MONITORING

Aromatic Hydrocarbons	Benzene
	Ethyl benzene
	Toluene
	Xylene
		Commercial solvents and intermediates for synthesis in the chemical and pharmaceutical industries.	Blood
			Bone marrow
			CNS[a]
			Eyes
			Respiratory system
			Skin
			Liver
			Kidney
				All cause:
				  CNS[a] depression: decreased alertness, headache, sleepiness, loss of consciousness.
				  Defatting dermatitis.
				Benzene suppresses bone-marrow function, causing blood changes. Chronic exposure can cause leukemia.
				Note: Because other aromatic hydrocarbons may be contaminated with benzene during distillation, benzene-related health effects should be considered when exposure to any of these agents is suspected.
					Occupational/general medical history emphasizing prior exposure to these or other toxic agents.
					Medical examination with focus on liver, kidney, nervous system, and skin.
					Laboratory testing:
					  CBC[b]
					  Platelet count
					  Measurement of kidney and liver function.

Asbestos (or asbestiform particles)		A variety of industrial uses, including:
		Building Construction
		Cement work
		Insulation
		Fireproofing
		Pipes and ducts for water, air, and chemicals
		Automobile brake pads and linings
			Lungs
			Gastrointestinal system
				Chronic effects:
				  Lung cancer
				  Mesothelioma
				  Asbestosis
				  Gastrointestinal malignancies
				Asbestos exposure coupled with cigarette smoking has been shown to have a synergistic effect in the development of lung cancer.
					History and physical examination should focus on the lungs and gastrointestinal system.
					Laboratory tests should include a stool test for occult blood evaluation as a check for possible hidden gastrointestinal malignancy.
					A high quality chest X-ray and pulmonary function test may help to identify long-term changes associated with asbestos diseases; however, early identification of low-dose exposure is unlikely.

Dioxin (see Herbicides)					

Halogenated Aliphatic Hydrocarbons	Carbon tetrachloride
	Chloroform
	Ethyl bromide
	Ethyl chloride
	Ethylene dibromide
	Ethylene dichloride
	Methyl chloride
	Methyl chloroform
	Methylene chloride
	Tetrachloroethane
	Tetrachloroethylene (perchloroethylene)
	Trichloroethylene
	Vinyl chloride
		Commercial solvents and intermediates in organic synthesis.	CNS[a]
			Kidney
			Liver
			Skin
				All cause:
				  CNS[a] depression: decreased alertness, headaches, sleepiness, loss of consciousness.
				  Kidney changes: decreased urine flow, swelling (especially around eyes), anemia.
				  Liver changes: fatigue, malaise, dark urine, liver enlargement, jaundice.
				Vinyl chloride is a known carcinogen; several others in this group are potential carcinogens.
					Occupational/general medical history emphasizing prior exposure to these or other toxic agents.
					Medical examination with focus on liver, kidney, nervous system, and skin.
					Laboratory testing for liver and kidney function; carboxyhemoglobin where relevant.

Heavy Metals	Arsenic Beryllium Cadmium Chromium Lead Mercury	Wide variety of industrial and commercial uses.	Multiple organs and systems including: Blood Cardiopulmonary Gastrointestinal Kidney Liver Lung	All are toxic to the kidneys. Each heavy metal has its own characteristic symptom cluster. For example, lead causes decreased mental ability, weakness (especially hands), headache, abdominal cramps, diarrhea, and anemia. Lead can also affect the blood-forming mechanism, kidneys, and the peripheral nervous system.	History-taking and physical exam: search for symptom clusters associated with specific metal exposure, e.g., for lead look for neurological deficit, anemia, and gastrointestinal symptoms. Laboratory testing:

Table 3-1. (*Continued*)

HAZARDOUS SUBSTANCE OR CHEMICAL GROUP	COMPOUNDS	USES	TARGET ORGANS	POTENTIAL HEALTH EFFECTS	MEDICAL MONITORING
Heavy Metals (*Continued*)			CNS[a] Skin	Long-term effects[c] also vary. Lead toxicity can cause permanent kidney and brain damage; cadmium can cause kidney or lung disease. Chromium, beryllium, arsenic, and cadmium have been implicated as human carcinogens.	Measurements of metallic content in blood, urine, and tissues (e.g., blood lead level; urine screen for arsenic, mercury, chromium, and cadmium). CBC[b] Measurement of kidney function, and liver function where relevant. Chest X-ray or pulmonary function testing where relevant.
Herbicides	Chlorophenoxy compounds: 2,4-dichlorophenoxyacetic acid (2,4-D) 2,4,5-trichlorophenoxyacetic acid (2,4,5-T) Dioxin (tetrachlorodibenzo-p-dioxin, TCDD), which occurs as a trace contaminant in these compounds, poses the most serious health risk.	Vegetation control.	Kidney Liver CNS[a] Skin	Chlorophenoxy compounds can cause chloracne, weakness or numbness of the arms or legs, and may result in long-term nerve damage. Dioxin causes chloracne and may aggravate pre-existing liver and kidney diseases.	History and physical exam should focus on the skin and nervous system. Laboratory tests include: Measurement of liver and kidney function, where relevant. Urinalysis.
Organochlorine Insecticides	Chlorinated ethanes: DDT Cyclodienes: Aldrin Chlordane Dieldrin Endrin Chlorocyclohexanes: Lindane	Pest control.	Kidney Liver CNS[a]	All cause acute symptoms of apprehension, irritability, dizziness, disturbed equilibrium, tremor, and convulsions. Cyclodienes may cause convulsions without any other initial symptoms. Chlorocyclohexanes can cause anemia. Cyclodienes and chlorocyclohexanes cause liver toxicity and can cause permanent kidney damage.	History and physical exam should focus on the nervous system. Laboratory tests include: Measurement of kidney and liver function. CBC[b] for exposure to chlorocyclohexanes.
Organophosphate and Carbamate Insecticides	Organophosphate: Diazinon Dichlorovos Dimethoate Trichlorfon Malathion Methyl parathion Parathion Carbamate: Aldicarb Baygon Zectran	Pest control.	CNS[a] Liver Kidney	All cause a chain of internal reactions leading to neuromuscular blockage. Depending on the extent of poisoning, acute symptoms range from headaches, fatigue, dizziness, increased salivation and crying, profuse sweating, nausea, vomiting, cramps, and diarrhea to tightness in the chest, muscle twitching, and slowing of the heartbeat. Severe cases may result in rapid onset of unconsciousness and seizures. A delayed effect may be weakness and numbness in the feet and hands. Long-term, permanent nerve damage is possible.	Physical exam should focus on the nervous system. Laboratory tests should include: RBC[d] cholinesterase levels for recent exposure (plasma cholinesterase for acute exposures). Measurement of delayed neurotoxicity and other effects.
Polychlorinated Biphenyls (PCBs)		Wide variety of industrial uses.	Liver CNS[a] (speculative)	Various skin ailments, including chloracne; may cause liver toxicity; carcinogenic to animals.	Physical exam should focus on the skin and liver.

Table 3-1. (*Continued*)

HAZARDOUS SUBSTANCE OR CHEMICAL GROUP	COMPOUNDS	USES	TARGET ORGANS	POTENTIAL HEALTH EFFECTS	MEDICAL MONITORING
			Respiratory system (speculative)		Laboratory tests include:
			Skin		Serum PCB levels.
					Triglycerides and cholesterol.
					Measurement of liver function.

[a]CNS = Central nervous system.
[b]CBC = Complete blood count.
[c]Long-term effects generally manifest in 10 to 30 years.
[d]RBC = Red blood count.
Source: NIOSH 1985.

- A thorough medical examination prior to employment.
- Periodic medical examinations during employment.
- Followup examinations and consultations, as needed.
- An examination upon termination of employment.
- An examination immediately after any injury or accidental overexposure.
- An examination as soon as possible after the employer is notified that an employee is experiencing symptoms which may be exposure-related.

Employers, employees, and medical professionals must be actively involved and communicate openly, and on a regular basis, if the medical surveillance program is to be effective. The advantages to both employers and employees of an effective medical surveillance program are many. For example, *employees* who participate in the program may benefit from advantages such as:

- Early warning of impending health problems.
- Ability to substantiate future claims of occupation-related illness (since occupational illness often mimics nonoccupational illness).
- Counseling from a physician.
- The establishment of baseline data with which future data can be compared.
- Appropriate employee task assignments (thus, safer working conditions).

Employers may benefit greatly, in the event of future litigation, from baseline data and other information generated through medical surveillance.

PRE-EMPLOYMENT MEDICAL EXAMINATION

This examination is designed to determine an individual's general fitness for work and to provide baseline data for comparison with future medical data. This examination should consist of the components discussed in this section.

Medical and Occupational History

Compilation of the applicant's medical and occupational history should be the major focus of the examination procedure. The medical and occupational history should be compiled prior to the physical examination of a prospective employee and should include the following:

- *Baseline information* such as name, address, social security number, next of kin, etc.
- *Past medical history*, as represented by existing medical records and the subject's input during questioning. The questioner should begin at the present and work back through time. Questions should focus on all body systems, from head to toe, of the subject, so that a complete *systems review* is conducted prior to physical examination. Special attention should be given to atopic disease (such as eczema), asthma, lung disease, and cardiovascular disease.
- *Family history*, as this may indicate a genetic predisposition to certain medical problems.
- *Personal and social data*, focusing on factors such as dietary habits, exercise habits, alcohol consumption, and use of tobacco.
- *Occupational history* should be compiled from present to past and cover all previous employers, all previous job duties, all previous work locations, and any military service. This part of the examination should emphasize any previous occupation-related injuries, illnesses, and/or symptoms (including heat injuries), and any previous exposures or reactions to hazardous substances.

Physical Examination

Physical examinations should involve a detailed examination of all body systems, with special attention to the pulmonary, cardiovascular, and skeletomuscular systems. The examination should cover the following items:

- *General appearance*, as this provides clues to personality, mental state, and general fitness, of the subject.
- *Vital signs*, including blood pressure, heart rate, respiration rate, temperature, weight, and height.
- Additional major areas of the body to examine include the *skin*, and the *head*.
- Also, *ears* should be examined, by audiometric testing as required by 29 CFR 1910.95. Note that perforated eardrums may preclude respirator use.
- *Eyes* should be examined with tests to measure refraction, depth perception, and color vision.

- A complete medical exam will also include examination of the *neck, chest and lungs, breasts, abdomen, rectum, and genitalia*.

More extensive analysis should involve the *heart* and *blood vessels*, and the *muscular, neurologic,* and *lymphatic* systems.

The *mental state* of the applicant should also be examined to determine ability to deal with job-related psychological stresses.

Laboratory Tests

Information from the physical examination should be supplemented with the following tests:

- Blood tests.
- Urinalysis.
- EKG.
- Chest x-ray.
- Pulmonary function.
- Other tests, (for example, allergy testing) as needed.

Examples of the application of specific tests are given in Table 3-2.

Analyses based on whole batteries of specific tests provide excellent baseline data, and sometimes provide early warnings of the tendency toward development of specific illnesses. The use of comprehensive analysis techniques as discussed in Appendix 3-1 is recommended, with specific tests (as listed in Table 3-2) to include:

- Alkaline phosphates.
- BUN.

Table 3-2. Tests Frequently Performed by Occupational Physicians.

FUNCTION	TEST	EXAMPLE
Liver:		
General	Blood tests	Total protein, albumin, globulin, total bilirubin (direct bilirubin if total is elevated).
Obstruction	Enzyme test	Alkaline phosphatase.
Cell injury	Enzyme tests	Gamma glutamyl transpeptidase (GGTP), lactic dehydrogenase (LDH), serum glutamic-oxaloacetic transaminase (SGOT), serum glutamic-pyruvic transaminase (SGPT).
Kidney:		
General	Blood tests	Blood urea nitrogen (BUN), creatinine, uric acid.
Multiple Systems and Organs	Urinalysis	Including color; appearance; specific gravity; pH; qualitative glucose, protein, bile, and acetone; occult blood; microscopic examination of centrifuged sediment.
Blood-Forming Function	Blood tests	Complete blood count (CBC) with differential and platelet evaluation, including white cell count (WBC), red blood count (RBC), hemoglobin (HGB), hematocrit or packed cell volume (HCT), and desired erythrocyte indices. Reticulocyte count may be appropriate if there is a likelihood of exposure to hemolytic chemicals.

Source: NIOSH 1985.

- SGOT.
- Bilirubin.
- SGPT.
- Other tests, as needed.

Specific blood and urine testing procedures are described in Appendix 3-1.

EXAMINATION FOLLOWUP/CONSULTATION

Based on medical examination results, health care professionals should be able to:

- Disqualify prospective employees who are physically unsuited for the demands of work on a hazardous waste site, or may pose a safety risk to fellow workers.
- Recognize employees with a history of vulnerability to specific substances.
- Recommend job assignments, or reassignments, based on the employee's state of fitness and the job demands or hazards.
- Assess an employee's capacity to perform while wearing a respirator, as required by 29 CFR Part 1910.134.
- Counsel employees on personal habits which may affect susceptibility to site hazards. For example, synergistic effects may occur between specific chemicals encountered on a site and alcohol consumed off the site.
- Provide employees with early warnings of impending medical problems, so that precautionary steps can be taken to avoid illness and disability.

PERIODIC MEDICAL EXAMINATION

A medical monitoring program should be developed for each employee based on factors such as:

- Medical and occupational history.
- Current state of health.
- Current and potential exposures on site.
- Routine job tasks of the employee.

Examinations must be conducted at least every two years during the employment of a worker, but it is strongly recommended that examinations be conducted at least annually.

Workers should be examined as soon as possible after an accidental overexposure or the appearance of symptoms of overexposure, heat stress, or other potential problems. As part of the program, workers should be trained to recognize signs and symptoms of chemical exposure and heat injury (as shown in Table 3-3).

Table 3-3. Signs and Symptoms of Chemical Exposure and Heat Stress that Indicate Potential Medical Emergencies.

TYPE OF HAZARD	SIGNS AND SYMPTOMS
Chemical Hazard	Behavioral changes
	Breathing difficulties
	Changes in complexion or skin color
	Coordination difficulties
	Coughing
	Dizziness
	Drooling
	Diarrhea
	Fatigue and/or weakness
	Irritability
	Irritation of eyes, nose, respiratory tract, skin, or throat
	Headache
	Light-headedness
	Nausea
	Sneezing
	Sweating
	Tearing
	Tightness in the chest
Heat Exhaustion	Clammy skin
	Confusion
	Dizziness
	Fainting
	Fatigue
	Heat rash
	Light-headedness
	Nausea
	Profuse sweating
	Slurred speech
	Weak pulse
Heat Stroke (may be fatal)	Confusion
	Convulsions
	Hot skin, high temperature (yet may feel chilled)
	Incoherent speech
	Convulsions
	Staggering gait
	Sweating stops (yet residual sweat may be present)
	Unconsciousness

Source: NIOSH 1985.

The periodic medical examination should basically follow the pre-employment examination format. However, this format should be appropriately modified according to changes in factors such as worker's symptoms, site hazards, and chemical exposures.

Periodic examinations should include, as a minimum:

- *A medical history update*, focusing on changes in health status, any illnesses present, and any symptoms which are possibly work related.
- *A physical examination*, as described above.
- *Specific medical tests*, as required based on the worker's potential for exposure, medical history, and examination results. These tests should include;
  — Pulmonary function tests for workers using respirators, exposed to irritating or toxic substances, or exhibiting breathing difficulties.
  — Audiometric tests for workers subjected to high noise levels.
  — Eye examination.
  — Blood and urine tests, as needed.

A comparison of periodic examination results with baseline data may provide early indications of adverse health trends and allow preventive measures to be taken. The same information can also be used in evaluating the effectiveness of protective measures used on site.

TERMINATION EXAMINATION

At the time of termination of employment or transfer to a job which does not require medical surveillance, all personnel should receive a medical examination unless all the following conditions are met:

- Less than 6 months has elapsed since the employee's last full medical examination was given.
- The employee has not been overexposed or injured since the previous examination.
- The employee has experienced no potentially exposure-related symptoms since the previous examination.

MEDICAL RECORDS

Hazardous waste site workers may work at a large number of different sites, and be exposed to a large number of hazardous substances, in the course of their careers. Thus, accurate records of previous potential exposures can be invaluable should the need for medical treatment arise.

OSHA regulations require that employers keep medical records on potentially exposed workers for a period of 30 years after their employment is terminated. These medical records must be made available to workers, their representatives, and authorized OSHA officials (29 CFR Part 1910.120). Employers are also required to post a yearly summary report of occupational illness and injuries

(29 CFR Part 1904). Other specific requirements regarding medical records are listed in 29 CFR Part 1910.120.

PROGRAM REVIEW

The medical surveillance program should be evaluated and updated at least annually to ensure its effectiveness. Also, reviews of medical records and test results should be conducted regularly as part of the safety and health plan evaluation procedure.

APPENDIX 3-1. SPECIFIC BLOOD AND URINE TESTING

The most important clues to the the presence of illness come from the medical and occupational history. Results of laboratory tests and of physical examinations often help to confirm the presence of occupational illness or help to raise suspicion that the illness is work-related. The initial measurements of the contents of the blood and urine are often performed through a battery of automated screening tests, for example the SMA 24 for blood (Sequential Multiple Analyzer) offers 24 blood screening tests. The test results are compared with known standard average values.

Blood Tests

Some of the substances tested for in the blood sample are calcium, phosphorus, total cholesterol, triglycerides, uric acid, creatinine, BUN (blood urea nitrogen), total bilirubin, alkaline phosphatase, SGOT (serum glutamic oxaloacetic transaminase), SGPT (serum glutamic pyruvic transaminase), LDH (lactic dehydrogenase) 5'-nucleotidase, total protein, albumin, potassium, chloride, total CO_2, glucose, etc.

Additional tests may be included depending on actual or potential workplace exposure, e.g., blood lead, ZPP (zinc protophorin), copper, iron, antimony ammonia, blood cholinesterase, PCB, etc.

Routine blood screening also includes a CBC (complete blood count) of red blood cells, platelets, hemoglobin, hematocrit, or packed cell volume. The CBC testing provides clues for the diagnosis of anemias, leukemias, and other blood diseases and malignancies. This is particularly useful where there is expected exposure to aromatic hydrocarbons, heavy metals, organochlorine insecticides, and ionizing radiation.

Interpretation of the Significance of Blood Testing.

Calcium (normal level 9–10.5 mg/dL). Low serum calcium levels are found in patients with hypothyroidism and dietary deficiencies; increased calcium levels may be associated with bone tumors and tumors of the parathyroid hormone where calcium has been mobilized from bone.

Phosphorus (normal level 2.5–4.5 mg/dL). Phosphorus and calcium have an inverse relationship. Increased phosphorus levels are found in Vitamin D intoxication and kidney failure. Decreased phosphorus levels occur in thyroid disease and during the menstrual cycle.

Total Cholesterol and Triglycerides (normal level 150–250 mg/dL for cholesterol and 40–150 mg/dL for triglycerides). Increased levels of each help to determine existence of increased risk for coronary heart disease.

Uric Acid (normal level 2.5–8.5 mg/dL for men and 2–6.6 mg/dL for women). Elevated levels usually indicate presence of gout, a painful disease of certain joints. Multiphasic screening tests have resulted in the early discovery of gout, which can be successfully treated. Low levels of uric acid may be associated with kidney failure.

BUN (Blood Urea Nitrogen) (normal level 5–20 mg/dL). BUN is one of the most commonly used tests to assess kidney function, together with creatinine and creatinine clearance. Urea is formed in the liver as an end product of protein breakdown and is excreted entirely by the kidneys. Blood concentration of urea is therefore directly related to the excretory function of the kidney.

Protein is digested in the gastrointestinal track into amino acids, which are absorbed into the blood stream and broken down into free ammonia in the liver. The ammonia molecule forms urea in the liver and is then transported to the kidney for excretion.

Nearly all major kidney diseases cause inadequate excretion of urea and the blood concentration (BUN) rises above normal. Kidney diseases, kidney stones and obstructions, and exposure to many toxic waste substances, such as aromatic hydrocarbons and chlorine insecticides, may cause elevated BUN levels.

Certain medications may decrease BUN, such as gentamycin and tetramycin. Dehydration, malnutrition, and combined liver and kidney disease also decrease BUN.

Creatinine (normal level 0.7–1.5 mg/dL). Creatinine is excreted entirely by the kidneys and therefore proportional to kidney excretory function. It is a breakdown product of creatine, which is used for skeletal muscular contraction. It is affected little by dehydration, malnutrition, or liver function. Only kidney disorders will cause abnormal elevation of creatinine. Normal BUN : creatinine ratio is about 20 : 1.

Serum Bilirubin (normal level 0.1–0.3 mg/dL). Bilirubin is the end product of hemoglobin breakdown. It normally exists in the red blood cell for about 120 days. The red blood cell is then destroyed by the spleen. The free hemoglobin is broken down by the liver to bilirubin. Bilirubin is excreted by the liver as a component of bile, which is excreted into the intestines. Bilirubin elevation in the blood causes skin and other tissue to become yellow (jaundice). Over 100 causes of jaundice exist, including obstruction of bile ducts by stones, tumors, or inflammation, liver infections (hepatitis), liver tumors, pancreatic tumors, and exposure to certain toxic chemicals such as aliphatic and aromatic hydrocarbons.

Liver Enzyme Tests:

- SGOT (serum glutamic oxalaocetic transaminase), normal level 5–40 IU/L (international units per liter).
- SGPT (serum glutamic pyruvic transaminase), normal level 5–351 IU/L.
- LDH (lactic dehydrogenase) normal level 90–200 mU/ml (micro units per millimeter).
- Alkaline phosphatase, normal level 30–85 mU/ml.
- 5'-nucleotidase, normal level 1.6–17.5 U/L (units per liter).

The liver houses many enzymes which speed up chemical reactions that occur within a cell. These five enzymes are stored and used within a liver cell. With cellular death, there is a release of these intracellular enzymes into the blood stream with tremendous elevations of the enzymes. Some of these enzymes are also produced in cells of other organs, such as heart, lung, and kidney. Injury or disease to these organs will cause a buildup of these enzymes in the blood. Therefore, although elevation of these enzymes is found in liver disease, it is not specific for liver disease.

SGOT, SGPT, and LDH values increase with diseases affecting the liver cell (hepatitis), as these cause very high serum levels. Alkaline phosphatase and 5'-nucleotidase levels are only minimally elevated.

LDH and SGOT also increase with diseases of the heart, lungs, and kidneys, SGPT, however, is made only in the liver. Unlike SGOT, then, SGPT when elevated, strongly incriminates the liver as the disease site. LDH and SGOT are much less specific.

Alkaline Phosphatase and 5'-Nucleotidase. With obstruction of bile ducts (by tumors, gallstones, or inflammation), the alkaline phosphatase and 5'-nucleotidase levels increases more than tenfold. The other three liver enzymes increase slightly. Other diseases or injuries, for example, bone fractures, or bone tumors, as well as normal bone development in children, can cause elevation of alkaline phosphatase. Like SGPT, however, 5'-nucleotidase is located only in liver cells. Its elevation incriminates only the liver. If only bone disease is involved, then only alkaline phosphatase is elevated. It is also possible to fractionate or separate parts or components of alkaline phosphate (and also parts of LDH). These different components are isoenzymes. Each isoenzyme comes from a specific organ. Therefore, fractionation of isoenzymes can differentiate liver disease from other organ diseases.

Serum Electrolytes:

- Sodium (Na$^+$): normal level 136–145 mEq/L (milliequivalents per liter).
- Potassium (K$^+$): normal level 3.5–5 mEq/L.
- Chloride (Cl$^-$): normal value 96–106 mEq/L.
- Carbon dioxide (CO$_2$): normal value 23–30 mEq/L.

Sodium (Na+). Sodium content of blood is the result of a balance between salt (sodium) intake and kidney excretion. Many factors assist in the sodium balance. Aldosterone, an adrenal hormone causes sodium retention by decreasing kidney losses. Water and sodium are closely interrelated. As free body water is increased, sodium is diluted in the blood and the concentration may decrease. The kidney compensates by removing (excreting) the water and conserving the sodium. If free body water decreases, serum sodium rises. The kidney responds by conserving water.

Some causes of increased sodium levels in blood are excessive dietary intake, excessive sodium in intravenous fluids, Cushing's Syndrome (overactive adrenal gland), excessive sweating, extensive burns, and diuretic medications.

Some causes of low level concentrations of sodium are deficient dietary sodium intake, deficient sodium in intravenous solutions, diarrhea, vomiting, kidney diseases, and exposure to a variety of hazardous waste substances, such as halogenated aliphatic hydrocarbons.

Chloride (Cl^-). The major function of chloride is to maintain electrical neutrality, mostly in combination with sodium, to follow sodium losses and accompany sodium excesses. It also serves as a buffer to assist the acid-base balance.

Potassium (K^+) is a major intracellular ion. Since its concentration within the cell is low, slight concentration changes are significant. Blood potassium depends on:

- Aldosterone, which tends to increase kidney losses of potassium.
- Sodium absorption (as sodium is reabsorbed, potassium is lost).
- Acid-base balance; acidotic states tend to raise blood potassium levels and basic states tend to lower potassium levels by causing a shift of potassium out of blood and into cells.

Some causes of increased potassium blood levels are: excessive dietary intake, kidney failure, and infection. This condition can be detected on an EKG reading.

Some causes of low potassium blood levels are: decreased dietary intake, diarrhea, drugs, and insulin, glucose, or calcium administration.

Other Blood Serum Constituents

Carbon Dioxide (CO_2). CO_2 content is a measure of bicarbonate (HCO_3^-) that exists in the blood. This anion is of secondary importance in electrical neutrality of cellular fluid. Its major role is in the acid-base balance. Increases occur with alkalosis and decreases occur with acidosis.

Serum Glucose (normal level 60–120 mg/dL). Serum glucose is most useful in helping to screen for diabetes mellitus, where true glucose elevations occur. In contrast, hypoglycemia results from decreased glucose levels. Causes of hypoglycemia include insulin overdose, tumors of the pancreatic islet cells, low functioning thyroid, and severe liver disease.

Other causes of hyperglycemia include: Acute stress response, over-functioning of adrenal glands, tumors of pancreas, tumor of adrenal medulla, pancreatitis, diuretic medications, hyperthyroidism.

Urine Tests—Urinalysis

Urinalysis is an informative, inexpensive test for kidney disease. Routine analysis includes:

pH (normal range 4.6–8.0 with 6.0 average). pH of a freshly voided urine specimen is an indication of the acid-base balance. Some diseases cause the kidney to secrete too much or too little acid. Certain types of kidney stones are formed in acid urine; others in alkaline urine.

Color. Urine color ranges from pale yellow to amber. Color varies with specific gravity. Abnormally colored urine can result from disease conditions or ingestion of certain medications and foods. Dark yellow may indicate bilirubin. Beets can cause red urine. Medications, e.g., phenazopyridine (pyridium) and phenytoin (dilantin) produce pink or red to reddish brown specimens.

Specific Gravity. Specific gravity is a measure of the concentration of particles (water and electrolytes) and indirectly is a measure of hydration: high specific gravity indicates a concentrated urine and low specific gravity indicates a dilute urine. Normal levels are 1.005–1.030 (usually 1.010–1.025).

Protein (Albumin). Normally protein is not present, but if the kidney filtering is injured, the spaces in the kidney filter become larger and allow protein to seep out into the urine. If the rate is high, the patient can loose much protein. Because protein keeps the water within blood cells, loss of protein will cause severe swelling of the skin (edema).

Glucose. Normally no glucose is detected in urine. In diabetics who are not well controlled by insulin, blood glucose levels become high. In most instances, when blood glucose level exceeds 180 mg/dL (the renal threshold) glucose spills over into the urine. As glucose level increases so does the urine level. The amount of glucose is measured as trace amounts to 4+ (read as "four plus").

Ketones. In poorly controlled diabetics, often young diabetics, massive fatty acid breakdown can occur. The purpose of this breakdown is to maintain an energy source at a time when glucose cannot be used by the cells because of a lack of insulin, which transports glucose into cells. Ketones are the end product of fatty acid breakdown. When ketones become increased in the blood, they spill over into the urine. Ketones in the urine can also be seen in nondiabetic persons with dehydration, starvation, and excessive aspirin ingestion.

Blood. In a number of kidney diseases (such as kidney tumors, stones, and kidney traumas), blood cells will enter the urine. Normally only 1–2 red blood cells or white blood cells are found in the urine under microscopic examination. Presence of more than 5 white blood cells indicates kidney infection. More than 5 red blood cells indicates hematuria (blood in urine).

Casts. Casts are clumps of materials or cells. They are in the shape of kidney collecting tubules. Microscopic examination of urinary sediment may detect white or red blood cell casts and may indicate kidney infection. Hyaline casts are conglomerations of protein and indicate the presence of protein in the urine. Fresh specimens will show casts which can break up rapidly.

Bacteria. Bacteria detected under microscopic examination indicates kidney infection.

Crystals. Crystals detected by microscopic examination usually indicate kidney stones are forming or have formed.

Routine Examination. The pH, presence of protein, glucose, ketones, or blood can be detected by using Multistix reagent strips for urinalysis. Multistix is a plastic stick to which several reagent strips are fixed for testing various substances.

- *Specific gravity* is measured by using a weighted instrument (urinometer) which is suspended or floated in a cylinder of urine. Concentration of urine determines the depth at which the urinometer will float. This depth is measured by a calibrated scale on the urinometer and is called the SPG.
- *Microscopic Examination.* A small amount of urine is placed in a test tube and spun around for several minutes. A drop of the sediment is then placed on a glass slide and examined under a microscope. Another drop of sediment is placed on a slide for staining. Cell casts, crystals, and bacteria are then identified.
- *Heavy Metal Screening.* The urine should also be screened for specific heavy metals when indicated. (Examples: arsenic, mercury, chromium and cadmium.)

Several followup clinical procedures are available for diagnosis of kidney disease. These include creatinine clearance, kidney biopsy, renal vein assay for renin, and split renal function tests. Cystoscopy affords direct visualization of the prostate, bladder, and urethra to determine the presence of tumors and assess the structure and function of the urethra, bladder, prostate, and ureter.

REFERENCES

David, W. D. 1988. Lecture to Author, January 1988.
NIOSH. 1985. *Occupational Safety and Health Guidance Manual for Hazardous Waste Site Activities.* NIOSH Publication No. 85-115. Washington, DC: U.S. Government Printing Office.
U.S. Department of Labor. Title 29 Part 1904. Washington, DC: U.S. Government Printing Office.
U.S. Department of Labor. Title 29 Part 1910. Washington, DC: U.S. Government Printing Office.

4
Hazard Recognition

INTRODUCTION

Hazardous waste sites are dangerous places because of the nature of the materials found there. The constraints of clothing and equipment which must be used to deal with these materials, and the difficulties encountered in this disorderly environment, create a number of health and safety hazards.

Three of the four phases of a hazardous waste site cleanup (preliminary investigation, remedial investigation, and construction) present a number of potential hazards. Chemicals encountered may be reactive, flammable or toxic; other substances which are radioactive or biologically active may be present on the site as well. Heavy equipment, heavy loads, and steep or slippery surfaces are only a few of the potential safety hazards there.

Methods have been developed to deal with all the potential hazards on hazardous waste sites once they have been identified; however, some difficulties in recognizing and identifying hazards may be encountered.

OBJECTIVES

The reader will:

- Have a basic understanding of chemistry and chemicals in the workplace.
- Be aware of the need for advance planning regarding hazard avoidance and correction.
- Know specific hazards and how to identify them.
- Understand the chemical and physiological basis for certain hazards.
- Consider both engineering and personal hazard controls.
- Know symptoms of human illness resulting from hazards.

RISK ASSESSMENT

The risks undertaken in performing each task on a hazardous waste site may be estimated in the same way as they are for any kind of work, where

$$\text{Risk} = \text{Probability of accident} \times \text{Danger}.$$

Probability of an accident on a task is difficult to quantify in many cases, as it depends upon knowledge of the number of times an accident has occurred previously with the same equipment or materials under the same circumstances. Relative probability or an "educated guess" may have to be used. Quantifying the danger presented by a particular accident is sometimes easier to do: for example, a spill of benzene, with a permissible exposure limit (PEL) of 1 ppm and vapor pressure of 75 mm is more dangerous than a spill of chlorobenzene, with a PEL of 75 ppm and a vapor pressure of 8 mm. The benzene is more toxic, and is more likely to volatilize into the air due to its greater vapor pressure.

In many cases economic factors enter into determination of acceptable risk. The Environmental Protection Agency allows a contractor to choose a cancer risk from materials remaining onsite after a cleanup of between 1 in 10,000 people and 1 in 10 million people. The figure of 1 in 1 million is commonly used when deciding on cleanup methods; however, at a Superfund site in Rhode Island a cancer risk of 1 in 100,000 was selected and the decision was described as a "compromise to reduce cleanup costs" by the Office of Technology Assessment of the United States Congress, who called the remedial plan "an excellent interpretation of cost-effectiveness for making technology choices." Cost in money is a consideration which is often balanced with cost to health.

CHEMICAL HAZARDS

Objectives

At the end of this section, the reader should:

- Understand the nature of atoms, molecules, ions and compounds.
- Be able to define and give examples of acids and bases.
- Understand and be able to apply the concept of pH.
- Comprehend the concept of chemical reactions.
- Comprehend atomic number and be able to use the periodic table.
- Understand the difference between ionic bonding and covalent bonding.
- Understand the nature of organic chemical compounds.
- Appreciate the special hazards presented by heavy metals, strong oxidizing agents, and organic monomers.
- Most importantly, be able to apply basic chemistry knowledge to safe practices in working with chemicals.

Basic Chemistry

Chemistry is the study of matter and its transformations. Scientists define matter as anything having mass and occupying space. The matter in the universe, and

HAZARD RECOGNITION   43

therefore on our planet, is in the form of different elements. Each *element* has a specific name and a specific organization into atoms. There are 92 naturally occurring elements and all the things on earth are made up of different arrangements of these 92 types of atom.

Atoms of all the various elements are constructed from fundamental particles: *protons, electrons,* and *neutrons*. Protons and neutrons are found in the nucleus at the center of the atom, while electrons occur in orbits around the nucleus. Table 4-1 summarizes the most important properties of these particles.

In Fig. 4-1, on the left is an atom of *hydrogen*. It has one proton in its nucleus and one electron (dark dot) orbiting around the nucleus at some distance. Notice that the number of positively charged particles in the nucleus (protons) equals the number of negatively charged particles (electrons) orbiting about the nucleus. The atom, therefore, is neutral overall, without charge.

A *helium* atom is shown on the right. In its nucleus are two protons and two neutrons, and there are two electrons in orbit. Since there are two positively charged particles and two negatively charged particles, this atom is neutral overall, too. The heavy neutrons in the nucleus add mass to the nucleus and therefore to the atom, but they contribute no charge.

Each element has an *atomic number* that is equal to the number of protons in the nucleus. Hydrogen has atomic number 1 and helium has atomic number 2.

The *symbol* for Hydrogen is H; helium = He. Every element has a symbol which serves as a shorthand way of identification, and recognizing some common symbols may help protect workers on a hazardous waste site from exposure to dangerous chemical. For example, a worker who recognizes the abbreviation Na (sodium) on a container should be alert for reactivity or perhaps for damage from skin exposure, depending on what other elements are combined with the sodium. The symbols for heavy metals (Hg for mercury, Pb for lead) are warnings of potential toxicity.

A look at elements having higher atomic numbers reveals that the electrons may be represented as occurring in several orbits surrounding the nucleus.

Notice that the examples in Fig. 4-2 reveal that the electron shell (orbit) nearest the nucleus becomes filled with two electrons. Each of the next two shells is considered full with eight electrons. In both these examples, these atoms have the same number of orbiting electrons as nuclear protons. Because

Table 4-1. Properties of Atomic Particles.

Particle	Mass ("Weight")	Charge	Location
Proton	Heavy (1)	Positive	Nucleus
Electron	Light (0)	Negative	Orbit
Neutron	Heavy (1)	Neutral	Nucleus

44 WORKER PROTECTION DURING HAZARDOUS WASTE REMEDIATION

Fig. 4-1. Structure of two atoms.

Cl (chlorine) has 17 protons its atomic number is 17 and Na (sodium), with 11 protons, has an atomic number of 11. (See the Periodic Table to confirm.)

Each element is in its most stable condition when its outermost electron shell is filled. This fact is the basis for chemical reactions. Considering the two atoms above, if sodium were to lose its outermost electron and if chlorine were to gain an electron in its outermost shell, each would be in stable configurations, as shown in Fig. 4-3.

But notice that the sodium atom's total negatively charged electrons will now be one less than the number of positively charged protons in the nucleus; consequently the resulting particle has a positive charge of +1 and is called a *sodium ion*. Similarly, the chlorine atom now has one more electron than protons and bears a charge of −1, and is a *chloride ion*. The positively charged sodium ions and negatively charged chloride ions arrange themselves into a structure; these are crystals of the *compound* NaCl, or common table salt. This example shows how *ionic chemical reactions* occur by the donation of electrons from atoms of one element to atoms of another element, forming chemical *compounds*.

Atoms are stable when their electron shells are full. Why does this matter to a hazardous waste worker? Because unfilled shells lead to unstable (we could say "unsatisfied") atoms on the prowl for another atom to share with or steal

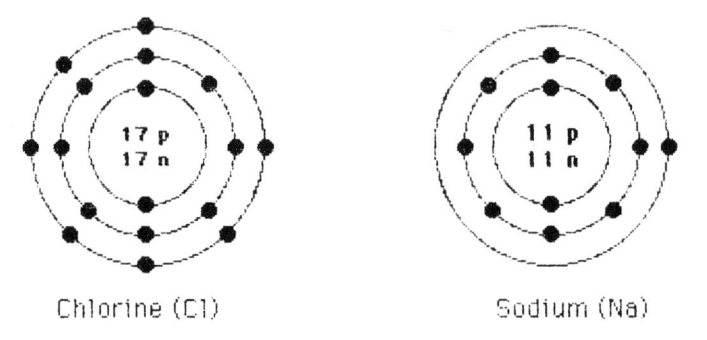

Fig. 4-2. Structure of the chlorine and sodium atoms.

HAZARD RECOGNITION   45

Chloride ion (Cl⁻)   Sodium ion (Na⁺)

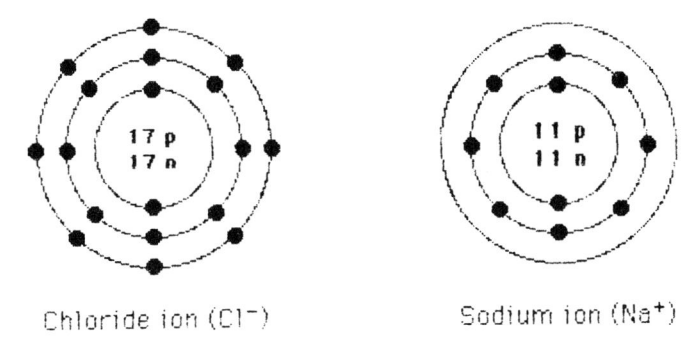

Fig. 4-3. Chlorine and sodium ion structures.

electrons. This unsatisfied atom may cause trouble in the form of a chemical reaction which releases heat or toxic chemicals, sometimes even violently.

The *physical properties* of compounds are often different from those of the elements from which they are made. NaCl, for example, is a solid, while chlorine is a gas and sodium is a metal. The toxic properties may also be very different: sodium metal can be dangerously reactive and chlorine gas is toxic, but table salt is relatively harmless. The resulting NaCl is called a *molecule* because it is composed of more than one atom. Molecules may be made up of *dissimilar* atoms, as in the case of NaCl, or of *similar* atoms, as in the case of a molecule of chlorine gas, Cl_2, composed of two atoms of chlorine.

Besides ionic chemical reactions there is one other major form of reaction between atoms. It is called *electron sharing*, or *covalent bonding*, and occurs in the compounds of carbon, forming *organic* chemicals.

The element carbon (atomic number = 6) has 4 electrons in its outermost (second) shell. Hydrogen, as you recall, has only one electron in its outermost (first) shell. If the single electron of each of four hydrogen atoms is shared with each of carbon's four orbital electrons, there is now a total of eight electrons orbiting about each carbon nucleus, and two electrons orbiting about each hydrogen nucleus. Each *pair* of shared electrons forms one *covalent bond*. Therefore, we could say that there are four covalent bonded hydrogen atoms in each methane molecule.

In Fig. 4-4, we can see that each atom is "satisfied" because its outer shell is filled. Each hydrogen shares a carbon atom, and so has 2 (one of its own and one of carbon's), which fills the only shell it has. Carbon needs 8 electrons to fill its shell, since it has 2 shells. It is sharing 4 with the hydrogen atoms, and so is full.

The *Periodic Table of the Elements* organizes the elements by their atomic number and identifies them by symbol. Vertical columns contain related elements that occur in the same group or "family."

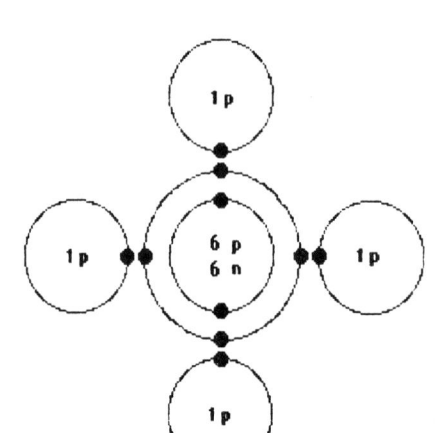

46 WORKER PROTECTION DURING HAZARDOUS WASTE REMEDIATION

The single electrons of four Hydrogen atoms are each shared with an electron in Carbon's outer orbit, forming four covalent bonds, and the molecule methane, CH$_4$.

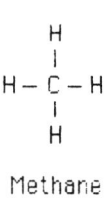

H
|
H – C – H
|
H

Methane

Fig. 4-4. Illustration of electrons in covalent bonds.

For example, the *halogen* group includes fluorine (F), chlorine (Cl), bromine (Br), and iodine (I). Each of these elements has seven electrons in its outermost orbit, so each reacts similarly with other atoms in forming compounds. You saw above that NaCl results when sodium (Na) contributes its single outermost electron to chlorine (Cl); NaBr (sodium bromide) and NaI (sodium iodide) are formed when bromine and iodine react with sodium (Na) in the same way. Much of the concern about hazardous chemicals deals with organic substances (halogenated organic compounds) containing atoms from this group. Since a halogen lacks 1 electron in its outer shell, it is very reactive with many other atoms and compounds.

The *alkali metals*, including lithium (Li), sodium (Na), and potassium (K) form another group of elements which react similarly in ionic chemical reactions. Thus, LiCl (lithium chloride) and KCl (potassium chloride) are compounds that are "salts" of the alkali metals, and are formed by the same sort of electron donations as in the example of NaCl described above. You have probably figured out that atoms of lithium and potassium, like sodium atoms, have one electron in the outermost shells.

Lithium, sodium, and potassium are elements in this group that are reactive, as pure metals, in air and water (see Table 4-2). For example, when sodium metal is dropped into water, a strong reaction occurs, releasing hydrogen:

$$2\,Na^0 + 2\,H_2O \rightarrow 2\,Na^+ + 2\,OH^- + H_2$$

This reaction produces enough heat to cause the hydrogen gas to react explosively with the oxygen in the air. On one completed cleanup, the first priority was to safely remove drums of sodium metal that were stored there, without allowing the contents to come into contact with water.

When sodium is combined with other atoms, however, as in NaCl (table salt), it is not reactive when dropped into water:

$$NaCl + H_2O \rightarrow Na^+ + Cl^- + H^+ + OH^-$$

Alkali metal elements may also be combined with hydrogen to form saltlike compounds called *hydrides*. These compounds are extremely reactive with water, releasing hydrogen:

$$LiH + H_2O \rightarrow Li^+ + OH^- + H_2$$

(lithium hydride) + (water) → (lithium hydroxide salt) + (hydrogen gas)

The word *hydride* will appear in the name of such a compound.

Any reaction which produces hydrogen gas is dangerous, as hydrogen gas is extremely flammable and explosive. Table 4-2 lists a number of chemicals which react with water to release hydrogen.

The *noble gases* (helium, neon, argon, krypton, etc.) are essentially non-reactive elements because they have filled outermost electron shells. Sometimes they are described as *inert*.

Heavy metals are elements that are frequently toxic to humans. Some, like copper (Cu) and zinc (Zn), are necessary in very low concentrations for life activities; most are industrial hazards. The heavy metals include: cobalt (Co), Nickel (Ni), silver (Ag), cadmium (Cd), chromium (Cr), and mercury (Hg). The last three, Cd, Cr, and Hg, are the most common toxic metals in industrial use.

Cleanup sites on former metal plating operations or battery plants may be contaminated with one or more heavy metals. Chemical plant operations which manufactured agricultural products may also.

Table 4-2. Water Reactive Chemicals.

Acetyl bromide	Phosphorus trichloride
Acetyl chloride	Potassium
Aluminum borohydride	Potassium peroxide
Calcium	Potassium hydroxide (solid)
Calcium oxide	Rubidium
Diborane	Sodium
Sodium peroxide	Sodium amide
Dimethyl sulfate	Sodium hydride
Lithium	Sodium hydroxide (solid)
Phosphorus oxychloride	Sulfur Chloride

Sources: (Meyer, 1979; Hammer 1985).

Acids and Bases

Acids. The formation of ions by gain or loss of electrons by an atom is related to the concept of acids and bases, too. An acid is defined as any substance that liberates *hydrogen ions* in solution.

A hydrogen atom, you recall, is composed of a single proton in the nucleus and a single electron orbiting around the nucleus. When hydrogen atoms *ionize* (give up an electron) what remains is the positively charged proton. Acids are simply solutions containing large numbers of protons (hydrogen ions). The higher the concentration of hydrogen ions, the more concentrated the acid.

Scientists use the pH scale (below) to describe the acidity or alkalinity (the term basicity is not commonly used) of a solution. Pure water is neutral (neither acidic nor basic) and has a pH of 7.0; the number of H^+ and OH^- ions is equal, since a molecule of HOH (water), contributes one of each ion when it dissociates.

Solutions with a pH of less than 7.0 are acidic; those with a pH greater than 7.0 are basic. The pH scale is logarithmic, so a solution with a pH of 3.0 is 10 times as acidic as one with a pH of 4.0; the 3.0 solution is 100× more concentrated in H^+ ions than a pH 5.0 solution. Fig. 4-5 shows the pH levels of some common solutions.

Organic acids like *acetic acid* tend to be *weak acids*. In water solution, acetic

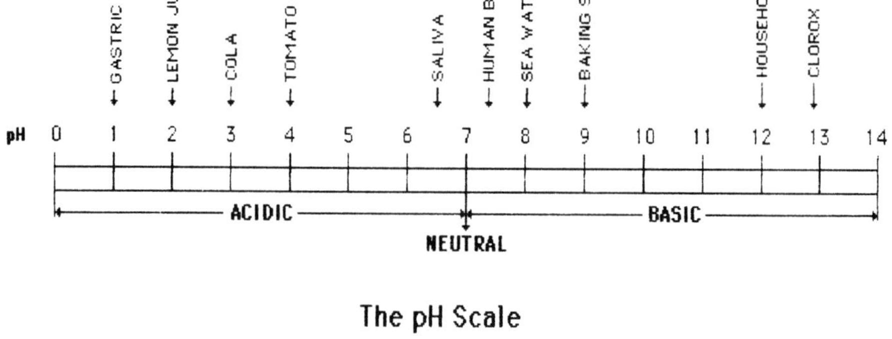

Ph of Common Solutions

The pH Scale

Fig. 4-5. Common solutions and their *p*H levels.

acid (CH_3COOH) *partially* ionizes into the acetate ion CH_3COO^- and hydrogen ions (H^+). Because they ionize to a lesser degree and contribute fewer hydrogen ions to the solution, organic acids are usually weak acids.

A *strong acid* such as *hydrochloric acid* (HCl) is actually a water solution of hydrogen ions (H^+) and chloride ions (Cl^-) *Sulfuric acid* (H_2SO_4) is, similarly, an aqueous solution of hydrogen ions (H^+) and sulfate ions (SO_4^{--}). These inorganic or *mineral acids* are strong acids because the HCl and H_2SO_4 molecules are completely ionized.

Hydrochloric acid, which has also been called *muriatic acid*, damages skin and eyes; concentrated solutions cause severe skin burns and permanent eye damage. It is used in many different industries such as ore refining, in some types of food production, and pickling and cleaning of metal products, and may remain at a cleanup site where one of these industries was located or at an abandoned disposal site.

Table 4-3 lists some common acids.

Experience has shown that it is possible to be badly injured by both weak and strong acids; special caution is called for when dealing with any acid. Concentrated solutions are highly *corrosive* and will attack materials and tissue; if spilled on skin, flush with lots of water.

Acid Pointers.   Sulfuric, Nitric, Hydrochloric, Acetic:

- These acids are highly soluble in water.
- When mixing with water, follow the AAA rule—Always Add Acid. Add the acid slowly to the water.
- Sulfuric and nitric acids are strong oxidizers; they should not be stored or mixed with *any* organic material.
- Concentrated (glacial) acetic acid is extremely flammable. Its vapors form explosive mixtures in air. It is dangerous when stored with any oxidizing material, such as nitric and sulfuric acids, peroxides, sodium hypochlorite, etc.
- The concentrated vapors of all the acids above are extremely harmful.

Table 4-3.   Acids.

Strong Acids	Weak Acids
Perchloric	Boric
Hydrochloric	Hydrocyanic
Sulfuric	Carbonic
Nitric	Acetic

Bases. Bases (also called *caustics*) are compounds which release hydroxyl ions (OH⁻) in solution. Potassium hydroxide (KOH), for example, dissociates (ionizes) to liberate potassium ions (K⁺) and hydroxyl ions (OH⁻) in a water solution. Bases like KOH and NaOH (sodium hydroxide) are powerful bases, or caustic compounds. Organic bases such as *amines* tend to ionize to a lesser degree and are generally weak bases.

Experience has shown that it is possible to be badly injured by both weak and strong bases; special caution is called for when dealing with any base solution.

Concentrated solutions of bases are highly *corrosive*. Bases penetrate the skin more deeply than most acids and are a greater hazard; if spilled on skin, flush with lots of water.

Base Pointers. Sodium Hydroxide, Ammonium Hydroxide, Calcium Hydroxide (Slaked Lime), Calcium Oxide (Quick Lime):

- These bases are highly soluble in water.
- When mixed with water, they generate large amounts of heat—especially true of sodium hydroxide and calcium oxide.
- Ammonium hydroxide can release ammonia gas when mixed with other strong bases; if mixed with chlorine compounds it can release deadly chlorine gas.

Chemical Reactions Involving Acids and Bases. Examples of some important chemical reactions of these substances include:

- Neutralization Reactions (Acid + Base). Example:

$$H^+ + Cl^- + Na^+ + OH^- \rightarrow H_2O + Cl^- + Na^+$$
$$(acid) + (base) \rightarrow (water) + (salt)$$

- Oxidation of a Substance by an Acid. Example:

$$Zn^0 + 2\,HCl \rightarrow Zn^{++} + 2\,Cl^- + H_2$$
$$(metallic\ zinc) + (acid) \rightarrow (salt) + (hydrogen\ gas)$$

- Reaction of a Base with a Metal. Example:

$$2\,Al + 6\,NaOH \rightarrow 2\,Na_3AlO_3 + H_2$$
$$(metallic\ aluminum) + (base) \rightarrow (salt) + (hydrogen\ gas)$$

Reactions Involving Strong Oxidizing Agents

Special mention should be made regarding molecules that are *strong oxidizing agents*. They include some acids, some elements, like fluorine and chlorine, and many compounds that contain oxygen. These substances participate in chemical reactions that usually release a large amount of energy, often violently. Because of this property, oxidizers should never be stored with any other substance. Many strong oxidizing agents can be recognized because the prefix "per" occurs in their name (*perchlorate, periodate, peroxides, permanganate*) and they often end in "ate" (*chlorate, dichromate*). Fig. 4-6 shows several examples of oxidizing agents and their relative strengths.

Practical Considerations

When dealing with reactive materials in the field, it is imperative to determine:

- How toxic is the corrosive material? Is it an irritant or does it cause severe burns?
- What kind of structural damage does it do, and what other hazards can it lead to? For example, will it destroy containers holding other hazardous materials, releasing them into the environment?
- What quantity of corrosive material is involved? What areas contain these materials?
- Is the corrosive concentrated? If so, how long has it been there? Corrosives with extreme pH levels will leave their containers given enough time.
- Are there potentially incompatible substances in the area?

Organic Chemicals

The compounds of carbon formed by covalent bonding make up a wide array of modern chemicals. Fig. 4-7 illustrates the magnitude of the organic chemical industry.

A few ordinary substances are represented below in their organic or covalently bonded form. Outer shell electrons are represented by x's and o's.

H⚬ₓH	Hydrogen Gas (H_2)	H—H
H⚬O⚬H	Water (H_2O)	H—O—H
O⚬ₓO	Oxygen Gas (O_2)	O=O
O⚬ₓC⚬ₓO	Carbon Dioxide (CO_2)	O=C=O
H⚬ₓC⚬ₓN	Hydrogen Cyanide (HCN)	H—C≡N
H⚬N⚬H with H	Ammonia (NH_3)	H—N—H with H

RELATIVE STRENGTHS OF OXIDIZING AGENTS

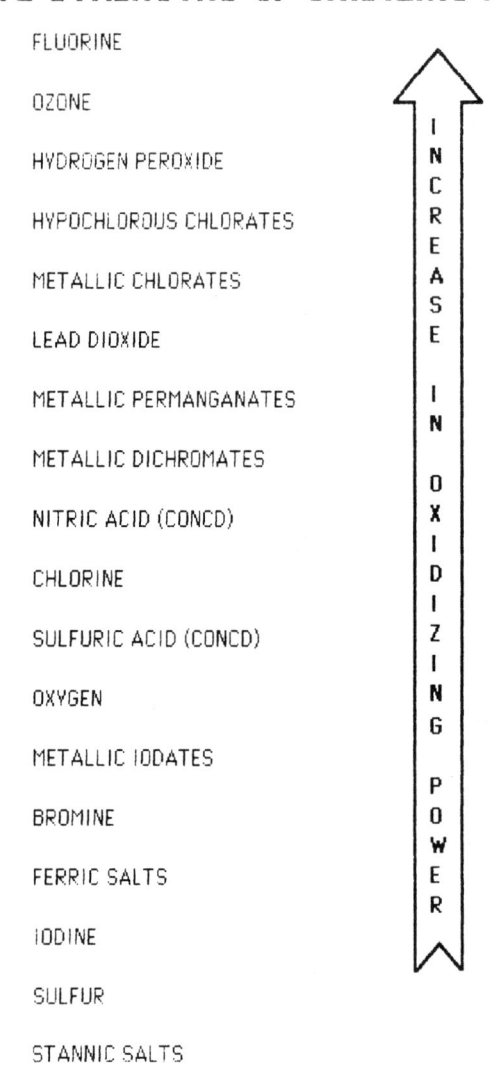

FLUORINE
OZONE
HYDROGEN PEROXIDE
HYPOCHLOROUS CHLORATES
METALLIC CHLORATES
LEAD DIOXIDE
METALLIC PERMANGANATES
METALLIC DICHROMATES
NITRIC ACID (CONCD)
CHLORINE
SULFURIC ACID (CONCD)
OXYGEN
METALLIC IODATES
BROMINE
FERRIC SALTS
IODINE
SULFUR
STANNIC SALTS

↑ INCREASE IN OXIDIZING POWER

Fig. 4-6. Examples of oxidizers and their relative strengths.

The simplest carbon-based organic chemicals are ones containing only a few carbons, together with hydrogen atoms. These are called *hydrocarbons*, and many of these compounds are familiar to us in everyday life. Some representative simple hydrocarbons are shown below:

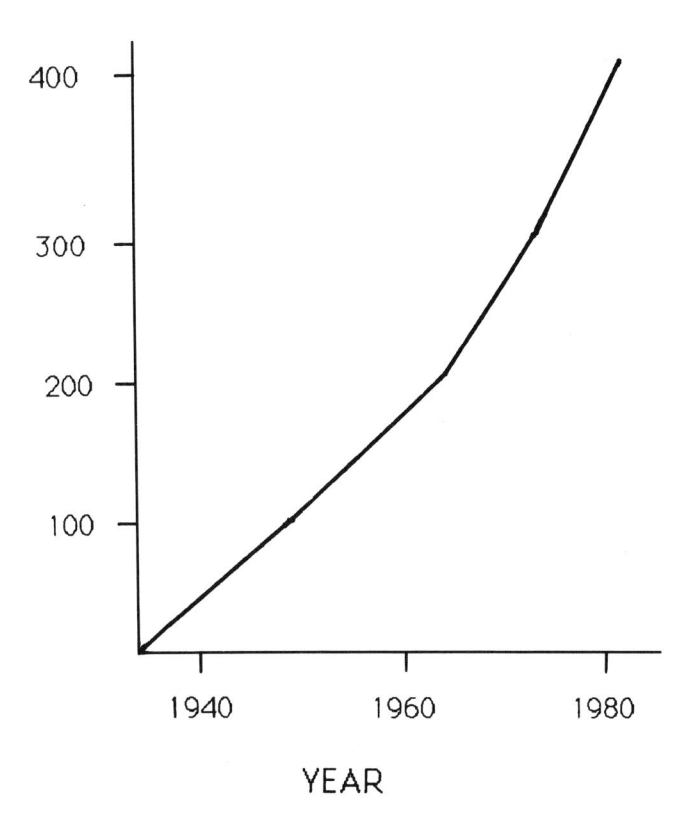

Fig. 4-7. History of organic chemical production.

```
        H              H   H           H   H   H           H   H   H   H
        |              |   |           |   |   |           |   |   |   |
      H—C—H         H—C—C—H         H—C—C—C—H         H—C—C—C—C—H
        |              |   |           |   |   |           |   |   |   |
        H              H   H           H   H   H           H   H   H   H
   Methane (CH₄)    Ethane (C₂H₆)     Propane (C₃H₈)         Butane (C₄H₁₀)
```

```
         H   H
         |   |
         C=C                           H—C≡C—H
         |   |
         H   H
     Ethylene (C₂H₄)                Acetylene (C₂H₂)
```

54 WORKER PROTECTION DURING HAZARDOUS WASTE REMEDIATION

In addition to the simple hydrocarbons, organic compounds exist in which one or more hydrogen atoms are replaced with certain *functional groups*. As can be seen in Table 4-4, similar compounds have similar functional groups present: $-OH$ in alcohols, $-COOH$ in acids, $-NH_2$ in amines, etc. These *classes* of organic compounds generally have similar properties and participate in the same kinds of chemical reactions.

Many organic compounds are "ring" compounds; a large group of these are called *aromatic* compounds (see Table 4-5), and they do frequently have a characteristic aroma, or odor.

Several groups of organic chemicals deserve special mention. They are discussed below.

Polymers. Polymers are very large molecules made up of many smaller, usually identical, units (monomers) that are chemically joined together. For example:

vinyl chloride molecules → polyvinyl chloride (PVC) molecules
(2 carbons each)   (thousands of carbons each)

Much of today's chemical industry is devoted to making polymers (polyethylene, polystyrene, polyvinyl chloride, nylon, orlon, etc.) and therefore large quantities of monomers (some extremely hazardous) are produced.

Table 4-4.
Some Organic Functional Groups

Name	Formula	Name	Formula
Methyl	$-CH_3$	Alcohol, or Hydroxy	$-OH$
Ethyl	$-C_2H_5$	Aldehyde	$-C(=O)H$
Propyl	C_8H_7	Ketone	$-C(=O)-$
Butyl	$-C_4H_9$	Acid	$-C(=O)OH$

Table 4-4. (Continued)
Some Organic Functional Groups

Name	Formula	Name	Formula
Phenyl	$-C_6H_5$	Ether	$-O-$
Chloro	$-Cl$	Ester	$-C(=O)-O-$
Bromo	$-Br$	Nitrile	$-C\equiv N$
Iodo	$-I$	Amino	$-NH_2$
Fluoro	$-F$	Nitro	$-NO_2$

Methyl Alcohol / Methanol / Wood Alcohol — CH_3OH

Ethyl Alcohol / Ethanol / Grain Alcohol — CH_3CH_2OH

n-Propyl Alcohol / Propanol — $CH_3CH_2CH_2OH$

n-Butyl Acohol / Butanol — $CH_3CH_2CH_2CH_2OH$

Formaldehyde

Acetaldehyde

Propionaldehyde

Butyraldehye

Formic Acid

Acetic Acid

Propionic Acid

Butyric Acid

Table 4-4. (Continued)
Some Organic Functional Groups

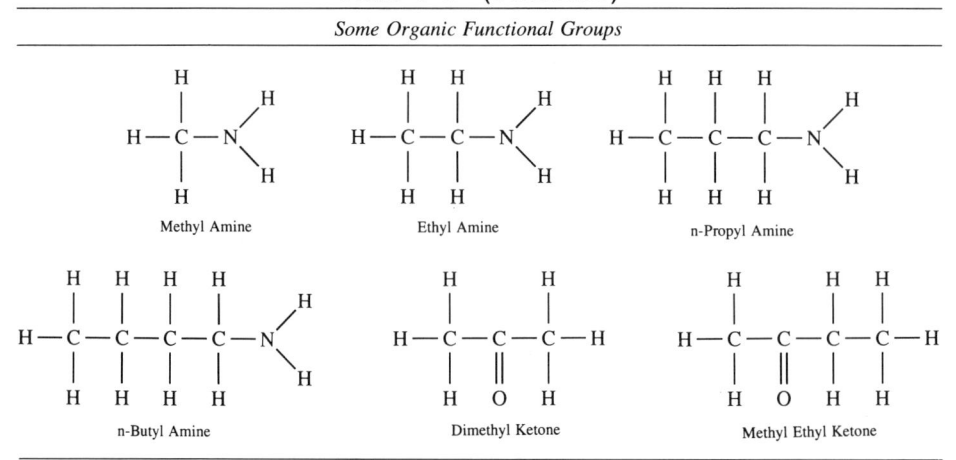

| Methyl Amine | Ethyl Amine | n-Propyl Amine |
| n-Butyl Amine | Dimethyl Ketone | Methyl Ethyl Ketone |

Pesticides. These include insecticides, herbicides, and fungicides. The *chlorinated hydrocarbon* pesticides (DDT, Chlordane, Mirex) are relatively insoluble compared to the *carbamate* (SEVIN) and *organophosphate* (Malathion, Parathion) pesticides. The organophosphates are strongly toxic. The most widely known herbicides are 2,4-D and 2,4,5-T; the latter is also known as *Agent Orange* because of its widespread use as a defoliant in Vietnam. Although the use of many of these substances has been banned or restricted, they have represented a major activity of the agricultural chemical industry in the past and their occurrence is not unusual at many hazardous waste sites.

Organic Peroxides. Like the inorganic strong oxidizing agents, this group includes substances that are serious fire and explosion hazards. They are typically sensitive to heat and shock; consequently, concentrated solutions are usually diluted before handling. Some representative organic peroxides are: methyl ethyl ketone peroxide; benzoyl peroxide; and peracetic acid. Note that, as with the inorganic oxidizers, the prefix "per" often occurs in the chemical name.

FIRE, EXPLOSION, AND CHEMICAL REACTION CONTROL

Objectives

- Introduce a basic chemistry vocabulary.
- Explain the physical properties which best characterize dangerous substances.

Table 4-5. Examples of Aromatic Compounds.

All three of the above represent the ring compound Benzene (C_6H_6)

Some other aromatic (ring) compounds:

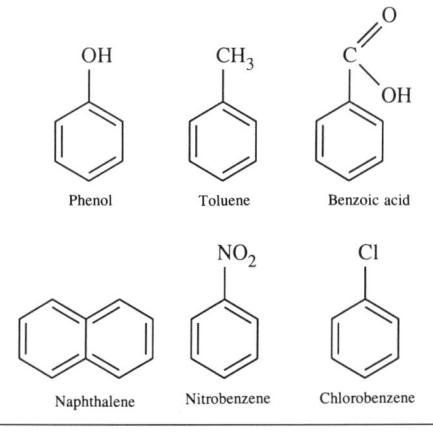

Phenol   Toluene   Benzoic acid

Naphthalene   Nitrobenzene   Chlorobenzene

- Classify different types of chemical reactions, fires, and explosions.
- List some of the effects of fire on personnel.
- Describe rationale and general methodology behind compatibility staging, fingerprinting, and bulking operations.

Working around chemicals requires knowledge of the "nuts and bolts" of chemicals' properties and reactivities to operate safely. This section of Hazard Recognition will introduce the basics of these specifics. For quick reference, a short list of definitions follows. A more extensive glossary of terms used throughout the book is located in the back. Knowing some of the following terms will greatly enhance an understanding of some basic concepts that will be detailed.

Definitions

Autoignition temperature (AIT);	The lowest temperature at which a flammable mixture will burn without the application of an outside spark or flame.
Brisance	The heat or shock of an explosion.
Catalyst	A chemical which, without changing itself, causes a chemical reaction to proceed faster.
Chemical Reaction	The conversion of one chemical substance to another.
Combustible	A liquid that becomes flammable when heated above 100°F (EPA).
Explosion	A rapid chemical transformation which suddenly undergoes a chemical reaction with the simultaneous production of large quantities of heat and gases (those gases being CO, CO_2, N_2, O_2 and steam).
Fire	Active burning from the combustion of fuel and oxygen in the presence of heat.
Flammability	A measure of combustibility.
Flammable or Explosive Limits (LEL & UEL)	For flammables, combustibles and explosives, the range of vapor concentrations (in air) in which they will burn is defined by their lower and upper limits (LEL and UEL, respectively). Below LEL a mixture is too "lean" to burn. Above the UEL a mixture is too "rich" to burn.
Flashpoint	The minimum temperature at which a flammable liquid gives off enough vapors to ignite.
Oxidizing Agent	A material which gives off oxygen in a chemical reaction (gains electrons).
Polymerization	A chemical reaction in which two or more molecules join to form a larger more complex molecule, frequently in a chain structure.
Reactivity	The ability of a material to undergo a reaction with the release of energy, usually heat.
Reducing Agent	A material which accepts oxygen in a chemical reaction (loses electrons).

Solubility	The tendency of a material to dissolve in water or other solvents.
Volatile Percent	The fraction (by weight or volume) of solvent or evaporable content in a mixture.
Vapor Pressure	The pressure exerted by a vapor at a given temperature.

Basic Concepts

At an uncontrolled hazardous waste site, or at an incident involving hazardous waste releases, personnel may be exposed to various dangers including smoke inhalation, oxygen deficiency, or the generation of toxic gases such as phosgene, hydrogen cyanide, hydrogen chloride, and metal fumes. In order to take fast, careful, and educated actions to reduce risks, it is important to know the fundamentals of these dangers and their capacity to interact. This section introduces the reader to facts about the control of fires, explosions, and chemical reactions that may take place at uncontrolled waste sites.

First, an understanding of the vocabulary of reactions will help us. The *explosive range* of a flammable liquid (as a vapor in air) and its *lower explosive limit* (LEL) are perhaps the two most important characteristics to check when trying to determine a substance's flammability. At what concentration of its vapors in air is a chemical likely to catch fire from a spark? The limit and range are known for all chemicals and can be looked up. Those substances with a low LEL and wide explosive range must be considered dangerous fire or explosion hazards.

Another critical physical characteristic of a substance is its *flashpoint*. Those substances with low flashpoint temperatures are susceptible to igniting readily, and appropriate caution must be exercised when dealing with them. Note that substances with *autoignition temperatures* are even more dangerous, because they don't need an ignition source to combust. *Combustion* is the chemical reaction which leads to burning and explosions; whether a substance is rated as combustible, flammable, or explosive simply refers to the speed, or rate of combustion. There are many classification schemes for flammability and explosivity. Generally they are based on the flashpoint and rate of combustion, respectively.

Two other physical properties help in understanding the potential hazards of substances. They are *solubility* (or miscibility), and *vapor pressure*. Solubility is important with hazardous substances in that a substance can remain concentrated and hazardous if it will not mix with water, and conversely a substance can be hard to detect or clean if it mixes with water and is dispersed by it. The formation of soluble compounds in aqueous solutions can also become the "driving force" for many chemical reactions. Knowledge of substances having high vapor pressure is essential at uncontrolled sites, as the potential exists for these substances to rapidly release and diffuse over a large area. This can cause a sudden choking, contamination, or combustion hazard.

Thinking about these physical properties and learning to understand them will help workers to prevent exposure to an uncontrolled fire, explosion, or release of hazardous vapors.

Fire and Explosion

We have seen what physical characteristics make substances susceptible to combustion: their explosive range, LEL, etc. Now we shall look at what kinds of fires there are, how they work, and what harmful effects can result.

The most common classification of fires is based on the material that burns, and is broken down into four main categories:

- *Type A* fires, with an "A" shown in a green triangle (see Fig. 4-8) refers to fires which use cellulose matter as fuel: wood, paper, cotton, etc.
- *Type B* fires, with a "B" shown in a red square (see Fig. 4-9) refers to fires which use flammable liquids (or their vapors) as fuel: gasoline, paint, waste solvents, etc.
- *Type C* fires, with a "C" shown in a blue circle (see Fig. 4-10) refers to fires which originate from electrical sources (even if other materials burn with them). They may be caused by faulty wiring, switches, burned out motors, etc.
- *Type D* fires, with a "D" shown in a yellow star (see Fig. 4-11) refers to fires which use reactive metals as fuel: sodium, potassium, magnesium, etc.

A fifth category involves highly reactive chemicals such as rocket propellants and strong oxidizers. These symbols can be seen on all common fire extinguishers, and tell a lot about how to fight a fire.

Regardless of what material is burning, all fires require three basic components for combustion to take place: *oxygen, fuel,* and *heat*. The relationship between these components is well illustrated by a phase diagram called the *fire triangle* (see Fig. 4-12). It shows that without a minimum amount of each component, a fire cannot be sustained (Hammer 1985; Meyer 1977).

Various Flammables Classifications

EPA Classifications (from RCRA).

Ignitable Wastes	Hazardous	Wastes with a flashpoint less than 140°F (40 CFR 261.21).

Fig. 4-8. Type A fire symbol.

Fig. 4-9. Type B fire symbol.

Fig. 4-10. Type C fire symbol.

Fig. 4-11. Type D fire symbol.

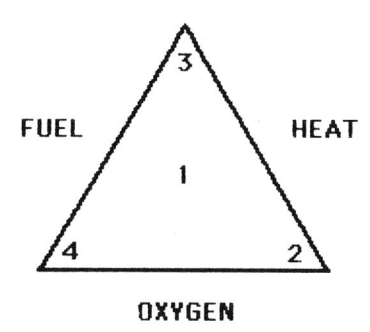

Each side of the triangle represents one of the necessary elements of a fire. The optimal situation, position number 1, is the best fuel-to-oxygen ratio, with sufficient heat to ignite the fuel and support its combustion. Each corner illustrates the removal of one component: in number 2 there is insufficient fuel (concentrations of vapors below the Lower Explosive Limit), in number 3 there is not enough oxygen (concentrations of vapors above the Upper Explosive Limit), and in number 4 the heat source is not adequate. A fire can be defined as a self-sustaining, flaming combustion (EPA 1985).

Fig. 4-12. The fire triangle.

DOT Classification of Flammable Liquids.

Flammables	Flashpoint below 100°F.
Combustibles	Flashpoints between 100°F and 200°F.
Nonflammables	Flashpoint greater than 200°F (49 CFR 173.115).

NFPA Classification of Flammable Liquids.

Class IA	Liquids with flashpoints below 73°F and boiling points below 100°F. An example of a Class IA flammable liquid is *n*-pentane (NFPA Diamond: 4).
Class IB	Liquids with flashpoints below 73°F and boiling points at or above 100°F. Examples of Class IB flammable liquids are benzene, gasoline, and acetone (NFPA Diamond: 3).
Class IC	Liquids with flashpoints at or above 73°F and below 100°F. Examples of Class IC flammable liquids are turpentine and *n*-butyl acetate (NFPA Diamond: 2).
Class II	Liquids with flashpoints at or above 100°F but below 140°F. Examples of Class II flammable liquids are kerosene and camphor oil (NFPA Diamond: 2).
Class III	Liquids with flashpoints at or above 140°F but below 200°F. Examples of Class III liquids are creosote oils, phenol, and naphthalene. Liquids in this category are generally termed combustible rather than flammable (NFPA Diamond: 2) (NFPA, 1986).

At a hazardous materials incident, the fuel and air (oxygen) are not easily controlled. Consequently, while working on site where a fire hazard may be present, the concentration of combustible gases in air must be monitored, and any potential ignition source must be kept out of the area. The properties that make flammable substances the most dangerous are:

- Low flashpoints or autoignition temperatures.
- Substances requiring little oxygen to support combustion.
- Substances with low LEL's and wide flammable (explosive) ranges.

Explosives are substances that combust at such a high rate that they pose hazard such as:

- Physical destruction due to shock waves, flying objects, and heat.
- Initiation of secondary fires or the creation of flammable conditions.
- Release of toxic or corrosive compounds into the surrounding environment.

There are separate classifications for explosives for chemists, the military, the D.O.T., and others. The most basic classification refers to the rate of combustion, and involves two types:

- High, or detonating explosives, are characterized by a shock wave which is faster than the speed of sound—as high as 4 miles per second. They may be detonated by shock, heat, or friction. High explosives are further classed into primary and secondary types, based on their ease of detonation.
- Low, or deflagrating explosives, which have slower combustion rates, but may be as dangerous. They generally need a booster to explode and are not as sensitive to shock, heat, or friction (Meyer 1985).

NFPA Hazard Identification System

Health Hazard (BLUE).

Rank Number	Description	Examples
4	Materials that on very short exposure could cause death or major residual injury even though prompt medical treatment was given.	Acrylonitrile, bromine, parathion
3	Materials that on short exposure could cause serious temporary or residual injury even though prompt medical treatment was given.	Aniline, sodium hydroxide, sulfuric acid
2	Materials that on intense or continued exposure could cause temporary incapacitation or possible residual injury unless prompt medical treatment was given.	Bromobenze, pyridine, styrene
1	Materials that on exposure would cause irritation but only minor residual injury even if no treatment was given.	Acetone, methanol
0	Materials that on exposure under fire conditions would offer no hazard beyond that of ordinary combustible material.	

Flammability Hazard (RED).

Rank Number	Description	Examples
4	Materials that (1) rapidly or completely vaporize at atmospheric pressure and normal ambient temperatures and burn readily or (2) are readily dispersed in air and burn readily.	1, 3-Butadiene, propane, ethylene oxide
3	Liquids and solids that can be ignited under almost all ambient temperature conditions.	Phosphorus, acrylonitrile
2	Materials that must be moderately heated or exposed to relatively high ambient temperatures before ignition can occur.	2-Butanone, kerosene
1	Materials that must be preheated before ignition can occur.	Sodium, red phosphorus
0	Materials that will not burn.	

Reactivity Hazard (YELLOW).

Rank Number	Description	Examples
4	Materials that in themselves are readily capable of detonation or of explosive decomposition or reaction at normal temperatures and pressures.	Benzoyl peroxide, picric acid, TNT
3	Materials that (1) in themselves are capable of detonation or explosive reaction but require a strong initiating source or (2) must be heated under confinement before initiation or (3) react explosively with water.	Diborane, ethylene oxide, 2-nitropropadene
2	Materials that (1) in themselves are normally unstable and readily undergo violent chemical change but do not deto-	Acetaldehyde, potassium

Rank Number	Description	Examples
	nate or (2) may react violently with water or (3) may form potentially explosive mixtures with water.	
1	Materials that in themselves are normally stable but which can (1) become unstable at elevated temperatures or (2) react with water with some release of energy but not violently.	Ethyl ether, sulfuric acid
0	Materials that in themselves are normally stable, even when exposed to fire, and that do not react with water.	

Special Information (WHITE). The white block is designated for special information about the chemical. For example, it may indicate that the material is radioactive by displaying the standard radioactive symbol, or unusually water-reactive by displaying a large W with a slash through it. For a more complete discussion of these various hazards, consult the NFPA Standard 704 M (NFPA, 1986).

Controlling Chemical Reactions

Considering the vast array of organic chemicals and the appreciable number of inorganics, there are a huge number of possible chemical reactions that may result in fires, explosions, and production of toxic substances. Storage of chemicals should follow rules that prohibit placing and housing incompatible substances in the same vicinity. Prevention of dangerous chemical reactions is one of the most important reasons for segregating apart incompatible materials during cleanup of a waste site.

Chemical reactions may be slow (the oxidation of iron or steel in air, resulting in rust) or rapid (the combustion, or oxidation, of ether, resulting in explosion). A number of factors determine the rate at which any chemical reaction takes place. The most important are:

- The *concentration* of reactants.
- The *temperature*. All chemical interactions are speeded up by an increase in temperature.

- *Pressure*. Just as potatoes cook faster in a pressure cooker, increased atmospheric pressure speeds up other chemical reactions.
- *Catalysts*. Some chemical reactions require the presence of a catalyst, a chemical substance that speeds up the interaction of reacting chemicals but which is not used up or changed in the reaction itself. Small quantities are often effective in promoting a chemical reaction because catalyst is not expended or changed in the reaction. Platinum acts as a catalyst in the oxidation of unburned hydrocarbons in an automobile's catalytic converter.
- *Surface area* of the reactants, at the reaction site. For example, in air a large chunk of coal is combustible; it burns (oxidizes) slowly, releasing heat energy slowly. But the same weight of coal dust particles has a much greater surface area for chemical interaction and is explosive; supported by the O_2 in the air, it oxidizes very rapidly, releasing energy very rapidly. Some catalysts act by increasing the reactive site surface area, thereby speeding the reaction.

Note: Some chemicals may be "shock-sensitive," and their oxidation may be initiated as a result of the energy received when a container is moved, shaken, or dropped. For example, picric acid containers with visible crystals around the cap or stopper are a definite explosive hazard when moved even slightly.

Table 4-6 lists several examples of hazardous chemical reactions.

Table 4-6. Some Hazards Due to Chemical Reactions (Incompatibilities).

Incompatible Substances	Consequences of Mixing
Acid and water	Generation of heat
Hydrogen sulfide and calcium hypochlorite	Fire
Picric acid and sodium hydroxide	Explosion
Sulfuric acid and plastic	Toxic gas or vapor production
Acid and metal	Flammable gas or vapor production
Chlorine and ammonia	Formation of a substance with a greater toxicity than the reactants
Fire extinguisher	Pressurization of closed vessels
Hydrochloric acid and chromium	Solubilization of toxic substances
Ammonia and acrylonitrile	Violent polymerization (EPA 40 CFR 300)

Categorization of Hazardous Wastes by Compatibility

The California Department of Health proposed a classification system based on the most commonly encountered types of wastes. Adverse human health and environmental effects occur upon mixing of certain materials. Injury may result from spontaneous generation of heat, explosion, release of toxic vapors upon mixing, release of toxic vapors during a fire, and release of flammable gases. Table 4-7 lists representative incompatibilities between reactive groups.

Compatibility Staging

Explosions and fires can result from reactions between chemicals not considered as explosives. The generation of toxic gases or the displacement of oxygen can also result. Some substances will react violently when simply exposed to air or water, or may change to become explosive with age or a temperature increase. Because of these hazards, one of the main jobs at uncontrolled waste sites involves the *staging of chemicals that are compatible* with one another. Once this is done, the chemicals can be stored until mixed (bulked) in large containers for removal (transport) to an appropriate disposal site.

Compatibility staging is a systematic set of field "spot" tests done on all

Table 4-7. Group A Must be Separated from Group B or the Indicated Consequence May Occur.

I Group A	*I Group B*
Acetylene sludge	Acidic sludge
Alkaline/caustic liquids	Acidic water
Alkaline cleaner	Battery acid
Alkaline corrosive liquids	Chemical cleaners ("Chromerge")
Alkaline battery fluid	Electrolyte, acid
Caustic waste waters	Etching acid, liquid, or solvent
Lime sludge	Liquid cleaning compounds
Lime waste water	(Muriatic acid)
Spent caustic	Pickling liquor and corrosive
	Acids
	Spent acid
	Spent acids, mixed
	Spent sulfuric acid

Imminent consequences upon mixing: generation of heat, violent reaction.

II Group A	*II Group B*
Asbestos waste	Cleaning solvents
Beryllium wastes	Data processing liquid
Unrinsed pesticide containers	Obsolete explosives
Waste pesticides	Petroleum waste
	Refinery waste
	Retrograde explosives
	Solvents
	Waste oil and other flammable explosive waste

Imminent consequence upon mixing: release of toxic substances during fire or explosion.

Table 4-7. (Continued)

III Group A	III Group B
Aluminum Beryllium Calcium Lithium Magnesium Potassium Sodium Zinc powder and other reactive materials	Any wastes in I Group A or B

Imminent consequence: fire or explosion due to release of hydrogen gas or intense exothermic reaction.

IV Group A	IV Group B
Alcohols Water	Any concentrated waste in I Group A or B Calcium Lithium Metal hydrides Potassium Sodium Thionyl chloride, $SOCl_2$ Sulfuryl chloride, SO_2Cl_2 Phosphorus trichloride Tricholor(methyl)silane and other water reactive wastes

Imminent consequence: contact may generate toxic or flammable gases or cause fire or explosion.

V Group A	V Group B
Alcohols Aldehydes Halogenated hydrocarbons Nitrated hydrocarbons and reactive organic compounds and solvents Unsaturated hydrocarbons	Concentrated wastes from I Group A or B and III Group A wastes

Imminent consequence: violent reaction, fire or explosion.

VI Group A	VI Group B
Spent cyanide and sulfide solutions	I Group B wastes

Imminent consequences: release of toxic cyanide or hydrogen sulfide gases.

Table 4-7. (*Continued*)

VII Group A	VII Group B
Chlorates and other oxidizers	Acetic acids and other organic acids
Chlorine	Concentrated mineral acids
Chlorites	II Group B wastes
Chromic acid	III Group A wastes
Hypochlorites	V Group A wastes and other flammable and combustible wastes
Nitrates	
Nitric acid, fuming	
Perchlorates	
Permanganates	
Peroxides	

Imminent consequence: reaction resulting in fire or explosion (EPA Title 40 CFR 300).

drums or containers of unknowns, and backed up by lab tests performed by chemists (see Table 4-8). These tests give what is called a "fingerprint" of the chemicals so that safe bulking operations can be performed. There are specific tests for radiation, flammability, corrosives, peroxides, sulfides, cyanides, PCBs, etc. Tests done in the field are not conclusive, and experience and a knowledge of chemistry are essential for safe bulking operations. Methodology varies from site to site and between companies doing the cleanup work, and any job requires specific tests. Note the flow chart in Fig. 4-13 describing the order and types of tests generally done. Table 4-8 lists detection methods for compatibility testing.

OXYGEN DEFICIENCY

Introduction

Oxygen deficient atmospheres are hazardous to workers without an air mask and an air supply, and may be present on sites. This condition cannot be per-

Table 4-8. Detection Methods For Compatibility Testing.

- Explosimeter or organic vapor analyzer
- Radioactivity scan
- Air reactivity test
- Water reactivity and solubility
- Halogenated/nonhalogenated
- Flammability test
- pH test
- Oxidation/reduction test

Source: Levine and Martin (1985).

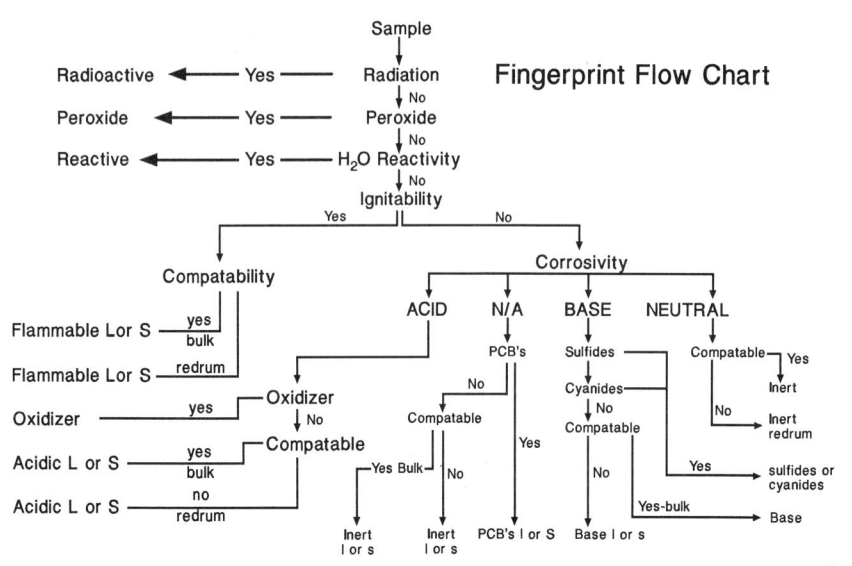

70 WORKER PROTECTION DURING HAZARDOUS WASTE REMEDIATION

Fig. 4-13. Test procedures for chemical compatibility.

ceived by human senses prior to the onset of symptoms; therefore, knowledge of potential locations, means of measurement, and protective equipment are important elements of safe work on hazardous waste sites.

Objectives

Readers will:

- Know normal and safe levels of oxygen in air.
- Recognize locations where oxygen may be expected to be deficient.
- Be aware of monitoring and air replacing equipment used in potentially hazardous locations.

Importance of Oxygen

Some living organisms can exist in environments where the percentage of oxygen is quite low, but a human being is not one of them. Each human cell requires the presence of oxygen to carry out life-sustaining activity, and cells quickly die if oxygen is not available to them.

The oxygen molecule in air, O_2, consists of two oxygen atoms bound together. It crosses the very thin layer of cells bounding the terminal alveoli in the lungs, the equally thin layer of cells forming the walls of the capillary network around the alveoli, and enters the blood within the capillaries. There it diffuses into the red blood cells where it bonds to iron atoms on hemoglobin molecules. As these red blood cells reach capillary beds in other tissues (muscles, brain, intestines, etc.) the oxygen molecules are released by hemoglobin, leave the blood across capillary walls, and enter working cells where they are used in cellular respiration. The amount of oxygen which is carried in the blood varies from person to person with individual differences based on health, fitness, and personal habits like smoking.

Percentage of Oxygen in Air

Oxygen is present in normal air sea level at a concentration of approximately 21%. Physiological effects of oxygen deficiency in humans are not apparent until the level becomes lower than 15%, but the OSHA standard for required use of an air-supplied respirator is 19.5%. Table 4-9 shows selected percentages of oxygen in air and expected corresponding symptoms.

Where May Oxygen Deficiency Be Likely?

Two occurrences may lead to oxygen deficient atmospheres:

- Displacement of oxygen by another gas. A heavier-than-air gas (like carbon monoxide) will flow into a low area and replace the normal air there.
- Consumption by chemical reaction, such as occurs in a fire in an enclosed area.

Knowing what situations may lead to oxygen deficiency enables the worker to anticipate where deficiency may occur. Suspect locations include:

- Low-lying areas such as natural valleys, ditches, trenches, pits, and basements of buildings.
- Confined spaces such as buildings and tanks.

Table 4-9. Oxygen Percentages and Effects.

% O_2	Effects
20.9%	Normal
19.5%	OSHA standard
16%	Less attention judgement, and coordination; increased breathing and heart rate.
<16%	Nausea and vomiting; brain damage; heart damage; unconsciousness and death
2–3%	Death within one minute

Adapted from Occupational Safety and Health Guidance Manual for Hazardous Waste Site Activities NIOSH/OSHA/USCG/EPA.

Measuring Oxygen Concentration

An oxygen meter should be used to determine the percentage of oxygen present before entering a location suspected of being deficient in oxygen. It is a portable, direct-reading instrument which you should learn to operate. An alarm on the machine sounds at oxygen levels below 19.5%. A worker who measures safe oxygen levels and enters a potentially suspect area without an air-supplied respirator should continue to monitor the air in case conditions change. If oxygen cannot be measured from outside the area before entering, an air supply apparatus should be used by the worker while monitoring and removed only after safe levels are determined to be present.

Dealing with oxygen deficient atmospheres requires conscientious adherence to four steps:

1. Assume the worst possible conditions in a suspect location.
2. Monitor the atmosphere before entering.
3. If step 2 is impossible, or measured oxygen is below 19.5%, use air-supplying equipment.
4. Continue monitoring during ongoing work if the conditions which originally made the location suspect continue to be present.

RADIATION

Introduction

Although not considered, the primary danger at a hazardous waste site, radioactive materials may be found there. These materials may be found in drums, including lab packs, as either a solid or liquid. Sites that have been used by hospitals and research facilities should be particularly suspect of having an ionizing radiation hazard. The best precaution for individuals working in hazardous waste sites is constant monitoring. Initial monitoring should be done to detect high levels of radiation. As drums are being opened, a second check for radioactive materials should be made. As new material is uncovered, the material should be checked for radioactivity. When in doubt, survey.

Objectives

The reader will become familiar with the type of radiation he may come in contact with at a hazard waste site and the potential health hazards associated with it. The reader will know the difference between acute and chronic effects; the basis of alpha, beta, and gamma radiation; the concept of radioisotope half-life; and how radiation is measured. The reader will have an understanding of the role of the regulatory agencies.

Radiation

Although we may be aware of other kinds of hazards through our senses (sight, smell), the detection of unlabeled or otherwise marked radiation hazards depends upon instrumental monitoring. Because we may or may not be aware of the presence of radiation hazards, initial site characterization must include a radiological survey.

By convention, radiation is classified into two types, *ionizing* and *non-ionizing radiation*.

Ionizing radiation has enough energy to cause the production of chemical ions which are often highly reactive with other chemicals in their vicinity. For this reason, ionizing radiation poses the greatest threat to biological tissues. It is known to break chromosomes, produce mutations in DNA, and induce cancer in living systems.

Gamma radiation. The most common ionizing radiations are X-rays, gamma rays, and cosmic rays. These forms of radiation represent the "high-energy" region of the continuum of energy wavelengths described by the electromagnetic spectrum, Fig. 4-14.

The earth is bombarded by a certain amount of gamma rays (which may be considered synonymous with cosmic rays) and X-rays. Certain isotopes of elements found in the rocks of the Earth's crust also emit gamma radiation when they undergo nuclear rearrangement ("decay"). There is thus a naturally occurring level (or "background") of high-energy radiation at the very short-wavelength end of the electromagnetic spectrum of energy.

Because they may be regarded as "pure electromagnetic energy" without mass, gamma rays and X-rays pass through many materials with ease, surrendering some of their energy in the process, and producing a high level of ionization. This property accounts for their ability to inflict serious tissue damage in biological systems, and is the basis for the use of X-rays in diagnostic radiography.

Protection from gamma rays and X-rays depends upon interposing some very dense material ("shielding") between the source and the target. Lead, for instance, is commonly used to shield against these radiations.

Two other forms of ionizing radiation exist. These result from the "decay" of *radioactive isotopes* of various chemical elements. *Alpha radiation* occurs when an atomic nuclear reaction ("radioactive decay") causes the ejection of a helium nucleus (two protons and two neutrons, you recall) from the nucleus of some heavier atom. These *alpha particles* have a high mass and as they pass through a material they surrender some of their energy to the material, ionizing atoms in the vicinity. Isotopes that undergo this nuclear rearrangement with its accompanying radiation are called *alpha emitters*.

Because these particles are relatively large, there is a high probability that

THE ELECTROMAGNETIC SPECTRUM

```
GAMMA
    X-RAYS    UV    INFRARED   MICROWAVE      RADIO
10⁻³    0.1    10          10⁵      10⁷     10⁹     10¹¹   nm

            VISIBLE SPECTRUM
         violet-blue-green-yellow-orange-red
         400    500    600    700  nm
```

Relationship between Wavelength and Energy (Arrows indicate increasing direction):

⟶ WAVELENGTH (nm)

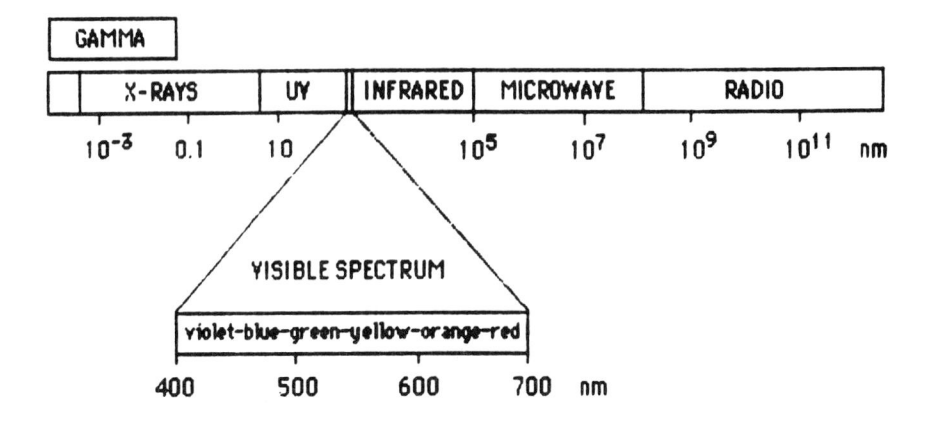

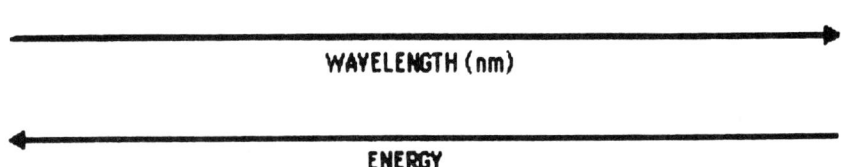

⟵ ENERGY

Fig. 4-14. The electromagnetic spectrum.

they will "bump into" atoms and lose their energy quickly. Consequently, they do not typically penetrate a material, including biological tissues, to a great extent. A sheet of paper is sufficient to absorb the energy released by alpha emitters; alpha radiation is absorbed by the epidermis (nonliving) layer of the skin. As you might expect, their greatest hazard results from inhalation or ingestion of the isotopic substance, followed by intracellular uptake of the isotopic atoms.

Beta radiation occurs when an atomic nuclear reaction ("radioactive decay") causes the ejection of a fast-traveling high-energy electron (negatron) or a similar positively charged particle (positron) from the nucleus of an atom. These *beta particles* have a low mass and as they pass through a material they surrender some of their energy to the material, ionizing atoms in the vicinity. Isotopes that undergo this sort of nuclear rearrangement with its accompanying radiation are called *beta emitters*.

Because of their low mass, these particles are less likely to "bump into" other atoms and they therefore penetrate materials, including biological tissues, to a greater extent than do alpha particles. Beta radiation may penetrate up to 15 mm of living tissue before being absorbed. Beta-emitting isotopes are thus a skin (contact), inhalation, and ingestion hazard.

Since all mineral elements and atmospheric gases consist of a small proportion of radioactive isotopes, these sources also contribute to the background of ionizing radiation occurring naturally.

It should be noted also that a number of isotopes "decay" by more than one process; an isotope may thus be both a beta and a gamma emitter. *Non-Ionizing radiation* includes all of the other forms of electromagnetic radiation except gamma (cosmic) and X-rays. (Actually, a very small proportion of ultraviolet radiation near the X-ray wavelengths is also weakly ionizing). Ultraviolet radiation, visible radiation, infrared (heat) radiation, microwaves, and radio waves are forms of non-ionizing radiation.

Important Concepts Associated with Radioactive Isotopes

Half-Life. This is the rate at which a radioactive substance ("radioisotope") loses its radioactivity and stabilizes. Each radioisotope (including different isotopes of the same element) is characterized by a specific period of time required for one-half of its atoms to *decay*. Therefore, after one half-life only half of the original isotopic atoms remain. After another equal length of time, half of these remaining atoms decay, leaving one-quarter of the initial amount of radioactivity in the sample. In this way radioactive elements give off smaller and smaller amounts of radiation per unit time. Table 4-10 shows this decline in radioactivity; Table 4-11 lists some representative half-lives.

Specific Activity. This is the proportion of radioactive atoms in a total sample. A sample having 30% of its atoms as radioactive isotope has twice the specific activity of another sample with 15% radioactive atoms. The emission characteristics (alpha, beta, or gamma), the amount, and the specific activity of a sample all contribute to its potential as a hazard.

Table 4-10. Table of Half-Life.

No. of Half-Lives Elapsed	*% Radioactivity Remaining*
0	100.0
1	50.0
2	25.0
3	12.5
4	6.25

Table 4-11. Half-Lives of Some Common Radioisotopes.

Type of Radiation	Isotope	Half-Life
Alpha	Radon-226	1622 years
	Uranium-238	4.5 billion years
Beta	Carbon-14	5730 years
	Hydrogen-3 (tritium)	12.26 years
	Strontium-90	27 years
Gamma	Cobalt-60	5.26 years
Beta and gamma	Iodine-131	8 days
	Cesium-137	33 years

Adapted from Meyer. *Chemistry of Hazardous Materials.*

Bioaccumulation. This is the concentration of some specific isotope in a specific organ or tissue as a result of the peculiar metabolism of that tissue. Iodine-131, for example, becomes concentrated in the thyroid gland because of that organ's role in the metabolism of iodine generally; high levels may induce thyroid cancer. Strontium-90 exchanges, atom for atom, for calcium in bone, where its decay may induce leukemia, a disease originating in the bone marrow.

General Cautions Regarding Radioactivity

- There is no way to tell by looking at an object how much radiation is present. It must be *measured* (see Table 4-12).
- There is no warning to the senses. If a radioactive substance is not properly labeled you may not know it is present.

Biological Effects of Radiation

Acute Effects. When the radiation source is external (outside the body), radiation penetrates and reacts with tissues locally, as a one time encounter. The result is destruction of some cells and failure of division of some types of cells.

Chronic Effects. More tissues are affected with passage of time. Radiation-linked diseases may be triggered (genetic effects, cancer, anemia).

Dose Rate Effects. The effects of received radiation are not the same for equivalent total doses (measured in roentgens or rads) if the time intervals for the doses are unequal. A given amount of radiation delivered over a long time (chronic exposure) causes less harm than the same amount delivered in a short time period (acute exposure).

HAZARD RECOGNITION 77

Table 4-12. Radiation Detecting Devices.

Detector	Types of Radiation Measured	Typical Full Scale Readings	Use	Advantages	Possible Disadvantages
Scintillation counter	Beta, x, gamma, neutrons	0.02 mR/h to 20 mR/h	Survey	1. High sensitivity 2. Rapid response	1. Fragile 2. Relatively expensive
Geiger-Muller counter	Beta, x, gamma	0.2 to 20 m R/h or 800 to 80,000 counts/min	Survey	Rapid response	1. Strong energy dependence 2. Possible paralysis of response at high count rates or exposure rates 3. Sensitive to microwave fields 4. May be affected by ultraviolet light
Ionization chamber	Beta, x, gamma	3 mR/h to 500 R/h	Survey	Low energy dependence	1. Relatively low sensitivity 2. May be slow to respond
Alpha counter	Alpha	100 to 10,000 alpha/min	Survey	Designed especially for alpha particles	1. Slow response 2. Fragile window
Film	Beta, x, gamma	10 mR and up	Survey and monitoring	1. Inexpensive 2. Gives estimate of integrated dose	1. False readings produced by heat, certain vapors and pressure

Table 4-12. (Continued)

Detector	Types of Radiation Measured	Typical Full Scale Readings	Use	Advantages	Possible Disadvantages
				3. Provides permanent record	2. Great variations with film type and batch 3. Strong energy dependence for low energy x rays
Pocket ionization chamber and dosimeter	X, gamma	200 mR to 200 R	Survey and monitoring	1. Relatively inexpensive 2. Gives estimate of intergrated dose 3. Small size	1. Subject to accidental discharge
BF3 Counter	Neutrons	0–100,000 c/min	Survey	Designed especially for neutrons	

How Radiation Is Measured

The unit of measure for radiation is the Roentgen-equivalent-man (rem). 1 rem = 1 rad × RBE. The rad is the measure of adsorbed ionizing radiation equal to the energy transfer of 100 ergs of energy per gram. RBE stands for the relative biological effectiveness.

Container or Ground Radiation. Any newly entered area should be checked for radiation, as should each drum and container. External surfaces, and internal spaces as drums are opened, should be monitored for emission of ionizing radiation before other tests are made. Several types of portable equipment are available for monitoring radiation; the most useful field monitor is probably the Geiger-Muller counter. Operation of this device is described in Chapter 8.

Personnel Monitoring. Any worker who expects exposure to any dose of ionizing radiation should be provided with a personal dosimeter of some kind to keep him informed of the cumulative dose to which he has been exposed. These include badges holding disposable film, and chamber-type dosimeters which may be reset and reused. (Chapter 8.)

Potential Sources of Radiation at Hazardous Waste Sites

Ionizing Radiation Sources.

- Marked sources (labeled as containing radioactive waste).
  A container marked with the symbol in Figure 4-15 should be dealt with in the following manner.
  — Measure radioactivity at the surface of the package.
  — If radiation is not detected, read and record any labels or markings on

Fig. 4-15. Symbol for radioactive materials. (*Source*: DOT.)

the package and try to contract the disposer to determine the nature of the contents.
— If radiation is detected, call the Nuclear Regulatory Commission for advice. Some states have a radiation safety agency which should be contacted.
* Incoming or stored material that is improperly marked or unmarked.

If radiation is detected at the surface of any container, or inside when the container is opened, contact the Nuclear Regulatory Commission or the state agency for radiation safety.

* Tools which have been, or are being, used on the site. It is possible that old thickness gauges, such as those used for inspecting welded pipes or depth gauges used in tank and bins, have been accidentally left behind on a site. Radiation sources from gauges may have been in disposal materials from other locations. Several pieces of equipment which might be used during Phase II (such as nuclear density meter to measure soil density) present a hazard if improperly used.

Non-Ionizing Radiation Sources.

* Lasers used for alignment, range finding and other construction measurements emit potentially damaging radiation.
* Welding equipment emits strong visible light + ultraviolet. Ultraviolet may cause damage to the retina of the eye at some distance, and UV welding "sunburns" may occur on areas of skin not protected by clothing.
* Sunlight (includes UV and infrared). UV exposure of the skin increases the risk of skin cancer. Infrared heating of the air increases ambient temperatures, thereby causing heat stress hazard.
* Incinerators give off infrared heat.

Regulatory Agencies

The Nuclear Regulatory Commission (NRC) is the Federal body which regulates radioactive materials. NRC has a 24-hour hotline telephone by which they can be contacted for information and advice if a radioactive source is found on a hazardous waste site. They can be contacted by calling (202) 951-0550.

Roughly half the states, referred to as "agreement states" because of their agreement with the Federal Government to take over all functions of the NRC, handle all radiologic emergencies within their borders. All the southeastern states are agreement states. In Alabama, for example, the Alabama Department of Radiologic Physics in Montgomery (205) 261-5315 is the agency to call if radioactivity is detected on a site in Alabama.

Local Civil Defense authorities in all areas can give information and additional help as needed on a site where radioactivity has been detected. These officials may, in some locations, be more accessible than state officials and will call the appropriate agency for you.

NOISE

Introduction

Although noise on a hazardous waste site is not considered to be the primary hazard, it still represents a danger to persons working on the site, particularly those working around vehicles and machinery. There is also the danger of an explosion which can lead to serious hearing damage for those in close proximity. Testing for noise levels should be done periodically if there is any doubt that levels are lower than OSHA standards.

Objectives

The reader will:

- Know the OSHA standard for noise level.
- Understand the dB system of noise measurement.
- Know hearing protection devices and their proper use.
- Identify potential noise hazards at a hazardous waste site.

Effects, Measurement, and Protection

Work involving heavy equipment and vehicles often creates excessive noise. The effects of noise can include:

- Workers being startled, annoyed or distracted.
- Physical damage to the ear, pain, and temporary and/or permanent hearing loss.
- Communication interference that may increase potential hazards due to the inability to warn of danger and the proper safety precautions to be taken.

If employees are subjected to noise exceeding an 8-hour time weighted average sound level of 90 dBA (decibels on the A-weighted scale), feasible administrative or engineering controls must be utilized. In addition, whenever employee noise exposures equal or exceed an 8-hour time weighted average sound level of 85 dBA, employers must administer a continuing effective hearing conservation program as described in OSHA regulation 29 CFR Part 1910.95.

It is important to note that the decibel scale is not linear, but exponential. A doubling of any sound pressure corresponds to a 6 dB increase. The decibel (dB) is used to express the sound level associated with noise measurements. The weakest sound that can be heard by a person with very good hearing in an extremely quiet location is assigned the value of 0 dB. At 140 dB the threshold of pain is reached.

Noise levels may be measured by various electronic devices, both for area and individual monitoring. Like most equipment, instruments must be calibrated according to manufacturers recommendations and interpreted by trained persons.

While loudness depends primarily on sound pressure, it is also affected by frequency. (Pitch is closely related to frequency.) The reason for this is that the human ear is more sensitive at high frequencies than it is at low frequencies. The upper limit of frequency at which airborne sounds can be heard depends primarily on the condition of a person's hearing and on the intensity of the sound. For young adults this upper limit is usually quoted as being somewhere between 16,000 and 20,000 Hertz (frequency). It is important to realize that most people lose sensitivity for the higher frequency sounds as they grow older. The aging effect is called *presbycusis*.

Three main techniques are employed to protect workers' hearing:

- *Engineering Controls*. Those physical means to lower the impact of sound damage, such as mufflers, sand baggings, and design innovations.
- *Ear Plugs*. Devices that fit in the ear, both disposable and reusable.
- *Ear Muffs*. Devices that fit on the head covering the ears.

The best approach is provided by engineering controls which eliminates the problem so that hearing protection is not needed.

ELECTRICITY

Introduction

Electrical wiring is often necessary on a hazardous waste site for operation of lighting, cooling, and equipment. At some locations, particularly those which include buildings, high and low voltage wiring is already on the site.

Objective

Upon completion the reader will:

- Understand the mechanism of electrical hazard.

- Know how to avoid being injured by electricity.
- Be able to list safe electrical work practices.

The Nature of Electricity

Electricity is defined as the flow of electrons, the charged particles around the nuclei of atoms. For electrical action to take place, the electrons must flow in a complete loop, or circuit. Any break in the circuit causes the electrical current to stop flowing, and electrical activity stops.

Electrical Hazards May Be Divided Into Five Categories

- Shock to personnel.
- Ignition of combustible or explosive materials.
- Overheating causing damage to equipment or burns to people.
- Electrical explosions.
- Inadvertent activation of equipment.

Resistance

Electrons flow more easily through some materials than others. Materials through which electron flow is impeded are glass, rubber, and plastic. These materials offer *resistance* (measured in ohms) to electron flow, and are used as *insulators* for protection. Materials through which electrons flow easily and rapidly are metals, water, and the human body (which is mostly water).

The resistance of the human body to current flow is confined almost entirely to the skin, particularly the dead, scaly cells of the outer layer. Different parts of the body differ in their resistance to current flow (see Table 4-13).

The result of exposure to electricity depends on the amount of current flow (see Table 4-14), the current path, and the frequency and duration of the flow. For example, relatively large currents can pass from one leg to another with only contact burns. A similar current from arm to arm or arm to leg may stop the heart or paralyze the respiratory muscles (see Table 4-14).

Table 4-13. Human Resistance to Electrical Current.

Dry skin	100,000–600,000 ohms
Wet skin	1,000 ohms
Hand to foot	400–600 ohms
Ear to ear	100 ohms

Reprinted with permission from the National Safety Council: *Protecting Workers Lives: Safety and Health Guidelines for Unions*. Chicago: National Safety Council, 1983.

84 WORKER PROTECTION DURING HAZARDOUS WASTE REMEDIATION

Table 4-14. Results of Exposure to Electrical Current.

	Current in milliamps			
	Males		Females	
Symptoms	DC	AC	DC	AC
Slight sensation	1	0.4	0.6	0.3
Perception threshold	5	1.1	3.5	0.7
Shock	9	1.8	6	1.2
Shock painful	70	9	40	6
Shock: muscle control lost	90	23	60	15
Shock: possible heart stop	500	100	500	100

Reprinted with permission from the National Safety Council: *Protecting Workers Lives: Safety and Health Guidelines for Unions.* Chicago: National Safety Council, 1983.

Prevention of Electrical Accidents

Prevention of electrical accidents requires planning and alertness. Common causes of such accidents are:

- Contact by raised equipment with a live overhead wire.
- Re-energizing of a circuit on which an electrician is working.
- Improper grounding of tools and equipment.
- Uncontained electrical discharge into flammable or explosive environments.
- Defective insulation of lines and equipment.

Hazardous waste sites offer many conditions which deteriorate or damage electrical line insulation, including rain and humidity, sunlight, chemicals, abrasion, crushing, or even biological factors such as rats and molds. All of these may operate over time to destroy insulation. In addition, temporary wiring may be hurriedly constructed and poorly done.

The following checklist may be useful in assessing electrical hazards.

- Are there any extraordinarily high voltage or amperage levels used which would require special safeguards? Have those safeguards been provided?
- Are all items which should be electrically grounded, grounded adequately? Are the grounds tested periodically?
- Is there any location where a live circuit is not insulated? Is adequate protection provided to keep personnel from contacting such circuits? Is protection also provided where insulation might have deteriorated to the point where it should be replaced?
- Is there any surface, other than a heating element, hot enough to burn a person or ignite a material?
- Are the voltage and amperage high enough to cause arcing or sparking which could cause ignition of a flammable gas or combustible material?

- Are there any points, such as motor brushes or open circuit breakers, where arcing or sparking can occur close to any fuel?
- Are fuses, circuit breakers, and cutouts sized to protect the circuits and equipment they are supposed to protect? Are fuses and circuit breakers in a readily accessible and safe location? Are accesses to them kept clear?
- Are all the electrical installations and systems on the site in accordance with the requirements of the OSHA standard and of the National Electrical Code?
- Is an interlock provided to cut off power upon access to any equipment interior where a person could receive a fatal shock?
- Are wires and cables protected against chafing, pinching, cutting, or other hazards which could damage the insulation so a person could get a shock, or which could cut the metal conductor?
- Are the locations of underground cables marked so that they will not be cut by excavating equipment?
- Are wires, cables, and conduits adequately secured to the structures along which they pass or to the chasses of the equipment on which they are installed?
- Are wires and cables kept off paths over which vehicles must pass? If they must be on the path, are they adequately protected against damage?
- Where batteries are used, is the location marked with the polarity, voltage, and type(s) of battery to be used?
- Where batteries may be "jumped" for engine starts, are they posted with instructions indicating the proper way it is to be done and the precautionary measures to be taken?
- Are materials and equipment which can generate static electricity grounded to prevent accumulations of static charges?
- Are tools to be used in opening containers made from nonsparking materials?

HEAT AND COLD

Introduction

Weather on a hazardous waste cleanup site may be hazardous because it cannot be controlled. The thermostat on most sites is regulated only by Mother Nature, and she is not always cooperative.

Objectives

Upon completion the reader, students will:

- Be able to recognize the weather conditions under which heat and cold stress can occur.

- Be aware of contributing factors to heat and cold stress.
- Know how to monitor conditions and workers subject to heat and cold stress.
- Know symptoms, prevention, and treatment of temperature stress response.

Normal Mechanisms

The human body has inherent thermoregulatory mechanisms which maintain temperature at 98.6°F ($\pm 1°$). A significant portion of the energy burned by all the body's cells is used to generate the heat necessary to maintain normal temperature. Since this metabolic heat is constantly being produced, a mechanism for losing heat from the surface of the body is necessary when metabolic heat plus environmental heat raise body temperature above normal.

A regulatory "thermostat" in the hypothalamus portion of the brain monitors body temperature and sets two heat-loss mechanisms into action when temperature is raised.

Radiant heat loss is increased by vasodilation (enlargement) of skin capillaries, which increases blood flow at the surface of the body. If the ambient temperature is lower then 98.6°F, heat in blood from the overheated body core is radiated into the air.

Evaporative heat loss occurs due to increased sweating. Liquid perspiration is converted to water vapor on the surface of the skin; the heat used in breaking the chemical bonds holding the liquid molecules together is derived from the body.

Conversely, if body temperature drops below normal, mechanisms to generate and conserve heat are triggered.

Vasoconstriction of skin capillaries reduces the amount of blood which transfers core heat to the body surface where radiant heat loss can occur.

Involuntary muscle contractions (shivering) are initiated by the nervous system; the metabolism involved in these contractions generates additional heat.

Contributing Factors

Several factors contribute to heat stress conditions. The most obvious are the initial cause, ambient air temperature, and humidity, or the amount of water vapor in the air.

Personal protective clothing interferes with both methods of thermoregulation by creating a microenvironment inside the suit which is hot and moist. No body heat can radiate into this small space if the temperature there is above 98.6°F, and no perspiration can evaporate since the air inside the suit quickly becomes saturated with water vapor and will accept no more. No cooling takes place at any body surface inside the clothing. In addition to enclosure, the dark coloration of many plastic and rubberized suits allows more heat absorption into the suit from the sun. Other considerations which result in individualized responses are acclimatization, overweight, food and fluid consumption, and smoking vs. nonsmoking.

Two factors influence the development of a cold injury: ambient temperature and the velocity of the wind. Wind chill is used to describe the chilling effect of moving air in combination with low temperature. For instance, 10°F with a wind of 15 miles per hour (mph) is equivalent in chilling effect to still air at −18°F.

As a general rule, the greatest incremental increase in wind chill occurs when a wind of 5 mph increases to 10 mph. Additionally, water conducts heat 240 times faster than air. Thus, the body cools suddenly when chemical-protective equipment is removed if the clothing underneath is perspiration soaked.

Symptoms

If the body's physiological processes fail to maintain a normal body temperature, a number of physical reactions can occur ranging from mild (such as fatigue, irritability, anxiety, and decreased concentration, dexterity, or movement) to fatal. Medical help must be obtained for the more serious conditions.

Heat-Related Problems. *Heat rash* is caused by continuous exposure to heat and humid air and aggravated by chafing clothes. As well as being a nuisance, heat rash decreases ability to tolerate heat.

Heat cramps are caused by profuse perspiration with inadequate fluid intake and electrolyte replacement. Signs: muscle spasm and pain in the extremities and abdomen.

Heat exhaustion is caused by increased stress on various organs to meet increased demands to cool the body. Signs: shallow breathing; pale, cool, moist skin; profuse sweating; dizziness and lassitude.

Heat stroke is the most severe form of heat stress. The body must be cooled immediately to prevent severe injury and/or death. Signs: red, hot, dry skin; no perspiration; nausea; dizziness and confusion; strong, rapid pulse; coma. *Medical help must be obtained immediately.*

Cold-Related Problems. Local injury resulting from cold is included in the generic term *frostbite*. There are several degrees of damage. Frostbite of the extremities can be categorized into:

- *Frost nip or incipient frostbite*, characterized by suddenly blanching or whitening of skin.
- *Superficial frostbite*, where the skin has a waxy or white appearance and is firm to the touch, but the tissue beneath is resilient.

- *Deep frostbite*, where tissues are cold, pale, and solid; this is an extremely serious injury.

Systemic hypothermia is caused by exposure to freezing or rapidly dropping temperature. Its symptoms are usually exhibited in five stages; (1) shivering; (2) apathy, listlessness, sleepiness, and (sometimes) rapid cooling of the body to less than 95 degrees Fahrenheit; (3) unconsciousness, glassy stare, slow pulse, and slow respiratory rate; (4) freezing of the extremities; and, finally, (5) death.

Prevention

Prevention of heat stress may be as simple as erecting a canopy over a work area. If no shade is present on the site, a canopy or beach umbrella can provide shade for the rest area. Large fans at a rest area will facilitate evaporative cooling.

For all workers engaged in field operations in hot weather, the following practices will help to maintain physical performance:

- Gradually increase field time and task difficulty for an unacclimatized worker over a period of two weeks.
- Drink one quart of water each morning, at each meal, and before beginning hard work; in addition, drink frequently during the day.
- Replace salt loss by eating three meals a day, supplemented during the day with an electrolyte replacement beverage like Gatorade. Many athletes find that diluting these drinks half-and-half with water prevents intestinal upsets the beverages may cause.
- As temperature and humidity increase, rest periods in a cool place must be more frequent and work rate lowered (see Tables 4-15 and 4-16).
- Body water loss due to sweating should be measured by weighing the worker nude, or at least in the same dry clothing, each morning and evening. If daily loss exceeds 1.5% of body weight, the worker should be instructed to increase his daily intake of fluids.

In impermeable clothing, all these steps should be taken; however, temperature and humidity will quickly become much higher inside the suits than that measured in the surrounding air. More information is available in Chapter 9, Personal Protective Equipment.

Preventing cold stress includes similar work/rest schedules, and the rest area should be warm and dry. The selection of clothing which can be layered to enhance maintenance of dead air space around the worker is extremely valuable, with materials such as wool and the new thermal synthetics polypropylene, capilene, and synchilla as the fibers of choice. An outer layer of woven nylon or other wind-breaking material completes the garb. Head, hands, and feet should be covered with warm, layered garments. Proper hydration is important in cold environments as well as hot, since a great deal of water is lost through the lung membranes to cold, dry air.

Monitoring

A number of devices are available for measuring temperature, wind speed, and humidity. Since temperature and humidity combinations are important in predicting heat stress, and temperature and wind speed act together to cause cold stress, single measures of one parameter are of less value than combinations of the critical ones (Table 4-18).

The *wet bulb globe temperature* (WBGT) index, used to monitor hot environments, combines the effects of humidity, air movement, air temperature, radiant heat, and solar radiation. WBGT instruments are battery operated and require calibration, and usually are set up at fixed locations. The wet globe thermometer (WGT) is a simpler device, portable and easy to read. Table 4-15 outlines limiting WBGT readings on the work site. Table 4-16 outlines use of WGT readings to adjust work patterns.

In cold weather, reading from an outdoor thermometer and a wind speed indicator should be used to determine wind chill conditions, with work periods adjusted accordingly. Table 4-17 details wind chill effects and relative danger of combined cold and wind conditions.

Treatment

Treatment of heat stress response varies according to the extent of the illness, but always includes cooling the victim if his oral temperature is above 102°F.

- Place the victim in the shade. Use a fan or air conditioner if one is available.
- Cool his body by bathing him in cool water or an easily evaporated fluid like rubbing alcohol.

Table 4-15. ACGIH Permissible Heat Exposure Threshold Limit Values in °C WBGT.

		Work Load	
Work/Rest Regimen	Light	Moderate	Heavy
Continuous work	30.0°	26.7°	25.0°
75%/25% each hour	30.6°	28.0°	25.9°
50%/50% each hour	31.4°	29.4°	27.9°
25%/75% each hour	32.2°	31.1°	30.0°

Source: American Council of Governmental Industrial Hygienists.

Table 4-16. Water Intake and Work/Rest Cycles for Heat Acclimated, Fit Workers.

Wgt, °F	Water Intake, qt/hr	Work/Rest Cycles, min
80–83°	0.5–1.0	50/10
83–86°	1.0–1.	45/15
86–90°	1.5–2.0	30/30
90° & above	2.0	20/40

Source: American Council of Governmental Industrial Hygienists.

- If he is conscious, give the victim cool water (or whatever potable water is available) to drink.
- Monitor oral temperature, pulse, respiration, and blood pressure to be sure they are returning to normal:

Temperature	98.6°F
Pulse	70–80 beats/minute
Respiration	16–20/minute
Blood Pressure	120/80

- Allow an extended recovery period of one half to several days, based on extent of illness and rate of recovery, before the victim returns to work under stressful conditions.
- If symptoms of heat stroke are evidenced, get medical attention immediately.

Treatment of cold stress response begins by warming the victim. Remove him to a warm indoor area, provide him with warm, nonstimulant beverages (no caffeine or alcohol), and treat frost-damaged areas as follows:

- Rewarm the frozen part quickly by immersing it in water maintained at 102–105°F (comfortably warm to the inner surface of an unchilled forearm); discontinue warming as soon as flushing indicates the return of blood.
- Do not allow the victim to walk on a frozen foot, but have him exercise a thawed part.
- Prevent contact between the injured part and any surface except a sterile bandage, and elevate it after warming.
- Seek medical attention.

BIOLOGICAL HAZARDS

Introduction

Living organisms which are hazardous to humans may be encountered on an abandoned hazardous waste site. These biohazards may have been disposed of

Table 4-17. Cooling Power of Wind on Exposed Flesh Expressed as an Equivalent Temperature (under calm conditions).

Estimated Wind Speed (in mph)	Actual Temperature Reading (°F)										
	50	40	30	20	10	0	−10	−20	−30	−40	−50	−60
	Equivalent Chill Temperature (°F)										
calm	50	40	30	20	10	0	−10	−20	−30	−40	−50	−60
5	48	37	27	16	6	−5	−15	−26	−36	−47	−57	−68
10	40	28	16	4	−9	−24	−33	−46	−58	−70	−83	−95
15	36	22	9	−5	−18	−32	−45	−58	−72	−85	−99	−112
20	32	18	4	−10	−25	−39	−53	−67	−82	−96	−110	−121
25	30	16	0	−15	−29	−44	−59	−74	−88	−104	−118	−133
30	28	13	−2	−18	−33	−48	−63	−79	−94	−109	−125	−140
35	27	11	−4	−20	−35	−51	−67	−82	−98	−113	−129	−145
40	26	10	−6	−21	−37	−53	−69	−85	−100	−116	−132	−148

(Wind speeds greater than 40 mph have little additional effect.)

LITTLE DANGER
In chr with dry skin Maximum danger of false sense of security.

INCREASING DANGER
Danger from freezing of exposed flesh within one minute.

GREAT DANGER
Flesh may freeze within 30 seconds.

Trenchfoot and immersion foot may occur at any point on this chart.

Developed by U.S. Army Research Institute of Environmental Medicine, Natick, MA.

on site by research laboratories or hospitals, or may be ordinary inhabitants of the area in which the site is located. These wastes may be classified as biomedical, biohazard, or etiologic agents.

Objectives

Upon completion of the material, the reader will know:

- What biohazards may be present on a site and how to recognize them.
- What responses to biohazard recognition will reduce or prevent exposure to them.

Microorganisms

Microorganisms are so named because they are too small to be seen without a microscope. They include bacteria, viruses, some fungi, and even tiny plants and animals. Microorganisms from the first three of these groups will be considered here, especially those which are pathogenic (capable of causing disease). Some bacteria and fungi cause illness in humans by actively damaging human cells (such as skin cells, nerve cells, or blood cells) and some by pro-

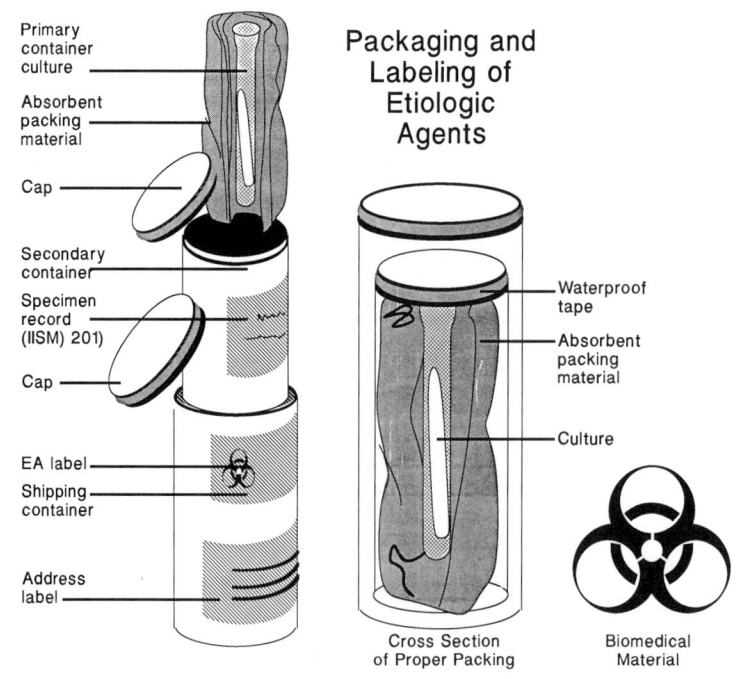

Fig. 4-16. Etiologic or biohazard shipping.

ducing natural chemicals which are toxic. Viruses also have several ways of causing disease: some are directly harmful in rupturing human cells, some inhabit cells and prevent their normal operation, and some rupture bacterial cells within our bodies (the bacterial parts are toxic when released).

Two methods of accidental or illegal disposal of microorganisms are possible, and attention to the packaging protocols for each will help workers avoid exposure.

- Disposal from a research facility or a hospital entails either subjecting materials to very high heat in an autoclave, usually in plastic bags clearly marked with the biohazard symbol, or burning them completely in an incinerator, for which they are packed in a solid red plastic bag. A red waste bag should never be sent to a landfill; if one is found there the generator should be called. On an abandoned waste site, this material should be burned unopened.
- Biohazards such as research bacterial cultures may be sent through the mail if they are packaged as shown in Fig. 4-16. Such packages, if located on an abandoned waste site, should be left in place until the responsible state agency has been contacted.

In May 1986, the Environmental Protection Agency issued a *Guide for Infectious Waste Management*, recommending categories of infectious wastes and practices for packaging, transportation, treatment, storage, and disposal of these wastes. Each state is responsible for regulation of infectious wastes.

If you have questions concerning infectious microorganisms on a site, call the Center for Disease Control, Atlanta, Georgia, at (404) 633-5313.

Bacterial decomposition of some types of landfilled materials will result in the production of dangerous compounds, especially methane gas, an explosion hazard, and vinyl chloride gas, which is carcinogenic. Exposure to these may occur unexpectedly during drilling operations on landfilled sites.

Plants and Animals

Many species of plants and animals which produce fluids toxic to humans may be encountered on abandoned, overgrown (or even indoors) waste sites:

- Plants such as poison ivy, poison oak, and poison sumac cause a severe allergic response in some people.
- Venomous insects, including wasps, hornets, and yellow jackets, may be expected to defend their nests by stinging those who come too close.
- Tick and spider bites can cause problems ranging from aggravation and irritation to severe illness and, rarely, death.

- Although the majority of snakes are nonpoisonous, poisonous species may be encountered.

The ability to identify and avoid biological hazards, both naturally occurring and improperly disposed of, is a worker's best protection against such hazards.

REFERENCES

American Conference of Governmental Industrial Hygienists. 1985. *Threshold Limit Values for Chemical Substances and Physical Agents in the Work Environment and Biological Exposure Indices with Intended Changes for 1984-85.* Cincinnati, OH: American Conference of Governmental Industrial Hygienists.

Hammer, Willie. 1985. *Occupational Safety Management and Engineering*, 3rd Ed., Chapters 20, 21. Englewood Cliffs, NJ: Prentice-Hall, Inc.

Levine, S., and Martin, W., Editors. 1985. *Protecting Personnel at Hazardous Waste Sites*, Chapters 5, 6. Stoneham, MA: Butterworth.

Meyer, Eugene. 1977. *Chemistry of Hazardous Materials.* Englewood Cliffs, NJ: Prentice Hall, Inc.

National Fire Protection Association. 1986. *Fire Protection Guide on Hazardous Materials*, 9th Ed. Quincy, MA: National Fire Protection Association.

NIOSH/OSHA/USCG/EPA. 1985. *Occupational Safety and Health Guidance Manual for Hazardous Waste Site Activities.* Washington, DC: U.S. Government Printing Office.

Supervisors Safety Manual, 6th Edition. Chicago, IL: National Safety Council.

U.S. Department of Transportation. Title 49 CFR 173.115. Washington, DC: U.S. Government Printing Office.

U.S. Environmental Protection Agency. Title 40 CFR 261.21. Washington, DC: U.S. Government Printing Office.

U.S. Environmental Protection Agency. 1985. *Syllabus—Hazardous Materials Response Operations.* Washington, DC: U.S. Government Printing Office.

U.S. EPA. Title 40 CFR 300 Appendix A. Washington, DC: U.S. Government Printing Office.

5
Toxicology

INTRODUCTION

Toxicity—The extent to which a chemical will cause harmful effects.

TOXIC SUBSTANCES = POISONS

"All things are toxic; the difference between a poison and a medicine is the dosage."—Paracelsus (c. 1500)

Toxicology is the science dealing with the effects, conditions, detection, etc., of poisons. Hazardous waste sites provide a fertile ground for individuals to be exposed to some of the most dangerous poisons known to man. In this chapter you will learn which substances cause harm, how these substances affect the human body, and how they enter your system. You will also learn how to use the *NIOSH Pocket Guide to Chemical Hazards* and to read a Material Safety Data Sheet to help protect yourself while working on a hazardous waste site.

OBJECTIVES

- The reader will associate toxic substances with *poisons*.
- The reader will understand *factors affecting toxicity*.
- The reader will be able to explain the concept of LD_{50}, and the relationship of LD_{50} tests to other tests.
- The reader will understand the concept: *exposure = dose × time*.
- The reader will understand that *exposure to more than one chemcial* can have several outcomes.
- The reader will understand that some toxic substances have delayed effects as *carcinogens, mutagens*, and/or *teratogens*.
- The reader will be able to distinguish between *acute* and *chronic* effects of exposure to toxic substances.
- The reader will be able to identify the three main *routes of entry* of toxic substances into the body.
- The reader will understand the concepts of *IDLH, PEL, REL*, and *TLV*.
- The reader will know the significance of *MSDS* (Materials Safety Data Sheets) and the right to examine them.

MEASUREMENT OF TOXICITY

All the information known today about the toxic effects of chemicals and other hazardous substances has been gained from two sources of information:

1. *Epidemiological studies*, in which numbers of affected people in one group are compared to numbers of affected people in another group. If one group shows a significantly higher incidence of an illness than the other group, a search for the reason begins. Sometimes the reason is exposure to a chemical in the workplace.
2. *Animal studies*. Most of what we know about the dangerous effects of toxic substances comes from animal studies. However, humans may react differently than animals to exposure to toxic materials. Also, higher doses than humans may be exposed to may be used in animal experiments. This is because it is often necessary to use higher doses to produce a measurable effect in an animal.

Factors Affecting Response of Humans and Laboratory Animals

The Chemical Itself. Some chemicals produce immediate and dramatic biological effects on animals. Others may produce no observable effects at all, or the effects may be delayed in their appearance.

Type of Contact. Certain chemicals appear harmless after one type of contact (skin, for example), but may have serious effects when contacted in another way (lungs, for example); carbon monoxide (CO) would be an example of this kind of contact-dependent effect.

Amount (Dose) of Chemical. The dose of a chemical exposure simply means how much of the substance is contacted. It might be expressed as milligrams (mg) if swallowed, or as parts per million (ppm) if it is in the air.

Individual Sensitivity. Humans and other animals vary in their response to any exposure to a chemical substance. For some, a certain dose may produce symptoms of serious illness; for others, only mild symptoms may appear, or there may be no noticeable effect at all. Often, a prior exposure to a chemical affects the way that an individual responds upon being exposed at a later time so there is not only variation between different individuals, there may be different responses in the same individual at different exposures.

Interaction With Other Chemicals. Chemistry is the study of the interaction of various chemicals with one another, for example, the reaction between acids and bases. Biological chemistry is much the same. Chemicals in combination can produce different biological responses than the responses seen when exposure is to one chemical alone.

Duration of Exposure. Some chemicals produce an effect after only one exposure (*acute*). Some produce symptoms only after exposure over a long period of time—say, days, weeks, or months (*chronic*). Some chemicals may have effects on humans from both kinds of exposure.

FORMS OF TOXIC SUBSTANCES

Workers need to know as much as they can about toxic substances in order to be able to protect themselves. An important first step in protecting the body from entry by hazardous chemicals is to know the route by which the chemical enters. The physical forms in which chemicals are encountered play a big role in their route of entry.

- SOLIDS. Principal hazard is usually from dusts or fumes produced when solids change form.
  Example: Polyurethane foam, when burned, gives off cyanide fumes.
  Principal Danger: Inhalation (lungs), ingestion (saliva), absorption (skin)
- DUSTS. Tiny particles of solids.
  Examples: Cement dust; metal dusts from grinding operations.
  Principal Danger: Inhalation of toxic materials into lungs.
- FUMES. Tiny particles from heating, volatilization, and condensation of metals.
  Examples: Zinc oxide fumes from welding of galvanized metal.
  Principal Danger: Inhalation (lungs), ingestion (saliva), absorption (skin).
  NOTE: TOXIC FUMES MAY RESULT FROM BURNING OF NON-TOXIC SUBSTANCES. (Example: Burning of polyurethene produces cyanide fumes.)
- LIQUIDS. Acids, organic solvents, chlorinated organic solvents.
  Examples: Benzene, sulfuric acid, 1,1,1-trichloroethane.
  Principal Danger. Absorption (skin), inhalation of vapors.
  NOTE: MANY DRUMS AT HAZARDOUS WASTE SITES CONTAIN ORGANIC SOLVENTS. THESE ARE ALSO A FIRE OR EXPLOSION HAZARD.
- VAPORS. Vapors are gases which may originate by evaporation of liquids, sublimation of solids.
  Examples: Phosgene.
  Principal Danger: Inhalation (lungs), absorption (skin).
- MISTS. Mists are liquid droplets in air.
  Examples: Acid mists from electroplating, solvent mists from spray painting.

Principal Danger: Inhalation (lungs), absorption (skin).
* GASES. A gas is a formless fluid occupying space.
Examples: Chlorine, carbon dioxide.
Principal Danger: Inhalation (lungs).

THE LD$_{50}$ CONCEPT

Once a chemical in any form has entered the body by any route, what will happen to the worker? Important considerations are the amount which enters, and the amount which will produce adverse effects. LD$_{50}$ = Lethal Dose$_{50}$. One way that biologists measure the toxic danger of compounds is to study what happens to laboratory animals (commonly mice or rats) when they are treated with chemicals. The LD$_{50}$ of a toxic chemical is the amount of the substance which, when administered to animals, causes the death of 50% of the animals within 14 days. LD$_{50}$ is therefore a measurement of toxicity, and the *lower* the LD$_{50}$ the *higher* the toxicity. For example, if 2 milligrams of Chemical A are injected into mice and half of them die in two weeks, while 4 milligrams of Compound B kill half the mice in two weeks when injected, then Chemical A is more toxic. It takes less to kill the same number of mice. Table 5-1 shows LD$_{50}$s for some chemicals and illustrates the different degrees of danger of different toxic materials. If you were on a job site containing each of these five chemicals, which would you be most careful to avoid?

It should be noted that LD$_{50}$ tests are frequently used to get first approximations of the dangerous levels of toxic chemicals. Further tests below this level are done to determine other effects of the chemical. A safety factor is included in arriving at TLV and PEL values.

BIOLOGICAL RESPONSE TO EXPOSURE TO MORE THAN ONE CHEMICAL

Why 2 + 2 Is Not Always 4

In testing chemicals in the laboratory, toxicologists have learned that many chemicals act together in certain ways on biological systems such as mice and men:

Additive Effect (2 + 2 = 4). Some toxic chemicals add their effects together in producing a biological effect. In this case, the effect is the same as being exposed to double the dose of either chemical alone. Example: Malathion + another organophosphate.

Synergistic Effect (2 + 2 = 6). Sometimes exposure to two different toxic chemicals produces a more severe effect than simply doubling the dose of either

Table 5-1. Examples of LD_{50}.

Class	LD_{50}, mg/kg	Example	Amount to Kill Average Adult
Super toxic	5	TCDD*	10 drops
Extremely toxic	5–50	Strychnine	1 teaspoon
Very toxic	50–500	DDT	1 teaspoon to 1 ounce
Moderately toxic	500–5,000	Morphine	1 ounce to 1 pint
Slightly toxic	5,000–15,000	Ethanol	1 pint to 1 quart
Relatively harmless	190,000	Water	3 gallons

*2,3,7,8-tetrachlorodibenzo-p-dioxin
Adapted from *Casarett and Doull* (1986).

one alone would have. Biologists call this *synergism*, and it really spells "watch out." Example: Isopropyl alcohol + chloroform. The alcohol ties up enzymes that would normally break down chloroform.

Potentiation (0 + 2 = 10). In some cases, a chemical without any known toxic effect (0 in the formula above) may act together with a known toxic substance (2 in the formula above) to make the toxic substance more potent, more dangerous. Example: Ethanol + chloroform (effect on liver).

Antagonism (4 + 6 = 8). The interaction of two toxic chemicals may be such that the effect produced is actually less than you would expect if the two added their effects together. But do not count on this at a toxic waste site; the chemicals there are more likely to interact to produce dangerous effects. Example: Phenobarbital + benzo(a)pyrene. Phenobarbitol increases the activity of enzymes that detoxify benzo(a)pyrene.

NOTE: A prescription drug being taken by a worker may interact with a hazardous chemical encountered in the workplace. Remember that the doctor may not have known that you would be exposed to toxic hazards, and you need to be conscious of any chemicals you bring to the job site yourself in this way.

TESTS FOR OTHER TOXIC EFFECTS

Tests for Carcinogenesis (Tumor Production) and Mutagenesis (Genetic Damage)

There is much concern today about the various forms of cancer and their prevention. It is well known now that certain chemicals can cause specific forms of cancer. A related worry is that many of the same chemicals are known to produce mutations in genes that are transmitted to our children. For this reason, tests of chemicals are done to try to measure their potential for producing tumors or mutations in the DNA of chromosomes.

One test is the *rabbit ear test*. Here the chemical to be tested is painted on the ears of rabbits; after a period of time, the ears are examined for the presence of tumors (cancer).

The *Ames test* is a laboratory test in which chemicals are scored for their ability to produce genetic mutations in bacteria. Many chemicals producing mutations also produce cancer, so the test indirectly is a measure of the ability of a chemical to produce tumors.

Tests For Teratogenesis (Embryo Damage)

Some chemicals are known to be especially dangerous in causing abnormal embryonic development. Chemicals are given to pregnant mice and abnormal effects on offspring are looked for.

ACUTE VERSUS CHRONIC TOXICITY

As noted before, *acute toxicity* results from a brief exposure to a toxic chemical. Effects (skin rash, throat or eye irritation, dizziness) usually occur soon after exposure. Acute toxic effects are often reversible; the symptoms disappear when the exposure stops.

Chronic toxicity is the result of repeated exposure over longer times (days, months, years). It may involve exposure to only small amounts of toxic substances, but the damage adds up. Symptoms of chronic toxicity may take some time to appear, even after the exposure to toxic substances has stopped. Many chronic effects are irreversible. An example is lung disease following long-term exposure to asbestos; cirrhosis of the liver is a well-known example of an effect of chronic intake of alcohol.

Exposure to the same chemical may produce both acute and chronic effects. For example, a brief exposure to benzene may result in dizziness or sleepiness, while long-term exposure to low levels of benzene may result in anemia and possibly leukemia.

Also, a single exposure to a toxic chemical may have both short-term (acute) and long-term (chronic) effects. For example, chlorine gas exposure results immediately in a stinging and irritation of the eyes and lungs. Later, lung scarring with reduced pulmonary function (ability to breathe) may occur as a result of the same exposure.

ROUTES OF ENTRY

Toxic substances enter the body in four main ways: through the *skin* by *absorption*; through the *eyes* by *absorption;* through the *mouth* by *ingestion* (swallowing); and via the *lungs* by *inhalation* (breathing). Actually, all toxic chemicals must be absorbed into the body's cells to be poisonous. In the cases of ingestion and inhalation, the absorption takes place at a point away from the external body surface. Many toxic substances are able to enter the cells of the body in more than one of these ways, sometimes all four ways.

Skin Exposure

The skin is a barrier between the environment and cells of body. Skin diseases are the most common occupational illnesses. This is not surprising, since it is common for workers to get chemicals on the skin. The skin is composed of an outer layer (the epidermis), containing the sweat glands, hair follicles, and blood vessels. Beneath this is the dermis, containing larger blood vessels, nerves, muscles, etc.

Exposure may result from skin contact with solids; liquids, and gases. Important responses to skin exposure are described below.

Corrosion. This is eating away of the skin tissue by strong chemicals like acids and caustics.

Contact Dermatitis. This results from direct chemical contact and appears at the site of contact. It is common on the hands. It usually goes away after exposure to the irritant is stopped. Some chemicals causing dermatitis include dilute acids and caustics, formaldehyde, ammonia, turpentine, metal dusts, and organic solvents.

Allergic Contact Dermatitis. This occurs when an individual becomes sensitized by prior exposure to a toxic substance. Poison ivy is a good example of this allergic response. A later exposure produces symptoms 2–3 days after exposure. Examples of chemicals frequently involved in allergic contact dermatitis include sodium bichromate, epoxies, aromatic amines, formaldehyde, and nickel metal.

More serious skin effects include acne and tumors, as described below.

Acne. Pimples at sites of exposure. A familiar form of this response is called *chloracne* and results from exposure to chlorine. Known acne-producing agents include: chlorine, oil, and tar.

Skin Cancers. Tumors of the skin may take 20–30 years to occur after toxic exposure. They may be caused by mineral oils, tars, and arsenic.

Eye Exposure

In general, eyes are affected by the same chemicals that affect skin. However, eyes are more sensitive than skin. Examples of chemicals with special toxicity for eyes include formaldehyde, ammonia, and chlorine gas.

Oral Exposure

This usually involves swallowing of chemicals, and is very serious since the path followed by ingested items is

Mouth → Stomach → Intestine → Blood → All Tissues

You can see that anything that gets into the bloodstream goes to all the body's organs and tissues: liver, pancreas, glands, muscles, nerves.

However, due to the blood-brain barrier, many chemicals in the circulation cannot pass into the brain cells; toxic substances that do can cause brain damage. Likewise, many chemicals in the circulation cannot pass into the cells of the testis. But toxic substances that do can cause testicular or scrotal cancer. In pregnant females, some chemicals in the circulation cannot cross the placenta into the tissues of the developing fetus. However, toxic substances that do can cause abnormalities of development and fetal death, so this is a special concern for female workers.

Oral exposure is usually the result of poor practices, including placing hands to mouth, eating, drinking, chewing gum, use of tobacco products, applying cosmetics. It may also result from involuntarily swallowing mucous from the respiratory system.

Lung Exposure

The route by which *inhaled* chemicals make their way to the body's cells is:

Nose or Mouth → Airways → Lungs → Blood → All Tissues

Inhalation is the most common way that toxic substances are absorbed in the workplace. The large volume of air (and dusts, fumes, vapors, mists, and gases) breathed in by humans—about 12,240 liters of air (432 cubic feet) each day—makes this route particularly dangerous.

Effects of Inhalation of Toxic Substances.
Direct—producing damage to the lungs themselves. An example is asbestos, which produces scarring of the lung tissue, and may result in a form of cancer called *mesothelioma*.

Absorption into the bloodstream and transport to the organs and tissues (brain, kidneys, liver, etc.). The lung contains little sacs, the *alveoli*, having a very large surface area (about the size of a tennis court) for absorption, and surrounded by tiny blood vessels.

A special danger to workers, particularly in confined spaces, is *asphyxiation*. Air is about 21% oxygen, the remainder being mostly nitrogen. A continuing supply of oxygen is necessary for life. Lack of oxygen results in asphyxiation.

In high concentrations, some substances (carbon dioxide, acetylene, argon) reduce the available oxygen in the air.

Some chemicals (carbon monoxide, hydrogen cyanide) combine with the hemoglobin in the red blood cells, preventing the hemoglobin from picking up oxygen and transporting it to the body's cells. These chemicals are toxic because they indirectly result in asphyxiation.

What Happens to Chemicals That Get Into the Bloodstream?

The human body deals with toxic chemicals in three main ways:

Metabolism. Chemicals absorbed into the bloodstream from the stomach or intestine make their way first to the body's largest organ, the liver. There useful nutrients like glucose are stored, sent on to other organs, or converted to other useful chemical compounds. This process of converting one chemical to another is called *metabolism*, and toxic compounds undergo metabolism too. Two main things may happen to a toxic compound as it is metabolized in the liver:

- *Detoxification*—"The Good News." The toxic substance is converted to a harmless substance. Sometimes this metabolite is water soluble and can be excreted from the body. Since the blood enters the liver first on its journey from the intestine, detoxification protects the cells further along from harmful effects of dangerous chemicals. The liver is the body's main detoxifying organ.
- *Formation of Reactive Intermediates*—"The Bad News." Sometimes the liver converts toxic chemicals to more toxic ones. These then leave the liver in the blood and make their way to various tissues of the body.

Excretion. The toxic chemical is given back to the environment in air breathed out of the lungs or in the urine formed by the kidneys. Often the body is not able to excrete all the substance quickly, so some remains in the tissues for a long time. Some toxic chemicals cannot be excreted.

Storage. Many chemicals, particularly those that are soluble in oil or fat, are not removed from the body but are stored instead. Usually this occurs in fat tissue. This is how human fat tissue has come to contain DDT, a chlorinated hydrocarbon pesticide; everyone who lived in the 1940s–1960s, when DDT was widely used on agricultural crops, has DDT in the fat cells of his body.

TOXIC CHEMICALS AND TARGET TISSUES

Certain toxic chemicals have effects on special cells of the body. Like a bullet, these chemical compounds are aimed at particular tissues. For example, it was discovered long ago that chimney sweeps exposed to chimney soot had a very high incidence of scrotal cancer. We know now that the organic chemical residues from coal in chimney soot are very potent cancer-producing agents. More recently, the compound DBCP (dibromochloropropane) was discovered to be associated with male infertility and sterility. Organophosphate insecticides used in agriculture directly affect nerve cells, leading to neurotoxicity, including tremors and spasms. Notice that all three of these examples are industrial diseases associated with particular kinds of jobs.

BIOLOGICAL TOXIC EFFECTS

Neurotoxicity—Brain and nerves
Chlorinated hydrocarbon pesticides (DDT, chlordane)
Organophosphate pesticides and nerve gases (malathion, parathion, GB, VX)

Metabolic Poisons—All tissues or target tissues
Cyanide, fluorine

Carcinogenesis (cancer)—All tissues or target tissues
Benzene, benzo(a)pyrene, DMAB (dimethylaminoazobenzene), DBCP (dibromochloropropane)

Teratogenesis—Damage or death to fetus
Thalidomide, dioxins

LIMITING EXPOSURE TO TOXICANTS

The worker has the ability, and the ultimate responsibility, to protect himself from chemical hazards through knowledge. Many public and private groups and federal agencies provide information about the hazards of chemicals and about the safe levels, based on research, of exposure to these chemicals. Abbreviations and acronyms are used in these publications, and workers must be familiar with these, as well as which information they may expect to find in each publication, in order to fully utilize these sources.

Know the Nature of the Chemical Substance
(*NIOSH Pocket Guide to Chemical Hazards*)

Concentrations.

IDLH   Immediately Dangerous to Life and Health (see Table 5-2)
PEL   Permissible Exposure Level (see Table 5-2)
REL   Recommended Exposure Level
TLV   Threshold Limit Value

Table 5-2. Some Toxic Compounds and Their IDLH and PEL Values.

	IDLH, ppm	PEL (TWA), ppm
Acrolein	5	0.1
Methyl isocyanate (MIC)	20	0.02
Phosgene	2	0.1
Toluene	2000	100
Toluene-2,4-diisocyanate (TDI)	10	0.005

Special Considerations.

Ca Identifies known or suscpected *carcinogens* (*NIOSH Guide*)

Know the Route(s) of Exposure of the Chemical

INH Inhalation
ABS Absorption
ING Ingestion

Use Appropriate Personal Protective Equipment (*NIOSH Guide*)

SCBA Self Contained Breathing Apparatus
PAPR Powered Air-Purifying Respirator
GM   Air-Purifying Respirator ("gas mask") with appropriate canister

Know the Nature of the Workplace

With training, workers can learn to read and interpret information about how to protect themselves from chemical exposure. Once sources of protection information are known, all that remains is to find out what chemicals are known or suspected to be present on a job site. An employer must furnish this information. One source of information is the Material Safety Data Sheet (see the end of this Chapter).

MSDS—Material Safety Data Sheets.

- All workers have access to the MSDS sheets on file at the job site.
- MSDS gives: Material identification (name, synonyms); ingredients; hazards (TLV, biologic effects); physical data; fire and explosion data; reactivity data; health hazard information; protection information (PPE, ventilation, etc.); and special precautions.

THRESHOLD LIMIT VALUES

Concentration limits for safe exposure have been set after research by the National Institute of Occupational Safety and Health, the research arm of OSHA. Specific terms are used to define certain limits.

TLV-C. Threshold Limit Value—Ceiling:

- The concentration that should not be exceeded even momentarily.
- Important for some irritant gases.

TLV-STEL. Threshold Limit Value—Short Term Exposure Limit:

- The maximum concentration to which workers can be exposed for up to 15 minutes continuously without suffering from:
  — Irradiation,
  — Chronic or irreversible tissue change, or
  — Narcosis of sufficient degree to increase accident proneness, impair self rescue, or materially reduce work efficiency, provided that:
  — No more than 4 excursions per day are permitted, that:
  — At least 60 minutes elapse between exposure periods, and that:
  — The daily TLV-TWA is not exceeded.

TLV-TWA. Threshold Limit Value—Time Weighted Average:

- The TWA concentration for a normal 8-hour workday or 40-hour workweek to which nearly all workers may be repeatedly exposed, day after day, without adverse effect.

CARCINOGENS

Carcinogens are cancer-causing chemicals. Cancer in humans is more difficult to predict than some other kinds of effects of exposure to toxic materials. There are two reasons for this:

- The vast majority of chemicals in use have not been tested on laboratory animals for cancer-causing effects. Of approximately 60,000 chemicals in commercial use only 284 have been tested. Of these, about half (144) have been shown to cause cancer in lab animals. Only 21 of these are regulated by OSHA.
- It can take 10–40 years to see the results of exposure to a cancer-causing chemical. You may be healthy for 20 years and get it the very next year. The time it takes for the cancer to show up after exposure is called the *latency period*. Table 5-3 shows some of the latency periods for different known carcinogens.

It may be too late to do anything about previous exposures, but any reduction in exposure now and in the future will reduce the risk of getting cancer. Because

Table 5-3. Average Number of Years After Exposure for Cancer to First Appear.

Substance	Years
Chromium & Chromates	21
Arsenic	25
Mustard Gas	17
Acrylonitrile	23
Benzene	10
Vinyl Chloride	15
Benzidine & its Salts	16
Napthylamine	22
Asbestos Fibers	—

IS THERE A SAFE LEVEL FOR EXPOSURE TO CARCINOGENS?

there is always new information on potential carcinogens it is important that workers keep records of their workplace exposures.

Many scientists agree that for some toxic chemicals there are "safe" levels of exposure. Below this "safe" level exposure is not supposed to cause any bad health effects. However, there is good reason to believe that this is not the case with carcinogens. Many scientists believe that exposure to any amount of a carcinogen is unsafe, and may lead to cancer. See Table 5-4 for a list of some common chemicals which are suspected carcinogens.

Table 5-4. Some Suspected Carcinogens.

Compound	TLV—TWA, ppm
Carbon tetrachloride	2
Chloroform	2
Ethylene dibromide	20
Formaldehyde	3
Hydrazine	0.1
o-Toluidine	5
Vinyl bromide	5

THE MATERIAL SAFETY DATA SHEET

A Material Safety Data Sheet is required by OSHA 1910.1200 (Hazard Communication Standard) to be available to all employees, members of the community, and the Local Emergency Planning Committee responsible for emergency response in the area of a site on which hazardous chemicals are located in certain designated quantities. As soon as a chemical is identified on a hazardous waste cleanup site, the MSDS sheet for that chemical can be ordered from a company which manufacturers the chemical. A great deal of useful information can be gained from reading a Material Safety Data Sheet. The MSDS is explained in detail below.

Identity: Chemical Name/Trade Name

The name appearing in this area will usually be a trade name. The name must be the same as that on the container label and chemical list. In some cases, the scientific or chemical name and synonyms will appear.

Section I. Name and Address

Material Safety Data Sheets will have the name and address of the manufacturer, importer, or other responsible party who can, if necessary, provide additional information about the chemical and appropriate emergency procedures.

Emergency and Information Telephone Numbers. The number(s) appearing here are usually intended for emergency use. The number(s) may be that of the manufacturer or importer, or to a service such as Chemtrec.

Date Prepared. The date the information found on the MSDS was finalized or approved.

Section II. Hazardous Ingredients/Identity Information

This section lists the various components of the material and, when established, the allowable exposure limits. The MSDS must list the following exposure limits:

- Permissible Exposure Limits (PEL), established by OSHA; these are legal limits.
- Threshold Limit Value (TLV), established by the American Conference of Governmental Industrial Hygienists, reviewed and published annually.
- Any other exposure limit used or recommended by the manufacturer.

Chemical Family/Formula. Many manufacturers will include this information. The chemical family is the general class of the chemical, such as acid, solvent, organic amine, etc. For simple substances the manufacturer may provide the chemical formula.

Section III. Physical/Chemical Characteristics

This section lists chemical and physical properties of the substance as determined by laboratory testing. Only those tests applicable to the product will be shown and can vary from substance to substance.

Boiling Point. This is the temperature at which a liquid changes to a vapor, generally at a pressure of one atmosphere. In general, the lower the boiling point of a flammable liquid the greater the fire hazard.

Vapor Density. The relative density or weight of a vapor or gas (with no air present) compared to an equal volume of air at ambient temperature. With air rated at 1.0, a measurement greater than 1.0 indicates a vapor or gas heavier than air, less than 1.0 indicates it is lighter than air.

Solubility in Water. The percentage of a material (by weight) that will dissolve in water at ambient temperature. Solubility information can be useful in determining spill procedures and fire extinguishing agents and methods.

Appearance and Odor. A brief description of the material under normal room temperature and atmospheric pressure.

Specific Gravity. The ratio of the weight of the product compared to an equal volume of water. This is an expression of the density of the product.

Insoluble materials with specific gravities of less than 1.0 will float in water while materials with a specific gravity of more than 1.0 will sink in water. Most flammable liquids will float on water, an important consideration in fire fighting.

Melting Point. The temperature at which a solid substance changes to a liquid state. For mixtures, the melting range may be given.

pH. The degree of acidity or alkalinity of a solution, in a range of 0–14 with neutrality indicated as 7. The lower the number below 7 the more acid the solution; the higher the number above 7, the more alkaline or basic the solution.

Section IV. Fire and Explosion Hazard Data

This section describes factors that should be considered when encountering a fire or the potential for ignition of the chemical.

Flash Point. The flash point of a material is the lowest temperature which will cause vapor to be given off in sufficient quantity to ignite in the presence of an ignition source. Since flash points vary with the test method, the method is shown. Tag Closed Cup (PMCC), and Setaflash (SETA) are some of the more common test methods.

Flammable or Explosive Limits. When flammable vapors are mixed with air in the proper proportions, the mixture can be ignited. The range of concentrations of vapors over which the flash will occur is designated by the Lower Explosive Limit (LEL) and the Upper Explosive Limit (UEL). Flammable limits (explosive limits) are expressed as percent volume of vapor in air.

Extinguishing Media. Flammable or combustible chemicals behave differently when burning dependent on their physical characteristics and flammable characteristics; therefore, the extinguishing medium must be selected for its ability to extinguish a fire or not to increase the problems associated with the fire.

Special Fire-Fighting Procedures. General fire-fighting methods are not described but special or "exception to the rule" procedures may be listed.

Unusual Fire and Explosion Hazards. Hazardous chemical reactions, changes in chemical composition, or byproducts produced during fire or high heat conditions shown. Hazards associated with the application of extinguishing media will be shown if applicable.

Autoignition Temperature. The approximate lowest temperature at which a flammable or vapor-air mixture will spontaneously ignite without spark or flame.

Section V. Reactivity Data

This section describes any tendency or potential of the material to undergo a chemical change and release energy. Undesirable effects, such as temperature increase, formation of toxic or corrosive byproducts due to heating or as a result of contact with other materials described.

Stability. An expression of the ability of the material to remain unchanged.

Incompatibility. An indication of the byproducts which may result from contact with other materials.

Hazardous Decomposition. An indication of the relative hazards associated with decomposition of the material.

Section VI. Health Hazard Data Routes of Entry

This section provides information on the ways the chemical may enter the body.

Health Hazards/Effects of Exposure. This section provides information on the health effects associated with overexposure. Both acute and chronic effects should be listed. Many times there are three listings, one each for eye, skin, and inhalation. In addition, toxicological information may also be given. This data is usually the result of research.

Carcinogenicity. This section reports as to whether NTP, IARC, or OSHA have listed the substance as a known or suspected cancer causing chemical.

Signs and Symptoms of Exposure. A summary of some general effects (dizziness, nausea, headache, etc.) which are associated with exposure are indicated here.

Medical Conditions Aggravated by Exposure. Based on generally recognized effects and known cause and effect.

Emergency and First Aid Procedures. Based on anticipated effects, emergency and first-aid procedures are recommended.

Section VII. Precautions for Safe Handling and Use

Steps to Be Taken in Case Material Is Released or Spilled. Information describes how to properly contain and handle the material in the event of spills or leaks that may damage the environment.

This may include recommended cleanup materials, equipment, and personal protective clothing.

Waste Disposal. The manufacturer's recommended method for disposing of excess, spent, used, leaked, or spilled material.

Special Precautions. This section will provide information regarding special measures for storage and/or handling which were not covered in other sections.

Section VIII. Control Measures

Recommendations are given regarding types of control measures and protective devices that may be necessary. Recommendations will include personal protection such as respirators, eye and face protection, and protective clothing such as gloves. Engineering controls (e.g., local exhaust ventilation) may also be listed.

Other. Many manufacturers will provide information which is not designated on the Material Safety Data Sheet. Some of the common categories include:

- DOT (Department of Transportation) information such as required hazard labeling and placards.
- EPA (Environmental Protection Agency) information including whether the material is considered a hazardous waste as defined by EPA.
- CAS (Chemical Abstract Service) number. The CAS number is a unique number given to a chemical. The CAS registry can provide additional information about the chemical. It should be noted that not all chemicals have CAS numbers.
- Issue information, that is, the date the sheet was issued or last updated.

REFERENCES

E. Hodgson, and Levi, P. E. 1987 *A Textbook of Modern Toxicology*.

C. D. Klaassen, M. O. Amdur, and J. Doull. *Casarett and Doull's Toxicology*, The Basic Science of Poisons. 3rd Ed., p. 13. New York: Macmillan.

International Chemical Workers Union. 1989 *Draft Hazardous Waste and Chemical Emergency Response Resource Manual*. Akron, OH.

U.S. Department of Health and Human Services, 1985 *NIOSH Pocket Guide to Chemical Hazards*. Washington, DC: U.S. Govt. Printing Office.

OSHA. 1989. 29 CFR 19010.1000.

6
Engineering Controls

INTRODUCTION

In almost all situations, it is the individual who has most control over the safety of a hazardous waste activity. Built-in protection measures should, however, be inherent in any phase of a Superfund or other hazardous waste project. A complete understanding of the circumstances surrounding an operation and its problems is required so that the best control methods can be chosen. Unfortunately, hazardous waste sites represent the ultimate in an uncontrolled workplace. Nonetheless, the health of workers in a hazardous environment must be protected by controlling exposure to chemicals, fires, explosions, and physical agents.

Administrative controls (such as limiting exposures to hazardous chemicals) and personal protective equipment must be considered secondary to the use of engineering control methods because they are difficult to implement and maintain. The best time to introduce engineering controls is during the design phase of a cleanup so that they can be integrated into the operation. The design of engineering controls is made more difficult at a remediation site because of multiple substances and activities in the same area. Close cooperation is required between those characterizing the site and those designing the cleanup plans. At times the operation may have to shut down as conditions change and new engineering controls need to be implemented (National Safety Council, 1979).

Traditional engineering controls such as ventilation and explosion- and flameproof designs are mandated by stringent OSHA and EPA regulations, although many site conditions limit the use of these methods. This chapter will address engineering controls which are most applicable for an uncontrolled hazardous waste site cleanup. They are:

- Zoning of a site.
- Site characterization.
- Trenching methods and hazards.
- The use and investigation of dikes.

OBJECTIVES

- The reader will understand the definition of engineering controls.
- The reader will understand the reasons for engineering controls.
- The reader will be able to identify engineering controls that are used on hazardous waste sites and why.

SITE CHARACTERIZATION

The characterization of an uncontrolled hazardous waste site is the first task in the process of cleaning up a site. It is an important job, because the safety and health of those who will become involved in the cleanup are dependent upon the accuracy and thoroughness of the site characterization. As mentioned in Chapter 2 on Rights and Responsibilities, the site characterization is an integral part of the safety and health plan section of the OSHA standard, 1910.120.

These legal requirements from the OSHA standard are to ensure that proper personal protective equipment will be used from the initial entry onward. OSHA requires that workers be protected from exposures above permissible exposure limits (PELs). Escape air supplies must be available during the initial entry, where respirators are indicated as a result of the off-site assessment, unless a self-contained breathing apparatus (SCBA) is utilized. Engineering controls must be integrated into this phase also. The standards for site characterization require that workers be made aware of the dangers involving the hazardous wastes to which they may be exposed, and that they understand how to use proper judgment and caution.

The site characterization is crucial to the success of a cleanup and therefore must be done accurately. Site characterization furnishes specific information vital to worker protection.

There are three broad stages in a site characterization:

- *The off-site characterization*, which consists of a historical search and a perimeter reconnaissance.
- *The on-site survey*, which begins with an initial entry.
- *The ongoing monitoring and hazard assessment program* to keep abreast of the hazards as activities change (NIOSH/OSHA/CG/EPA, 1985).

Some preliminary considerations will direct the investigator towards completing these three phases successfully. The same concepts of recognition, evaluation, and control that guide the industrial hygienist in a normal industrial setting also apply here. There are differences in assessing a hazardous waste site, however. These include:

- The wide range of safety and health concerns.
- The considerations for the community as well as the worker.
- The realization that contaminants may be unique, unknown, multiple, and hard to identify.

These considerations force extensive preplanning and information gathering. Therefore, the first phase of a site characterization is off site, and consists of a historical search and a perimeter reconnaissance.

Historical Research

As with most of the steps in a site characterization, the information gathered during off-site characterization provides the foundation for the remainder of the project. Questions concerning the waste which the preliminary assessment must answer include:

- What is the waste composed of?
- What are the characteristics of the waste?
- Are the wastes mobile?
- By what routes might they migrate?
- What effects could result through discharge to air, water?
- What initial remediation steps are there?

Other information to be gathered should answer:

- Does an emergency exist?
- What is the severity of the problem?
- What specific areas need focus?
- What hazards and precautions are necessary for field personnel?
- What priorities should therefore be placed on further investigations?
- What previous studies can supplement the information?
- What resources are needed for the investigation?
- Who is responsible for documenting information?
- What is the expected duration of the clean-up? (Levine and Martin, 1985; OECD, 1983)

There are three sources to be tapped in answering these questions: people, files, and maps. A title search can determine who owned the site, and perhaps what was done there. It is very helpful to have statements made by previously involved owners, employees, or adjacent landowners verified by other individuals, and, if possible, to prepare written statements supporting those statements. If current or previous employees of a site can be located, they should be informed of the employee protection provisions under RCRA, section 7001, so that they may respond to questions without feeling intimidated.

If personal injury or property damage was claimed by anyone at or regarding the site, find out who was the physician or insurance adjuster involved. Local EPA branch office personnel or health department personnel with expertise in toxic substances, drinking water, solid waste, or enforcement may be aware of the site and be able to provide help. Government officials will know whether a RCRA or CERCLA notification file was processed for the site, and whether the site operator had a National Pollution Discharge Elimination System (NPDES) wastewater permit. If so, these permits will provide pertinent information.

If surface impoundments exist on the site, state inventories may describe them in SDWA (Safe Drinking Water Act), RCRA, or USGS (United States Geological Survey) groundwater investigations.

The historical search can also be aided through the use of computerized databases and other automated information systems.

Examples of Databases (Levine, Martin, 1985).

- Corporate information:
  — Subsidiaries.
  — Profit/loss statements.
  — Boards of directors.
  — History of company.
  — Products
- Information on specific chemicals:
  — Toxicity.
  — Physical or chemical properties.
  — Manufacturers.
  — Locations.
- Ownership of property, operations, employees, leasee, operator.
- Chemical Regulations and Guidelines System.
- Congressional Information Service—(800) 227-1960.
- SDC search service—(800) 421-7229.
- Westlaw (legal search)—(800) 328-9833.
- National Library of Medicine (source for health effects)—(800) 638-8480.
- NIH–EPA Chemical Info System (4 databases)—(800) 368-3432.
- Dun and Bradstreet—(212) 285-7000.
- Hazardline—(808) 223-8978.
- Records of Generators (since 1976), manifests and biannual summaries.

We recommend that (because of cost considerations) only experienced computer users log onto the premium data services. It can be less expensive to hire someone to do computer searches for you. To gain experience, novice computer users should consider familiarizing themselves with the more general computer database services such as the Compuserve network or Dialog. From this point, if they are convinced they can use the information listed in these services, they can then move onto more specific computer services online.

Site Map

Finally, a historical search must be supplemented with geologic, climatic, topographic, and environmental data which can be organized on a detailed site map. The type of soil or overburden must be determined, as well as its depth, permeability, and vegetative cover. This information can often be obtained from the county soil conservation service. The type of bedrock, its depth, its structure, and its ability to allow contaminants to migrate must be understood.

Surface waters must be located. Knowledge of local aquifers, their gradient, use, nature (confined or artesian, perched, etc.), direction, and rate of movement should be determined. For many areas nationwide, USGS water quality data may be found at NAWDEX—(703) 860-6031.

Environmental data should include local wells, floodplains, wetlands, sinkhole-prone areas, or other ecologically sensitive areas, and local land use (Desmarais and Exner, 1984). Topography, climate, prevailing winds, and population density must be considered. Topographic maps, aerial photos, even infrared or side-scan radar mosaic imagery (SLAR) may be of assistance. By means of all these sources a comprehensive understanding of a site can be developed in order to answer most questions for maximum safety during the next step, the perimeter reconnaissance.

Perimeter Reconnaissance

Armed with the information obtained from a thorough historical search, the investigator now conducts a perimeter reconnaissance. This will complete the site map and determine the personal protective equipment (PPE) needed for the initial entry. In walking around the site, careful observation (with binoculars and camera) should focus on (NIOSH/OSHA/CG/EPA, 1985):

- Buildings.
- Tanks, drums, or other containers.
- Labels, signs, placards, etc.
- Visible deterioration or unusual conditions.
- Biological indicators.
- Impoundments, location and size.

- Surface water/liquids and their color.
- Wind direction and barriers.

Constant monitoring of the air for radiation, combustible gases, toxic substances, and oxygen deficiency is essential. Sample the soil, surface waters passing through the site, and groundwater around the perimeter (Bixler and Hamson, 1984; Desmarais and Exner, 1984). Geophysical investigations of the subsurface, such as resistivity or ground penetrating radar may be warranted. Leave the area immediately if the presence of hazards above safe levels is indicated. Maintain constant communications with others. Document all observations thoroughly. Note the extent of field documentation listed in Table 6-1.

On-Site Survey and Hazard Assessment

The on-site survey is a more cautious and thorough repetition of the perimeter reconnaissance. Unless site hazards are positively identified during off-site assessment, personnel making the initial entry must use at least level B ensembles of personal protective equipment. Constant monitoring and acute observation for IDLH conditions are required. Verify the condition of the terrain, containers, impoundments, and any indicators of contamination. Note any safety hazards: confined spaces, cluttered or irregular surfaces, etc.

As with the perimeter reconnaissance, use remote sensing for the subsurface, sample the ambient air, water, and soil, and document procedures and label samples thoroughly; see Chapter 10 on Safe Sampling Procedures. Stress must be placed on cautious, conservative actions and careful observation. A mini-

Table 6-1. Example of Field Logbook Entries to Describe Sampling.

- Date and time of entry.
- Purpose of sampling.
- Name, address, and affiliation of personnel performing sampling.
- Name and address of the material's producer, if known.
- Type of material, e.g., sludge or wastewater.
- Description of material container.
- Description of sample.
- Chemical components and concentrations, if known.
- Number and size of samples taken.
- Description and location of the sampling point.
- Date and time of sample collection.
- Difficulties experienced in obtaining sample (e.g., is it representative of the bulk material?).
- Visual references, such as maps or photographs of the sampling site.
- Field observations, such as weather conditions during sampling periods.
- Field measurements of the materials, e.g., explosiveness, flammability, or pH.
- Whether chain-of-custody forms have been filled out for the samples.

mum of two people go on the site and two others, with similar PPE and radio communications, will remain at the perimeter as support. The individuals involved in this operation will consist of the project team leader and experienced laborers or technicians. Backup support should come from chemists, industrial hygienists, geologists/hydrogeologists, health physicists, and toxicologists (NIOSH/OSHA/USCG/EPA, 1985).

After all of the accumulated data have been gleaned from the samples and observations, in combination with information from the historical search, a site safety plan can be developed for the actual cleanup. Ongoing monitoring and hazard assessment will continue as the job progresses, so that changes in exposure potential are realized and taken into account.

ZONING

Zoning of an uncontrolled hazardous waste site is another form of engineering control. Zoning minimizes the spreading of chemical contamination out of the contaminated areas of the site.

The area of maximum contamination is called the *exclusion zone* or *hot zone*. It is where the highest level of personal protective equipment is needed and where the actual remediation/removal work takes place. Its boundary is called the *hotline*, and is often a physical barrier such as a chain link fence. Several hot zones may exist at a site, each with a different level of protection required. Therefore, zones and levels of protection must be identified on the site map and posted for review by all workers.

Entrance to the hot zone is through *access control points* where supplies, rest areas, emergency response personnel, and decontamination exist or occur. One corridor exiting the hot zone through the access control points is called the *contamination reduction corridor* (CRC). It leads into the next lower level of contamination. This buffer zone, which ideally surrounds the hot zone, is called the *contamination reduction zone* (CRZ). At the outer boundary of the CRZ is the contamination control line, through which no contaminants should pass. Outside of this is the *support zone*, where the command post exists, preferably upwind of the hot zone. The administrative, transportation, communications, weather center, and (if necessary) lodging facilities are placed here also.

In ideal conditions, these zones would be arranged in the shape of a target, or concentrically, so that access to and migration from the center is buffered by these zonations (see Fig. 6-1). Many real sites, however, cannot conform to these ideals due to topography, urban setting, or other impediments (see Figs. 6-2 and 6-3). Further, there may be more then one exclusion zone at a site or there may be areas within the exclusion zone which are "hotter" than their immediate surroundings. This will require different levels of personal protective equipment.

In delineating the size and locations of these zones at a site many factors

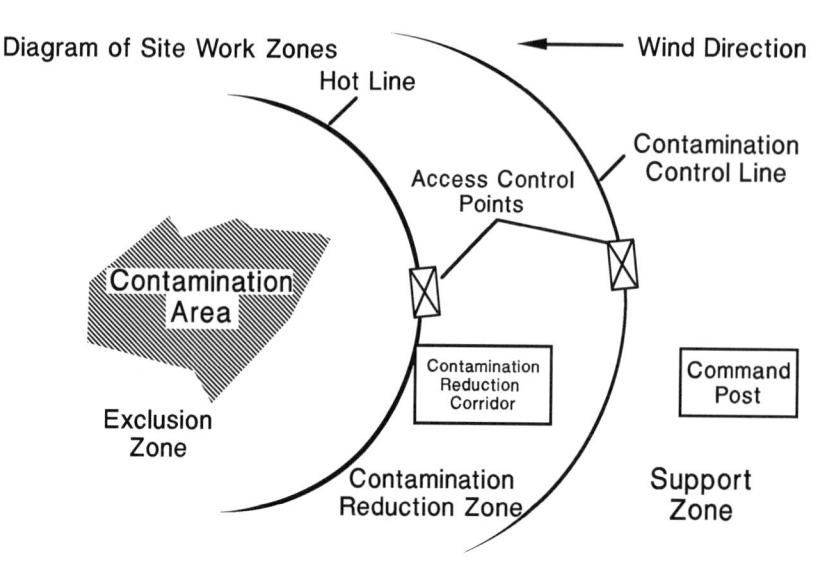

Fig. 6-1. Diagram of site work zones. (NIOSH, 1985.)

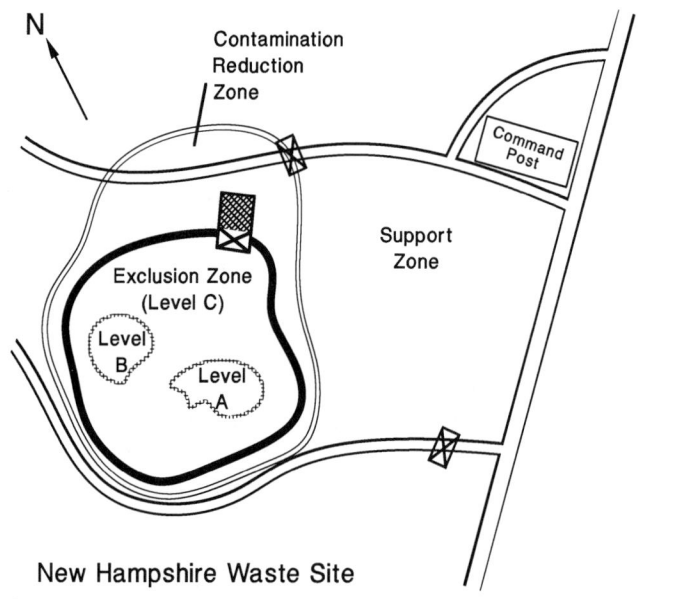

Fig. 6-2. New Hampshire Hazardous Waste Site Zones. (NIOSH, 1985).

ENGINEERING CONTROLS 121

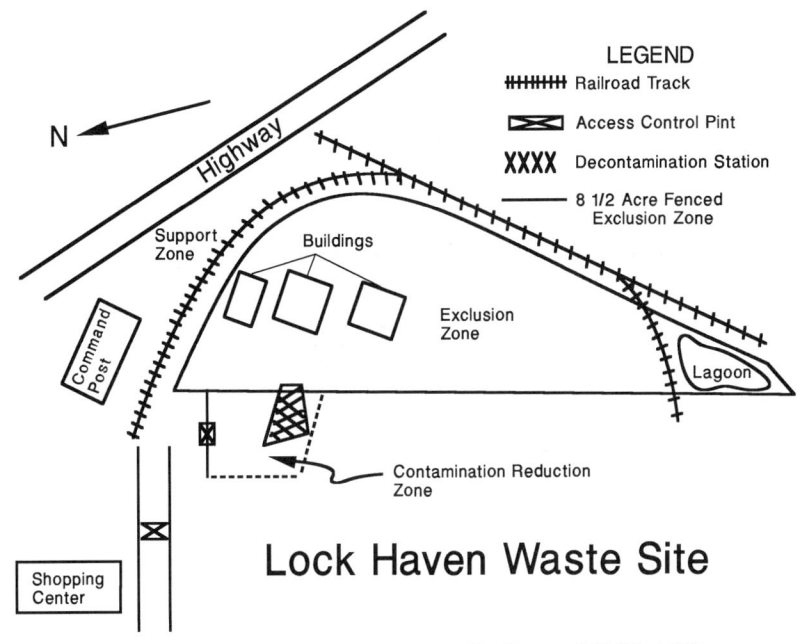

Fig. 6-3. Lock Haven Hazardous Waste Site Zones. (NIOSH, 1985).

come into play. Establishing the exclusion zone and its hotline is done by visually surveying the area and determining where the contaminant is, where streams or water bodies intersect the contaminant, and by determining through monitoring activities where airborne concentrations of contaminants exceed safe levels. The established hotline may move as conditions change. Other considerations for exclusion zone parameters involve fire/explosion buffer distances and room necessary for equipment and operations to take place. Understanding the physical, chemical, and toxicological characteristics of the substances present will increase and further delineate boundaries as a job progresses.

People and equipment must pass through access control points to enter or exit the contamination reduction zone. There are usually separate contamination reduction corridors for each. Long-term operations should involve methods (such as air surveillance, swipe testing, and visual examination) to determine if material is being transferred between zones. Site zones must be given thoughtful consideration based on all available information due to their site specific nature. In conclusion, zoning criteria and methodology provide essential engineering controls for the safety of the workplace/site (NIOSH/OSHA/USCG/EPA, 1985).

TRENCHING

Another engineering control that is commonly found on hazardous waste site cleanups are trenches (Fig. 6-4). Trenches are narrow excavations made below the surface of ground, generally deeper than they are wide and less than 15 feet in width. They are used in the risk assessment/feasibility study phase of a project and in the construction or cleanup phase. Activities in the trenches include sampling buried containers; sampling subsurface soil, sludges, and other materials; and establishing control areas around extremely hot areas to prevent further contamination.

Employees working in trenches may be exposed to many hazards other than the chemicals that are to be removed in the cleanup operation. Workers may come in contact with underground utility lines or pipes which can cause a potentially fatal electrical or fire hazard. Typical hazards include cave-in of trench walls, oxygen-deficient atmospheres, accumulations of heavier-than-air gases or vapors, and objects falling on workers from a higher elevation. Slips and falls are also prevalent inside the trench or from ground level into the trench. Heavy equipment in operation around the trench can cause serious accidents if all employees are not observant of heavy equipment safety rules (see Chapter 7).

Fig. 6-4. Typical existing trench at a hazardous waste site (EPA, 1987).

Injuries associated with trenching on hazardous waste sites can be minimized by following OSHA Construction Standards contained in Title 29 CFR 1926 Subpart P, as outlined below. (See also Table 6-2 concerning specific trench shoring requirements.)

Trenching Safety Precautions

Prior to excavation, observe the following precautions:

- Check the area to be excavated for any underground pipelines, transmission lines, etc. Consult with utility companies, as needed.
- Determine soil composition (e.g., through soil sampling, soil maps, etc.), and other relevant site conditions, with special emphasis on conditions conducive to cave-ins.
- Formulate a site-specific trenching safety plan for dealing with trench-related hazards. (Not required by OSHA, but recommended). Update the plan as required by changing conditions on site throughout the duration of site activities.
- Train all employees involved in safe trenching practices, with emphasis on factors such as (Melton, 1988):
  — Utility line locations.
  — Cave-in prevention measures.
  — Recognition of conditions which may cause cave-in.
  — Clues to impending cave-in (e.g., tension cracks, bulging walls, etc.).
  — Means of egress from trench.

During excavation and work in trenches, observe the following precautions:

- Follow standard construction safety procedures:
  — Heavy equipment safety.
  — Good housekeeping (e.g., keep tools and equipment clear of tops of trench walls).
  — Wear hardhats and other required protective equipment.
- Utilize ditches, dikes, pumps, or other means to keep surface water out of trenches.
- Water should not be allowed to accumulate in any excavation.
- Monitor the atmosphere in and around trenches on a regular basis to check for explosive, toxic, or otherwise dangerous gases and vapors:
  — Bear in mind that trenches represent a confined space hazard, as well as a low-lying area hazard.
  — Be especially cautious if heavier-than-air gases (i.e., gases having a vapor density in excess of one) are encountered.
  — Utilize appropriate engineering controls (e.g., ventilation), work practices, and personal protective equipment as needed.

- Trenches in excess of 4 feet deep must have steps or ladders located so that all workers within the trench are within 25 feet of a place of exit.
- Excavated material (e.g., "back dirt") shall be placed at least 2 feet from the edges of excavations, unless effective barriers are in place to prevent the excavated material from falling into the excavation.
- Precautions to prevent cave-in (as described below) should be strictly followed.

Cave-in Hazards

The following conditions increase the likelihood of cave-in:

- Soil materials composed of unconsolidated, uncompacted, and/or rounded particles (see Table 6-2 for relative stabilities of various materials). Special care must be used when trenching in areas which have previously been excavated and backfilled.
- Soils which have a high water content, or have been subjected to freeze-thaw or frost-heaving.
- Loading of trench walls by adjacent equipment, supplies, structures, "back-dirt" piles, etc.
- Vibration due to equipment operating near excavations.
- Trench walls which are steeper than the angle of repose of the material composing the walls (see Fig. 6-5).
- Deep trenches (i.e., high trench walls).

The following precautions should be used to prevent cave-ins in all trenches in excess of 5 feet deep. These precautions should also be used in trenches less than 5 feet deep whenever those site conditions just listed indicate the likelihood of a cave-in:

- *Sloping*. Trench walls should be sloped to the correct angle of repose, as shown in Fig. 6-5.
- *Shoring*. Vertical trench walls (unlesss composed of solid rock) must be shored and braced, or restrained with movable trench boxes, to prevent cave-ins (see Figs. 6-6 and 6-7).
  - Shoring systems must be designed by a qualified person and meet accepted engineering requirements (see Fig. 6-6 and Table 6-2).
  - Shoring must be installed from the top down, and removed from the bottom up.
- Additional precautions (e.g., added shoring bracing or a flatter slope angle) should be utilized whenever site conditions indicate that they are needed to prevent cave-ins.
- Excavations should be inspected by a competent person daily, and after

Table 6-2. Trench Shoring—Minimum Requirements.

Depth of trench	Kind of condition of earth	Uprights Minimum dimension (Inches)	Uprights Maximum spacing (Feet)	Stringers Minimum dimension (Inches)	Stringers Maximum spacing (Feet)	Cross braces: Width of trench Up to 3 feet (Inches)	3 to 6 feet (Inches)	6 to 9 feet (Inches)	9 to 12 feet (Inches)	12 to 15 feet (Inches)	Maximum spacing Vertical (Feet)	Maximum spacing Horizontal (Feet)
Feet 5 to 10	Hard, compact	3 × 4 or 2 × 6	6	—	—	2 × 6	4 × 4	4 × 6	6 × 6	6 × 8	4	6
	Likely to crack	3 × 4 or 2 × 6	3	4 × 6	4	2 × 6	4 × 4	4 × 6	6 × 6	6 × 8	4	6
	Soft, sandy, or filled	3 × 4 or 2 × 6	Close sheeting	4 × 6	4	4 × 4	4 × 6	6 × 6	6 × 8	8 × 8	4	6
	Hydrostatic pressure	3 × 4 or 2 × 6	Close sheeting	6 × 8	4	4 × 4	4 × 6	6 × 6	6 × 8	8 × 8	4	6
10 to 15	Hard	3 × 4 or 2 × 6	4	4 × 6	4	4 × 4	4 × 6	6 × 6	6 × 8	8 × 8	4	6
	Likely to crack	3 × 4 or 2 × 6	2	4 × 6	4	4 × 4	4 × 6	6 × 6	6 × 8	8 × 8	4	6
	Soft, sandy, or filled	3 × 4 or 2 × 6	Close sheeting	4 × 6	4	4 × 6	6 × 6	6 × 8	8 × 8	8 × 10	4	6
	Hydrostatic pressure	3 × 6	Close sheeting	6 × 10	4	4 × 6	6 × 6	6 × 8	8 × 8	8 × 10	4	6
15 to 20	All kinds or conditions	3 × 6	Close sheeting	4 × 12	4	4 × 12	6 × 8	8 × 8	8 × 10	10 × 10	4	6
Over 20	All kinds or conditions	3 × 6	Close sheeting	6 × 8	4	4 × 12	8 × 8	8 × 10	10 × 10	10 × 12	4	6

Trench jacks may be used in lieu of, or in combination with, cross braces.
Shoring is not required in solid rock, hard shale, or hard slag.
Where desirable, steel sheet piling and bracing of equal strength may be substituted for wood.
Source: 29 CFR 1926.652.

Approximate Angle of Repose
For Sloping of Sides of Excavations

Solid Rock Shale or Cemented
Sand and Gravels (90°)

Compacted Angular Gravels
1/2 : 1 (63°)

Recommended Slope for Average Soils
1 : 1 (45°)

Well Rounded Loose Sand
2 : 1 (26°)

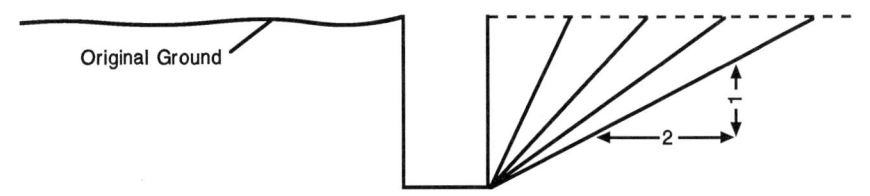

Fig. 6-5. Angle of repose (OSHA, 1975).

any event (e.g., rainfall) which may increase the likelihood of cave-in. If inspection indicates the potential for a cave-in, all work should cease until appropriate precautionary measures are taken.

Note: Trenches which (1) will not be entered by workers and (2) will be immediately backfilled (e.g., for permeable treatment walls) are exempted from the requirements listed here.

When the procedures outlined above are followed, trenching can be a useful control on a hazardous waste site. It should be emphasized that construction hazards must be avoided as much as chemical exposure hazards; therefore, site safety training should include an in-depth trench hazard awareness if trenching will be used in site remediation.

DIKING

Dikes and Diking Systems

Dikes are relatively impermeable barriers used to block the flow of liquid contaminants. Dikes are frequently used for containing liquids on hazardous waste sites due to EPA requirements under RCRA to prevent any rainwater runon or runoff at the contaminated site. Dike-related activities may involve working

ENGINEERING CONTROLS 127

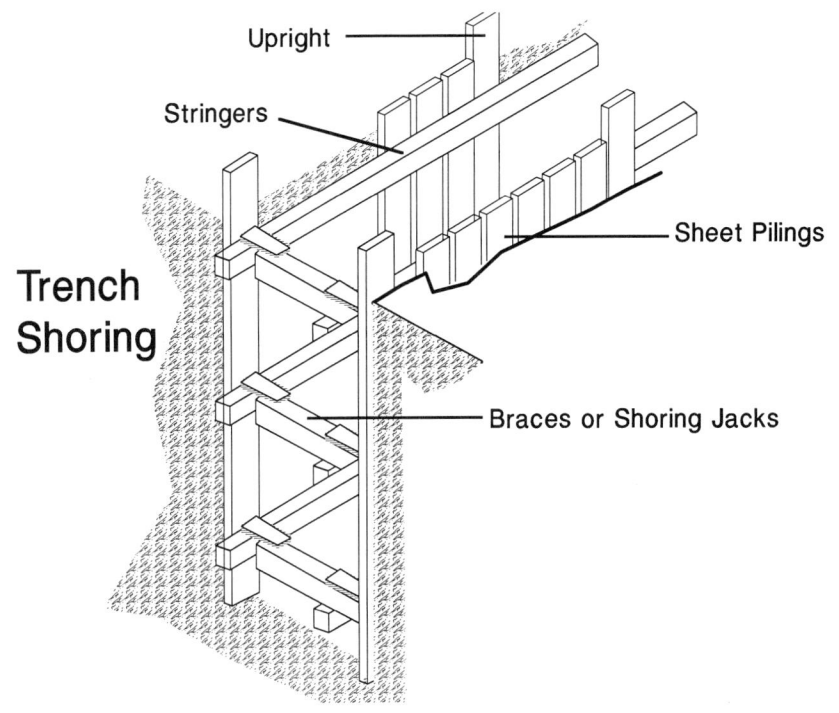

Fig. 6-6. Trench shoring details (OSHA, 1975).

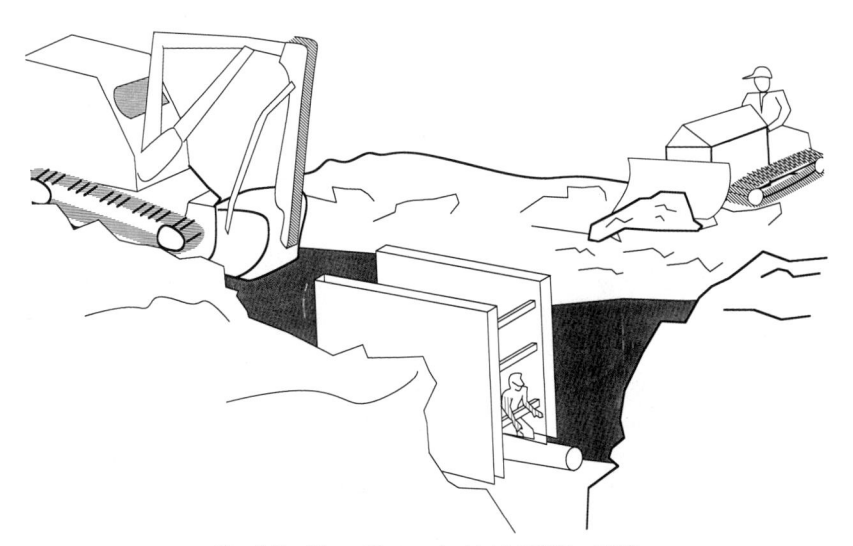

Fig. 6-7. Moveable trench shield (OSHA, 1975).

around existing dikes (e.g., around ponds and lagoons) and construction of dike systems (for spill containment). Thus, a basic knowledge of diking is desirable during site operations as it is an important engineering control to limit the spread of contamination.

Assessing the Stability of Existing Dikes

The failure of a diking systems may result in loss of life, environmental damage, and/or property damage. Therefore, appropriate caution must be used when working around existing dikes on hazardous waste sites. Factors to consider when assessing dike stability include foundation condition, material used in dike construction, type of liner (if any) used, and type of waste material impounded.

Information pertaining to dike stability may be made available as follows:

- *Historical Information* may be gathered through sources such as as-built construction plans and specifications, inspection reports, and existing geotechnical data. The location of the existing impoundment, lagoon, or pond should be plotted on a map showing the failure impact zone or leaking areas.
- *Reconnaisance Investigation* should be conducted through a detailed site investigation by a team with appropriate expertise, such as engineers, hydrogeologists, and other environmental specialists.
- *Geotechnical Investigation* will be required if available historical information is insufficient to allow for a complete analysis of dike stability. Geotechnical investigation may require compaction studies, seismic investigation, bedrock mapping, etc.
- *Engineering analysis* is required before dike stability can be positively determined. Stability analysis must be site-specific and requires a systematic technical approach and the professional judgment of experienced engineers. *Stability analysis cannot be based on visual inspection alone.*

Note: Dikes may require *periodic inspection* as site conditions change (e.g., heavy rainfall, freezing weather, and heavy equipment operation, etc. may affect stability).

Visible indicators of potential instability of dikes include areas of seepage or leakage in or around dikes, settling of dikes, cracks in dikes, and bulging or slumping of dikes. Sinkholes in or around dikes, erosion of dikes, undercutting of the toe of slope of the dike, growth of vegetation (e.g., tree roots) on dikes, and animal burrows in dikes are other obvious visual indicators of instability of diking systems. The absence of any or all of these features does not indicate stability of the dike.

Dikes and Spill Containment

Diking can be used for spill containment as a prevention measure and as an emergency response measure. As a prevention measure, staging areas on hazardous waste sites should be designed so that all liquids spilled will flow to the center for containment and ease of collection. All drainageways or low-lying areas sloping away from the staging area should be blocked with dikes. The staging area should be graded to slope toward the center (i.e., the center should be the lowest point). In some instances (e.g., sites requiring the handling of large numbers of full drums), a plastic liner may be required to limit the contamination of soil underlying the staging area. See Chapter 12 on Emergency Procedures for accidental spill handling procedures.

Construction of Dike Systems

On hazardous waste sites dikes are typically composed of earthen materials, specifically high density clays with geo-membrane double liners (made from high-density polyethylene). Lining systems as a method of containment for waste liquids and leachate are used on RCRA permitted hazardous waste disposal facilities and are not considered within scope of this chapter. Dikes and dike systems should be designed and constructed so as to adequately contain potential spills onsite. Fig. 6-8 shows a general spill containment system plan. Ramps are used to provide access for vehicles (see Section A-A in Fig. 6-8).

Staging Area Dike System

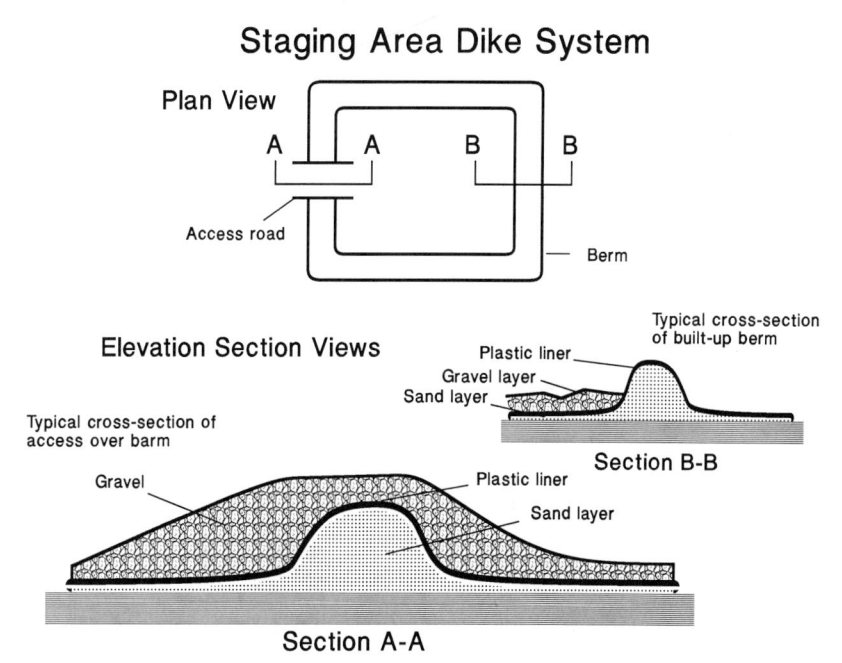

Fig. 6-8. Dike design for spill containment.

EPA regulations contained in Title 40 CFR Parts 264 and 265 require that areas used for the storage of hazardous wastes be surrounded by a dike system. The dike system must be able to contain the greater of the following volumes: 10% of the total volume stored in the diked area or the total volume of the largest single container in the area. Also, Title 40 CFR 112.7 presents the standards for the preparation and implementation of a spill prevention control and countermeasure plan (SPCC plan) and the requirements for storage of petroleum-based products.

REFERENCES

Bixler, D.B., and Hanson, J.B. 1984. Selecting Superfund Remedial Actions. In *Proceedings of the 5th National Conference on Management of Uncontrolled Hazardous Waste Sites*, pp. 493–497. Silver Spring, MD: Hazardous Materials Control Research Institute.

Desmarais, A.M.C., and Exner, P.J. 1984. The Importance of Endangerment Assessment in Superfund Feasibility Studies. In *Proceedings of the 5th National Conference on Management of Uncontrolled Hazardous Waste Sites*, pp. 226–229. Silver Spring, MD: Hazardous Materials Control Research Institute.

Levine, S.P., and Martin, W. F. 1985. *Protecting Personnel at Hazardous Waste Sites*. Stoneham, MA: Butterworth Publishers.

Melton, M. 1988. Through Proper Training, Education, Trenching Hazards can be Reduced. *News Digest*, February, pp. 5–6.

National Safety Council. 1979. *Fundamentals of Industrial Hygiene*, 2nd Ed., J. B. Olishifski, Editor, pp. 614–634. Chicago, IL: National Safety Council.

OECD. 1983. Hazardous Waste "Problem" Sites, Chapter 2. Paris, France: Organization for Economic Co-operation and Development.

U.S. OSHA. 1987. Title 29 CFR 1926 Subpart P. Washington, DC: U.S. Govt. Printing Office.

U.S. NIOSH/OSHA/CG/EPA. 1985. *Occupational Safety and Health Guidance Manual for Hazardous Waste Site Activities*, Chapter 6, NIOSH Publication no. 85-115. Washington, DC: U.S. Govt. Printing Office.

U.S. OSHA. 1975. *Excavating and Trenching Operations*. OSHA Publication no. 2226. Washington, DC: U.S. Govt. Printing Office.

U.S. EPA. 1983. *Training Manual for Hazardous Waste Workers* (Draft Copy).

U.S. EPA. 1984. *Standard Operating Safety Guides*. Washington, DC: The Office of Solid Waste and Emergency Response.

7
Safe Work Practices

INTRODUCTION

Safety considerations are frequently a subject that people do not want to hear about. Some people naturally think safety is just a matter of common sense; preaching safety at people is sometimes equated with insulting their intelligence. Others think that all the safety talk is equivalent to "big brother" watching over them. Some OSHA or employer's safety requirements may appear to be picky or useless, yet they are mandatory for employment.

There is another way to look at safety in the workplace. Safety can be an attitude of cooperation with your co-workers that can keep each of you alive; an attitude of positive social reinforcement that will not only work better, but work easier, too. To use this approach requires a little psychology. Once you think about safety as learned behavior used by a working group, the old "big brother" or "don't insult my intelligence" perspectives will seem outdated. This chapter begins, therefore, with a look at what makes people tick, and how positive social reinforcement works. The remainder of the chapter will focus on ways to minimize hazards commonly associated with hazardous waste site cleanups.

OBJECTIVES

- The reader will have an overall awareness of what causes unsafe acts and conditions.
- The reader will understand the use of positive social reinforcement in maintaining safe work practices.
- General rules for machinery, heavy equipment, and drill rig safety are presented for reference.
- The reader will be able to understand and safely carry out a confined spaces operation.
- Safe drum handling will be understood in a step-by-step approach.

UNSAFE ACTS

Why do people do the things they do? Behavior is controlled by cognition (the act of knowing) and the environment, according to social-learning theory. In other words, experience and its influences make us organize and select from the stimuli in our world so that we can affect and control what happens (Latham and Sari, 1979; Goldberg, 1975). Of course, we are not perfect, things are forgotten, and things are not learned correctly. Bad habits and attitudes develop in all of us when we do not keep goals in mind and concentrate. Additionally, physical impairments or plain bad judgment can cause events to get out of control (see Table 7-1).

People studying means of safe behavior recognize these facts. It is known that even accidents and injuries may only temporarily reinforce safe behavior. Without *direct*, *immediate*, and *consistent* reinforcement no one will continually follow any pursuit, whether it is safety rules or anything else. The consequences of unsafe acts are often not immediate and are never consistent. For example, not wearing a respirator when advised to may not cause illness for a long time. Also, not every unsafe act leads to injuries (NSC, 1978). For example, cigarette smoking around combustibles does not always lead to an explosion.

There are three excellent ways to correct unsafe acts: *concentration*, *training*, and developing good *habits*. Many factors may contribute to a lack of concentration. (Grimaldi, and Simonds, 1975):

- Too many interests.
- Lack of interest.
- Worry and fear.
- Narcotic effect from contaminants.

Thinking about the job at hand is crucial. After all, *you* are in control of events more than anyone or anything else.

Training organizes and develops patterns of activity by remembrance and repetition. Remembrance is a process by which someone receives an impression, retains it through an act of mental association, and can therefore recall it later.

Table 7-1. Human Factors That Create Unsafe Acts.

Physical	*Developmental*
Bad vision	Inexperience
Bad hearing	Emotional instability
Fatigue	Poor perception (visual discrimination)
Discomfort	
Youth	Immaturity
Poor reaction time	Bad attitude
Apathy	
Low intelligence	

The training sequence is as follows (Atec and Associates, n.d.):

- Tell the worker what he must do.
- Show him how to do it.
- Watch closely as he does it.
- Correct him (without ridicule) when he is wrong.
- Commend him when he does a job well.
- Warn him of dangers.
- Do not allow him to work alone until you are sure he is able to do so.
- Repeat any action above if necessary to assure proper training.

A good habit is developed (Grimaldi and Simonds, 1975) by:

- Defining the desired habit.
- Practicing repeatedly, and creating opportunities to do so.
- Allowing *no* exceptions to occur—never lapsing into the old habit.

Concentration, training, and the development of good habits can be used in a program where you work. It is a program of *positive social reinforcement* (see Table 7-2). That means keeping aware of the workers around you, and always telling them they are doing well when their actions are safety minded. It develops communication, trust, and friendship. It is the reverse of punishment (negative social reinforcement). Although punishment has a place, it says nothing about correct behavior, and usually affects behavior in other ways than those intended (such as creating resentment).

Judge this example for yourself. Alan, a fairly new employee at a cleanup site, is working with Ralph testing some unknown liquids during an initial staging operation. The zero knob on his oxidation meter is very stiff, and Alan cannot seem to set it correctly wearing two pairs of gloves. He has found it much easier to take off one of the outer gloves to do this. In a positive reinforcement response, Ralph sees what Alan is doing and says: "Look Alan, I've

Table 7-2. Social Reinforcement for Safe Procedures (after Goldberg, 1975).

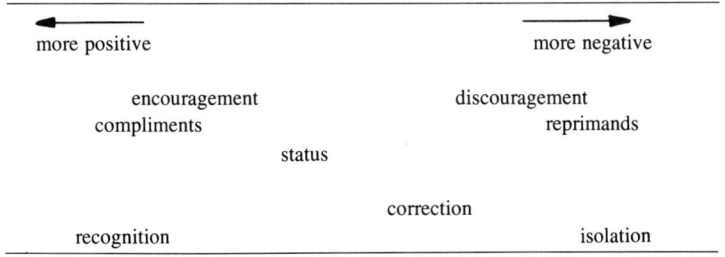

← more positive	more negative →
encouragement	discouragement
compliments	reprimands
status	
	correction
recognition	isolation

had trouble with that contraption too, so I figured out how to do it this way. It takes longer, but it sure beats getting that creeping crud on your hand." Then he lets Alan try it himself, and compliments him. Later on, everybody was changing into street clothes and Ralph said to Pete and Fred: "Alan really picks up on some of the hassles with our PPE well. You guys will appreciate working with him." Alan, of course, overheard this and resolved to maintain the two glove procedure, regardless of who he was working with. In a negative reinforcement response, Ralph sees what Alan is doing and says: "You idiot! It is such a hassle dealing with you new people. Use both gloves or I'll talk to the supervisor." Alan, of course, put his outer glove back on.

Working at positive social reinforcement at a job site requires concentration and awareness. It is training in action; reinforced behavior by your peers will be remembered long after you have read this book. The consistent repetition of well placed compliments and encouragement develops good safety habits in the workplace. When your friends are looking out for you, it is a lot more meaningful than a boss looking over you. A successful safety program based on positive social reinforcement will blend education with practical knowledge and fellowship.

UNSAFE CONDITIONS

The Site

Safe attitudes and behavior at a hazardous waste site must go hand in hand with an understanding of the unsafe conditions. The most obvious unsafe condition is the physical state of the site itself. When first encountered, a site may be nothing more than a vacant lot or field. On the other hand, it may be a chaotic jumble of leaking containers, twisted, rusting steel and machinery, black, stinking lagoons, and anything else you can imagine. As work begins the site changes, and new unsafe conditions present themselves. Excavation creates muddy, irregular terrain, and exposes those working there to new contaminants. The staging, testing, and bulking operations may clutter a site and further expose workers to contaminants and combustibles (see Fig. 7-1). Electrical lines, sharp or loud noise, improper illumination, climate, and the rigors of operating in personal protective equipment may singly or collectively compound unsafe conditions further. Many unsafe conditions with specific machines or procedures can be pinpointed and listed. The site itself, however, is a constantly changing condition. No general set of rules will apply here. A worker has to be aware and careful. Safe operation begins with the individual. Each person must be a responsible *housekeeper* on the work site to minimize accident potential. The principle reads: "A proper place for everything and everything in its proper place."

SAFE WORK PRACTICES 135

Fig. 7-1. Poor housekeeping increases the hazards at sites (EPA, 1987).

A listing of specific safety rules which will reduce unsafe conditions at a hazardous waste site follows (ATEC and Associates, Inc., n.d.).

Clothing

- Never wear *loose or ragged clothing* near moving machinery.
- *Tuck overalls* into boot tops or bind them at the ankles.
- Wear *PPE* specified for the job, properly donned and adjusted.
- If hazardous or flammable liquids are spilled on clothing, *decontaminate or remove* and don new PPE.
- Remove *rings, watches, and jewelry* when working near moving machinery.
- Wear *safety footwear*.
- Wear *gloves* when handling cable, rods, or sharp or splintery material. Protect hands from corrosives.
- Wear a *hard hat* in all job sites specifying hard hats.
- Wear a *safety belt* where required.
- Wear a *lifeline* in confined spaces.
- If *glasses* are worn, they must have safety lenses.
- Wear *goggles* where specified.

- *Observe signage* for clothing or PPE requirements in job area and comply accordingly.

Use of Tools

- *Store* tools safely so that they cannot fall on someone, or so that someone will not trip over them, causing a fall.
- Follow *grounding* requirements when using electrical power tools.
- Use *non-sparking* tools in environments where flammable or explosive substances are present or suspected.
- Follow procedures for *decontamination* of tools.

Fuel

- *Store* fuel in a safe manner (containers, marking, venting).
- Do not use gasoline to start a fire or wash clothing or hands.
- Shut off engines before *refueling*.
- Keep all *ignition sources* (matches, cigarettes), away from fuels.
- Prevent *fuel spills*; clean them up if they do occur. Use a funnel when dispensing fuel from containers; make sure containers are grounded.

Fire Prevention and Response

- Keep *ignition sources* (matches, smoldering cigarette butts, sparks from machinery exhaust) away from work site.
- *Store* properly all combustible substances (e.g., fuel).
- Never check a *battery* with a match.
- Know location of *fire extinguishers* and how to use them.
- Know procedures for summoning *firefighters and EMT personnel*.

Fire and explosion hazards will be covered in greater detail in Chapter 12.

HEAVY EQUIPMENT AND DRILL RIG SAFETY
(NIOSH, 1982; National Drilling Federation, n.d.)

Beyond general conditions that create unsafe acts are specific situations. Hazardous waste sites utilize all of the mechanical equipment used on any major construction site. Typical machinery to be found includes pumps, compressors, generators, portable lighting systems, pneumatic tools (drum openers), hydraulic drum crushers, pug mills, fork lifts, trucks, dozers, back hoes, and drill rigs. From a safety standpoint, it is important to be continually aware of this equipment around you. It poses a serious hazard if not operated properly, or if personnel near machinery cannot be seen by operators.

When working around light equipment and machinery:

- Keep loose clothing away from moving parts.
- Never pump flammable material with gas or electric pumps (use hand or diaphragm pumps).
- Be aware of the types of fittings on pumps/hoses. For example, acid and caustic will rapidly corrode aluminum.
- Ground equipment and containers near flammables.
- Do not fuel running equipment.
- Use non-sparking tools.

General *common sense* rules for heavy equipment operation:

- Inspect equipment before operation to be sure it is in good operating condition (no leaks of flammable substances, backup warning operable, steering, brakes, wheel lugs, exhaust muffler, tow hitches).
- Before moving machine, make sure all people are clear.
- Slow down when backing up, or when on ramps and curves.
- Allow for safe stopping distances.
- Do not drive through dust clouds.
- Exercise caution on slopes or near banks.
- Watch for overhead and unearthed obstructions or lines.
- Follow decontamination procedures for equipment.
- Use designated parking areas and shut down procedures.

For working around heavy equipment:

- Try to make eye contact with operators you are near.
- Always wear a hard hat and foot protection.
- Be aware of the location of the equipment near you.
- Do not operate anything you are not qualified to.
- Never walk under suspended loads.
- Never walk in front or back of moving heavy equipment.
- Be aware that equipment can be a source of ignition for flammable/explosive materials.

Drill crews are confronted with all of these heavy equipment hazards. They must be responsible for housekeeping around the rig because of all of the rods, auger sections, rope, and hand tools cluttering the operation. Maintenance is a constant requirement. Overhead and buried utilities require special precautions, because of electrical and natural gas hazards. Electrical storms may seek out a standing derrick. The hoist or cathead rope poses specific hazards that must be respected. Always use a clean, dry, sound rope. Keep hands away from the test hammer! Hearing loss, while not an immediate danger, is considerable over time. Use hearing protection.

Drilling operations at hazardous waste sites are subject to additional dangers. See Fig. 7-2. Intercepting contaminants and combustibles, and the impairments of PPE (personal protective equipment) during strenuous work are the most significant problems. Following is a list of some typical problems and various solutions that have been adopted.

Problems with Drilling Operations at Hazardous Waste Sites

Problem. Intercepting buried lagoons, drums, tanks, etc. which may expose drill crews to volatile, toxic, or flammable substances.

Solutions. First, site the rig in a crosswind configuration if possible. Keep your face away from the hole, especially when removing rods or augers. Monitor constantly for combustible gases, at the breathing zone *and* at the hole. If hollow stem augers are used (or hollow rods), place sniffer through the center of the auger when possible. If methane or other combustible gases are abundant, it may be possible (or necessary) to vent drill cuttings, fluids, and escaping gases away from the rig. When unknown liquids/sludges are encountered, wear gloves and splash protective clothing.

Fig. 7-2. Drill rig operations have special hazards (EPA, 1987).

Problem. Dealing with heat stress due to strenuous work in personal protective clothing during warm weather.

Solutions. Pace your work, and take frequent rest breaks. (Make a shaded area.) Force fluid intake—at least one quart of liquid an hour.

Problems. Cross contamination with split spoon, or transmission of contaminants to the surface or through the subsurface.

Solutions. Split spoons can be triple rinsed between uses, under the assumption that one of the decontaminating solutions will be effective on a given unknown. Another method is to make a disposable split spoon from PVC well casing pipe. When augering into contaminated soils a surface liner (plywood, plastic, etc.) should be placed around the hole to keep from contaminating the surface. To prevent subsurface transmission of contaminants the driller needs the experience to be able to "feel" when the auger encounters impermeable strata and stop.

Problem. Evacuation situation due to high concentration of contaminants or combustible vapors at hole.

Solutions. Based on monitoring, try to pull auger out of hole. Otherwise, exit *immediately* and wait for diffusion or upgrade PPE.

Problem. Inexperienced worker uses steam jet to decontaminate the augers and split spoon. The jet goes out of control and accidentally shoots steam into his rubber boots, causing second- and third-degree burns.

Solution. Proper training of workers designated to do decontamination of equipment would have avoided this accident. The knowledge that steam jets cause extensive burns would have alerted this worker to use more caution around the jet.

CONFINED SPACES

Definitions

What defines a confined space in the workplace? Those workspaces which are *enclosed* by design or nature are confined spaces. Storage tanks and trenches are examples. Why are confined spaces hazardous to workers? There are many reasons. The principal reason is limited egress, which means these spaces are not easy to leave.

Any place that is hard to get out of hinders escape, rescue, and ventilation. Poor ventilation may lead to low oxygen levels or high levels of gases which are explosive or poisonous. When working in a confined space you often cannot be seen from the outside. The ability to move freely is restricted because of limited space and your personal protective equipment. In addition, machinery or electrical lines in a confined space are usually controlled from the outside. There may be danger from falling equipment or materials to consider. Many types of confined spaces occur at hazardous waste sites, including:

- Tanks.
- Vats.
- Reaction vessels.
- Sumps.
- Pits.
- Unventilated buildings.
- Trenches.
- Ravines.
- Excavations.

What can happen to you if unprepared in a confined space? Exposure to:

- Oxygen deficiency.
- Fire or explosion.
- Chemical gases or vapors.
- Radiation.
- Very high temperatures.
- Very loud noise.
- Electrical shock.
- Accidental machinery startups.
- Falls or falling objects.

Safe Entry Procedures

One might say that since confined spaces can be so dangerous, we should just stay out of them. Sometimes, though, people have to work there. Well considered entry procedures can make these jobs safe. First, be *trained* in safe entry and rescue procedures. Be able to recognize hazards. Second, use a *plan*. A written entry permit issued by a responsible supervisor will make sure that a checklist of precautions is reviewed before entry. *Never* go in a confined space without an authorized written permit!

Entry Permit

Let us look at the things a good entry permit will cover. Overall, the permit must evaluate the hazards and address the measures which will protect workers. A permit is only good for a given time period (usually 8 hours). An entry permit should also address each of the items detailed below.

Preparation of Work Area. *Isolation.* Disconnect or cap all process lines connected to work area. Merely closing valves is not adequate.

Lockout. Any electrically operated equipment (e.g., pumps, mixers, conveyers, etc.). or electrical lines which are connected with the work area should have their main electrical switches locked in the "off" position.

Cleaning. Confined spaces should be cleaned, flushed, or purged of hazardous materials to the maximum extent possible prior to entry.

Ventilation. Fans, portavents, air movers, or natural drafts should be used to provide positive ventilation, prior to and during the entire entry period.

Atmospheric Testing. This will be conducted prior to and during entry in order to monitor for the following conditions:

Oxygen Content. The oxygen content should be *no less than 19.5%* and *no greater than 25.0%* in any part of the confined area.

Flammability. The atmosphere should be nonexplosive (i.e., less than 10% of the lower flammable limit).

Toxicity. Toxic concentrations of vapors should *not* be present.

Personal Protective Equipment. Required PPE should be specified. It should be used only in situations in which engineering controls and safe work practices cannot adequately control the hazards present.

Area Safety Equipment.

- A *safety harness and lifeline* should be worn by any person entering a confined area.
- For entries through top openings, a *hoisting device* should be provided for lifting out the worker.
- *Ladders* are required when entry or exit involves a drop or climb of more than 3 feet.

Observer (Buddy System).

- A second worker (or standby) should be on hand at the entrance to the confined area throughout the entry time.
- The worker within the confined space should remain in sight of the observer at all times, or else remain in communication with the observer (e.g., by walkie talkie or by signaling through tugs on the lifeline.)
- The standby worker should wear all personal protective equipment required for the worker inside the confined space.
- The standby worker will be equipped with some means of communications to be used in summoning help if needed.

- Under no conditions should the standby worker enter the confined area unless other workers are standing by.
- Should the standby worker be required to leave his post, the worker inside must leave the confined area.

Rescue Procedures.

- A rescue plan should be specifically designed for each entry.
- In the event of an emergency, the standby worker should call for help, then try to rescue the worker by manipulating the lifeline from outside of the confined area.
- *Under no circumstances* should the standby worker enter the area prior to the arrival of other personnel.

Tools.

- All tools should meet general safety guidelines (e.g., ground fault interrupters on electrical circuits).
- All tools used must be safe for conditions within the confined area (e.g., non-sparking tools in potentially flammable atmospheres).

Posting.

- Entrances to confined areas should be posted to indicate that:
  — The area is a confined space,
  — A permit is required for entry.
- During work, it is advisable to post signs at all entrances indicating that workers are inside.

Hot Work. If the oxygen level of a confined space is *greater than 25%*, *no hot work* should be allowed until the oxygen level has been *reduced* to approximately 21%.

HANDLING DRUMS AND CONTAINERS

The OSHA Standard, 29 CFR 1910.120, requires that cleanup operations involving drums and containers be done safely. This means that the handling, sampling, testing, staging, transport, decon, evacuation, excavation, and bulking of drums and containers must be done with minimum risk. When new containers are used, they must meet minimum standards according to the DOT, OSHA, and EPA regulations. But what about buried, abandoned, or old, leaking containers? *Millions* of old 55 gallon drums alone are out there, half-full of unknowns, rusting and leaking their contents. With this in mind, a set of safe work practices must be *understood* and *used*. First, the requirements for container handling will be addressed with respect to OSHA 1910.120. Second, some general rules will be considered; then the handling and sampling methods which are safest. Special types of waste will be mentioned with precautions for them. Finally, procedures for staging and bulking operations will be covered.

Handling Drums and Containers—The Law: Requirements under 29 CFR 1910.120

General Rules.

- Drums and containers used must meet minimum DOT, OSHA, and EPA regulations for the wastes they contain.
- If practical, drums and containers will be inspected to insure their integrity prior to being moved. If drums or containers are stored or stacked so that inspection is impossible, they should be moved to an accessible location for inspection prior to further handling.
- Unlabeled drums and containers will be assumed to contain hazardous substances and treated accordingly until contents are positively characterized.
- Site operations shall be organized so as to minimize the amount of drum or container movement required.
- All employees exposed to a transfer operation shall be warned of potential hazards associated with contents of any drums or containers involved.
- DOT specified salvage drums or containers and suitable sorbent materials shall be available in areas where spills may occur.
- Where major spills are possible, a spill containment program shall be implemented as part of the employer's safety and health plan. The spill containment program shall allow for the containment and isolation of the entire volume being transferred.
- Drums and containers that can't be moved without rupture or leakage will be emptied into a sound container.
- Some type of detection system (such as ground-penetrating radar) shall be used to estimate the location and depth of buried drums or containers.
- Buried drums shall be excavated carefully to prevent rupture.
- Suitable fire extinguishing equipment will be kept on hand and ready for use.

Opening Drums and Containers.
These procedures are to be followed in areas where drums or containers are being opened:

- If airline respirators are used, air cylinder connections must be protected from contamination and the entire system shall be protected from physical damage.

- Employees who must work near drums or containers being opened must be provided protective shielding in case of explosion.
- Employees not directly involved in the opening procedures will be kept at a safe distance.
- Controls for opening equipment, monitoring equipment, and fire suppression equipment shall be located behind the shield.
- Non-sparking tools and equipment will be used when flammable atmospheres are a reasonable possibility.
- Drums and containers shall be opened so as to safely relieve excess pressure. Either:
  — Relieve pressure from a remote location, or
  — Place appropriate shielding between the employee and the drums or containers.
- Employees shall not stand on, or work from drums or containers.

Material Handling Equipment. Material handling equipment shall be selected, located, and operated so as to prevent ignition of vapors released during opening procedures. There are hazards associated with gas powered or electrically powered units.

Radioactive waste. Drums and containers containing radioactive waste shall not be handled until their hazard to employees has been properly assessed.

Shock-Sensitive Waste. When handling drums or containers containing or suspected of containing shock-sensitive wastes, the following spectral precautions should be followed:

- All nonessential employees shall be removed from the area of transfer.
- Material handling equipment shall be fitted with explosion containment devices or protective shields to protect operators.
- An alarm system will be used to signal the beginning and end of the procedure.
- Continuous communication will be maintained between the employee in charge of the handling operation and the site safety and health supervisor or command post during the operation.
- Pressurized drums shall not be moved until the cause of the excessive pressure is determined and appropriate measures are implemented.
- All drums and containers containing packaged laboratory wastes lab packs shall be considered shock-sensitive until proven otherwise.

Lab Packs. Lab packs shall be opened only by a person who is sufficiently knowledgeable to inspect, classify, and segregate the containers within the pack according to the hazards involved. If crystalline material is noted on any container, the contents shall be treated as shock-sensitive until positively identified.

Sampling Procedures. Sampling will be performed in accordance with a written sampling procedure, which is part of the site safety and health plan, see Chapters 10 and 15 for actual requirements.

Shipping and Transport. Drums and containers shall be identified and classified prior to packaging for shipment. Staging areas shall be kept to the minimum number necessary and shall be provided adequate entrance and exit routes.

Bulking of hazardous wastes shall be permitted only after a thorough characterization has been completed.

Tank/Vault Entry Procedures. Tanks and vaults will be handled in a manner similar to that for drums and containers, taking the size of the tank or vault into consideration. Provisions for safe confined space entry (see previous section in this chapter), as included in the site safety and health plan, must be followed for entering tanks and vaults.

Occasions for Container Handling

Waste containers of various types on a hazardous waste site may need to be handled during sampling, waste characterization, preparation of material for disposal, in addition to other reasons.

Reasons for Concern

In the process of moving and/or opening drums, there is an increased potential for accidents. Containers may leak, rupture, detonate, or release vapors, resulting in fires, explosions, vapor generation, cause direct contact of workers with chemicals due to splashing or spraying, or contamination of soil or water. There is additional potential for worker injury due to muscular stress or strain during container handling, due to working around stacked drums, and due to working around heavy equipment involved in container handling.

Minimization of Danger

There are two main elements to keep in mind in an effort to minimize the dangers involved handling of containers: keep container handling to a minimum, and when moving or opening of containers is necessary, use equipment and procedures which isolate workers from the containers being handled to the maximum extent possible.

The Container Handling Process

Visual Inspection. Prior to handling, visually inspect the containers for the following: to determine if the contacts might show whether the materials may be radioactive, explosive, corrosive, toxic, flammable, or lab-packed:

- Symbols, works, or markings.
- Signs of deterioration such as corrosion, rust, or leaks.
- Indications the container is under pressure, such as swelling or bulging.
- Drum type. Polyethylene or PVC-lined drums often contain strong acids or bases. If the lining is punctured, the substance usually quickly corrodes the steel, resulting in a leak or spill. Exotic metal drums such as aluminum, nickel, stainless steel, or other unusual metal are very expensive drums which usually contain an extremely dangerous material. It is not always obvious that the metal of a drum is one of these exotic metals. Single-walled drums are used as pressure vessels. Such drums have fittings for both product filling and also placement of an inert gas, such as nitrogen. These drums may contain reactive, flammable, or explosive substances. Lab packs are used for disposal of expired chemicals and process samples from university labs, hospitals, and similar institutions. Individual containers within the lab pack are often not packed in absorbent material. They may contain incompatible materials, radioisotopes, shock-sensitive, highly volatile, highly corrosive, or very toxic exotic chemicals. Lab packs can be an ignition source for fires at hazardous waste sites.
- Configuration of drumhead. Open top drums were originally designed to contain solids. Bunged drums were originally designed for liquids; however, in a hazardous waste setting, the material inside may consist of solidified sediments.
- Conditions in immediate vicinity of containers. Crystalline material on or around the containers could indicate shock sensitive material. In addition, there may be other material leaked or spilled from the containers onto the ground which might give a clue as to what may be in the drum.

Monitoring. Before any moving or opening of containers takes place, direct-reading instruments should be used to detect the presence of organic vapors, combustible gases, or above-background levels of radiation (see Chapter 8).

Subsurface Investigation. If there is any reason to suspect the presence of buried containers, some type of non-destructive ground penetrating system should be used to determine the approximate location and depth of such containers.

Preliminary Classification

As a precautionary measure, any unlabeled containers should be assumed hazardous until it is learned otherwise. Using the information gathered by visual inspection, monitoring, and subsurface investigations, preliminarily classify any containers thought to be radioactive, leaking/deteriorated, under pressure, explosive/shock-sensitive, lab packs, or buried.

Planning

Based on inspection and preliminary classification, decide if any hazards are present and the appropriate response activity. Determine which drums need to be moved in order to be opened and/or sampled. A preliminary handling plan should be developed dealing with the extent of any necessary container moving or handling, the personnel designated for the job, and the most appropriate procedures based on the particular hazards revealed during preliminary inspection. The handling plan should be revised as new information comes to light during operations at the site.

Moving of Containers

There are a number of reasons for which it might be necessary to move containers. Such reasons include the need to respond to any obvious and immediate problems that could impair worker safety, such as precariously stacked drums, leaking containers, or incompatible wastes in too close proximity to each other. Other reasons might include the need to orient drums for sampling by unstacking or providing aisle space for movement of personnel. In addition, it might be necessary to organize the drums into different areas to facilitate staging and remedial action.

Container moving may not be necessary, depending on how the containers are positioned on site. Since accidents frequently occur during moving of drums, drums should only be moved if absolutely necessary.

Equipment frequently used in drum moving includes a drum grappler attached to a hydraulic excavator, a small front-end loader which can be loaded manually or equipped with bucket sling, a rough terrain forklift, a roller conveyor equipped with solid rollers, and drum carts. However, the drum grappler is the preferred piece of drum handling equipment. It keeps the operator removed from the drums so there is less chance of injury if a drum detonates or ruptures.

Guidelines for Container Moving. *Radioactive.* If instrument readings or container labeling indicate the possibility of radioactive waste, do not handle the containers in any way. The first step which should be taken is to notify the worker's supervisor immediately. The supervisor should immediately contact the Nuclear Regulatory Commission and/or a radiation specialist for further guidance.

Explosive/Shock-Sensitive. Special assistance should be sought before any handling of any containers for which there is any indication that explosive or shock-sensitive waste may be present. If handling is necessary, use extreme caution. Before handling, move nonessential personnel away. Use a grappler unit constructed for explosive containment. Before transporting drums of shock-sensitive material, palletize the drums and secure them to the pallet. An audible siren signal system should be used to signal beginning and ending of handling of explosive wastes. Finally, continuous communication should be maintained with the Site Safety Officer during the handling procedure.

Bulging/Swelling Containers. Bulging or swelling containers indicate the container's contents may be under pressure. Pressurized drums are very hazardous, and if they must be moved, should be moved with a drum grappler equipped for explosive containment.

Lab packs. Lab packs contain individual containers of lab materials. They can be an ignition source or they can contain explosive or shock-sensitive waste. When handling is necessary, it should be assumed that the lab pack contains shock-sensitive waste until it is known otherwise, and handling procedures should follow those for shock-sensitive waste. Once a lab pack is opened, a chemist should inspect, classify and segregate the bottles within the pack (without opening them). After the materials have been segregated, they should be packed with absorbent. Repacked lab pack drums should be palletized and secured to the pallet.

Leaking, Open, and Deteriorated Drums. If such containers cannot be moved at all without rupture, the contents should be pumped to a sound container. If the container can be moved without rupture, it can be placed in an overpack drum by use of a drum grappler or other device. These procedures and other spill containment procedures are discussed in detail in Chapter 12.

Buried Drums. Before beginning excavation, use ground penetrating systems to estimate location and depth. Soil should be removed with great caution to minimize potential for drum rupture. A dry chemical fire extinguisher should be on hand to control small fires.

Opening Containers

If supplied air respiratory protection is used, place a bank of air cylinders outside the work area and supply air to the operators via airlines and escape SCBAs. Keep personnel at a safe distance from the drums being opened. If possible, monitor for radiation and combustible gas during opening.

Remotely Controlled Opening Devices.
If possible, use remotely controlled devices for opening drums. Some of these devices are discussed here.

The backhoe spike is a metal (bronze) spike attached or welded to a backhoe bucket. It is very efficient and thus is advisable for large-scale operations. The drums should be in rows with adequate aisle space to allow ease of backhoe movement. Once in rows, drums can be quickly opened by punching holes in the drum tops with the spike. To prevent cross contamination, the spike should be decontaminated after each drum is opened. Although some splash or spray may occur, the operator of the backhoe can be protected by mounting a large shatter-resistant shield in front of the operator's cage. This precaution, combined with normal PPE, should be enough to protect the operator. Additional respiratory protection can be provided by equipping the operator with an on-board airline system.

A hydraulically operated drum piercer consists of a manually operated pump which pressurizes oil through a hydraulic line. A piercing device with a metal point is attached to the end of the line and pushed into the drum by the hydraulic pressure. The piercing device can be attached so that the hole is made in the side or top of the drum.

A pneumatically operated bung remover operates by means of compressed air delivered through a high-pressure airline to a pneumatic drill which is adapted to turn a bung fitting which fits the bung to be removed. An adjustable bracket has to be attached to the drum before the drill can be operated and must be removed before the sample can be taken. This bracketing procedure is time consuming and is a major drawback. This opening process does not permit slow venting. Also, it cannot remove bungs that are rusted shut. The drums have to be upright and fairly level to use this process.

Manually Operated Opening Devices.
The risks are greater when manually opening drums than when using remotely operated means. When using manual devices the drums must position to allow easy worker access to the drums.

A bung wrench, when used, should be of the non-sparking kind, and should be marked as such. However, although a non-sparking wrench should prevent sparking between the wrench and the bung, it will not prevent sparking between the bung and the threads on the drum. The bung should be turned very slowly to allow pressure to dissipate. The small bung should be opened first, as a pressure release. Eye protection should be worn in case of spraying of liquid contents under pressure. Also, try to avoid leaning over the drum while opening it.

A drum deheader can be used when the bung is not removable with a bung wrench. It can be used only with closed-head drums, not on open-top drums. It is used by first positioning the cutting edge just inside the top chime and then tightening the adjustment screw so the deheader is held against the side of the drum. Move the handle up and down while sliding the deheader against the side of the drum, then make initial cut very slowly in case the contents are under pressure.

Hand picks, pickaxes, and spikes are not recommended for opening drums because the drum usually must be struck with much force, creating great potential for splashing or spraying. Also, drums cannot be opened slowly with this method, so opening drums under pressure can cause spraying. In addition, there is great hazard using this method on containers of shock-sensitive material. Neither chisels nor firearms should be used to open drums.

Staging of Containers

Fires, explosions, and dangerous reactions may result from the comingling of chemicals which are incompatible with one another. Some substances, if indiscriminantly mixed, may generate gases which are toxic or displace oxygen. To prevent this a two step process of staging of containers is done at uncontrolled hazardous waste sites.

Initial staging involves the movement of containers with similar labels and appearance in an organized manner to predesignated areas. Drums are usually laid out in rows of two with lanes in between. This makes it easy to sample the drums and provides fire and equipment lanes.

Once this is done, the second step, compatability staging, is performed. Compatability staging is a systematic set of field "spot" tests done on *all* drums or containers of unknowns, and backed up by lab tests performed by chemists. These tests give what is called a "fingerprint" of the chemicals so that safe bulking operations can be performed. There are specific tests for radiation, flammability, corrosives, perioxides, sulfides, cyanides, PCB's, etc. Tests done in the field are *not* conclusive and an intimate knowledge of chemistry and experience are essential for safe bulking operations. An example of specific field tests is shown in Fig. 6-1; however, methodology varies from site to site and between companies doing the cleanup work, and any job requires specific training. Note the flow chart (Fig. 4-13) from one company describing the order and types of tests generally done.

The extent of staging needed depends on site-specific circumstances such as the number of drums involved, the number of waste streams, the extent of any on-site treatment of waste, the number of potential disposal site destinations, and the perceived hazards regarding proximity of the various waste streams to each other. Staging of containers should be kept to a minimum in order to minimize drum movement.

Up to five separate staging areas have been used at hazardous waste sites: the initial staging area, the opening area, sampling area, holding area, and the final staging area (bulking area). The initial area is where drums are organized according to type, size, and suspected contents and where the drums can be stored prior to sampling. The opening area is where drums are opened, sampled, and resealed. This area should be separated from other areas in case of explosion. The sampling area may be used in large-scale or emergency operations, where it may be advisable to sample drums in a different area from where they are opened in order to minimize the number of people present in the opening area. The holding area is where drums are temporarily stored after sampling pending characterization of their contents. Unsealed drums with unknown contents should not be placed in the holding area, in case they contain incompatible materials; the contents should either be placed in a sound container or the drums should be overpacked. The bulking area is where substances that have been characterized are bulked for transport to treatment or disposal facilities. The bulking area should be located as close as possible to the site exit. It should be graded and covered with plastic sheeting, and should have a one-foot-high dike around it. A separate bulking area should be created for each type of waste present in order to prevent possible mixing of incompatible wastes based on characterization resulting from sample analysis, the drums should be segregated according to their basic chemical categories (acids, metals, pesticides, etc).

In all staging areas, the drums should be arranged in rows two drums wide. These rows should be spaced at least 7–8 feet apart to enable movement of drum handling equipment, and to provide adequate fire lanes.

Bulking of Waste

Wastes that have been characterized are often mixed together and placed in bulk containers such as tanks or vacuum trucks for shipment to treatment of disposal facilities. Bulking increases efficiency of transport and it reduces disposal costs. It should be done only after thorough waste characterization. Preliminary field tests may give only a general indication of the nature of the individual wastes. In many cases additional sampling and analysis is needed to further characterize the wastes. However, even after thorough characterization of wastes, no bulking should take place until compatibility tests have been run on the material. Compatibility testing is done by mixing small quantities of different wastes together under controlled conditions and observing for signs of incompatibility such as vapor generation and heat of reaction.

Several cautions should be observed during bulking. When pumping hazardous liquids, use pumps that are properly rated according to NFPA standards. Pumps should have a safety relief valve with a splash shield. Pump hoses, casings, fittings, and gaskets should be compatible with the material being pumped. Hose lines should be inspected before beginning work to ensure that all lines, fittings and valves are intact with no weak spots. Take special precautions when handling hoses, as they often contain residual material that can splash or spill on operating personnel. Personnel should be protected against accidental splashing and hose lines should be protected from vehicular or pedestrian traffic. Flammable liquids should be stored in approved containers. Before transferring bulked materials to a tank truck, inspect each tank trailer and remove any residual materials from the trailer; this will prevent reactions between incompatible materials.

Special Case Problems

Tanks/Vaults. Tanks and vaults are often found on hazardous waste sites. When opening such containers, vent excess pressure if the container holds volatile substances. Place a deflecting shield between workers and the opening to prevent direct contamination of workers by materials forced out by pressure when tank is opened. Guard manholes to prevent personnel from falling into the tank. The contents of the tank/vault should be identified through sampling/analysis. If characterization indicates that the contents can be safely moved with the available equipment, vacuum contents into a trailer for transportation to disposal or recycling facility. Empty/decontaminate the tank or vault before disposal. If necessary to enter tank or vault for any reason (for example, to clean off solid materials or sludge on bottom or sides of tank or vault), observe precautions associated with confined spaces. This should include ventilating thoroughly prior to entry, disconnecting any pipelines entering vessel, and monitoring for combustible or toxic gases and for adequate oxygen levels.

Vacuum Trucks. Appropriate protective clothing/equipment should be worn when opening the hatch. If possible, use mobile steps of suitable scaffolding. Avoid climbing up the ladder and walking across the tank catwalk. If the truck must be climbed, raise and lower equipment and samples in carriers to enable workers to use two hands while climbing. If possible, sample from the top of the vehicle. If necessary to sample from the drain spigot, take steps to prevent spraying of excessive substances.

Elevated Tanks. In general, the same safety precautions described for vacuum trucks should be observed for elevated tanks. In addition, safety lines and harness should be used. Ladders and railings should be maintained in accordance with OSHA regulations.

Compressed Gas Cylinder. Before and during disposal of compressed gas cylinders, expert assistance should be obtained. Such cylinders should be handled with great caution because rupture of a cylinder can result in an explosion and the cylinder can become a dangerous projectile. Record the ID numbers on the cylinders to aid in characterizing their contents. Manufacturers keep records on the ID numbers and corresponding contents of cylinders.

Ponds/Lagoons. Drowning is a very real danger for personnel suited in protective equipment. Thus, where there is potential for drowning due to work at a pond or lagoon, wear water safety gear. Wherever possible, stay on shore—avoid going out on the water. The appearance of solid, cracked mud along shorelines can be deceptive, so use caution. The self-contained breathing apparatus, SCBA, used as a part of Level A and B protection *does not work underwater*. Do not be mislead by the familiar name "scuba" given to these respirators.

REFERENCES

ASTM. D4276-84. *Standard Practices for Confined Space Entry*.

ATEC and Associates, Inc. n.d. *Employee Safety Manual*, Appendix A, Drilling Division.

Goldberg, R. 1975. In *The Human Side of Accident Prevention*, Margolis, C., and Kroes, W., Eds. Springfield, IL: Charles Thomas.

Grimaldi, J. V., and Simonds, R. H. 1975. *Safety Management*, 3rd Edition. Homewood, IL: Richard D. Irwin, Inc.,

Latham, G. P., and Sari, L. M. 1979. Application of Social Learning Theory to Training Supervisors through Behavior Modeling. *Journal of Applied Psychology*, **64**(3): 239–246.

National Drilling Federation. n.d. *Drilling Safety Guide*. Diamond Core Drill Manufacturer's Association and National Drilling Contractors Association.

NSC. 1978. *Supervisor's Safety Manual*, 5th Edition, Chapter 2 and 3. Chicago, IL: National Safety Council.

U.S. NIOSH/OSHA/CG/EPA. 1985. *Occupational Safety and Health Guidance Manual for Hazardous Waste Site Activities*. NIOSH Publication No. 85-115, Chapter 2. Washington, DC: U.S. Govt. Printing Office.

U.S. NIOSH. *Working in Confined Spaces, Criteria for a Recommended Standard*. NIOSH Publication No. 80-106. Washington, DC: U.S. Govt. Printing Office.

U.S. NIOSH. *Working on Hazardous Waste Sites*. (Draft Copy).

U.S. NIOSH. 1982. *Hazardous Waste Sites and Hazardous Substance Emergencies*. Worker Bulletin. NIOSH Publication No. 83-100. Washington, DC: U.S. Govt. Printing Office.

8
Safe Use of Field Equipment

PART ONE. DIRECT-READING AIR MONITORING INSTRUMENTS

INTRODUCTION

As a worker on a hazardous waste site you will be working in environments that can be detrimental to your health. One means of protecting personnel on such a site is to monitor the air for harmful contaminants. Various types of direct-reading air monitoring instruments are used on sites to sample for airborne contaminants.

Response to an environmental incident requires careful preparation and prompt action to reduce the hazards. Concurrently, the health and safety of response personnel and the general public must be protected. Air monitoring instruments provide an integral portion of the information necessary to determine how these requirements are being met. The purpose of this section is to:

- List commonly used air monitoring instruments useful for hazards incident response.
- Describe the operating theories and principles of these instruments.
- Illustrate the proper interpretation and limitations of the data obtained.

Used correctly, these instruments provide data that help response personnel determine:

- Potential or real *effects on the environment*.
- Immediate and long-term *risks to public health*, including the health of response workers.

- Appropriate *personnel protection*, including respiratory protective equipment to be used on-site.
- Actions to *mitigate the hazard(s)* safely and effectively.

OBJECTIVES

- The reader will know the reasons for air monitoring.
- The reader will be familiar with the *principles of operation and use* of the following air monitoring devices.
  — Oxygen meter.
  — Combustible gas meter.
  — Flame ionization detector (FID).
  — Ultraviolet (UV) photoionization meter (PIM).
  — Direct reading colorimetric indicator tubes.
- The reader will know the *limitations* of each instrument.
- The reader will know how to *safely* use the equipment in a hazardous waste setting.

PURPOSE OF AIR MONITORING AT HAZARDOUS WASTE SITES

Airborne contaminants can present a significant threat to human health, thus, identifying and quantifying these contaminants by air monitoring is an essential component of the health and safety program at hazardous waste sites.

Air monitoring data are useful for assessing health risks to the public and to response workers, for selecting personal protective equipment, for delineating areas where protection is needed, for determining potential effects on the environment, and for selecting actions to mitigate hazards safely and effectively.

Direct-reading instruments provide information at the time of monitoring and do not require sending samples to a lab for analysis. This characteristic of direct-reading instruments enables decision making. Direct-reading instruments are used in preliminary site survey and in periodic monitoring, once the preliminary survey is done.

Preliminary Site Survey

This is considered a relatively rapid screening process for collecting preliminary data on site hazards. The time needed to conduct the preliminary survey depends on the urgency of the situation, information needed, size of the site, availability of resources, and level of protection required for initial entry personnel. Thus, the initial survey may need hours or days to complete and consist of more than one entry.

Priority for preliminary site monitoring is broken down into two sections:

poorly ventilated spaces and open, well-ventilated areas. In general, for poorly ventilated spaces—buildings, ships' holds, boxcars, or bulk tanks—which must be entered, combustible vapors/gases and oxygen-deficient atmospheres should be monitored first with team members wearing, as a minimum, Level B protective equipment. Toxic gases/vapors and radiation, unless known not to be present, should be measured next. For open, well ventilated areas, combustible gases and oxygen deficiency are lesser hazards. However, areas of lower elevation on site (such as ditches and gulleys) and downwind areas may have combustible gas mixtures, in addition to toxic vapors or gases, and lack sufficient oxygen to sustain life. Entry teams should approach and monitor from upwind, whenever possible.

Periodic Monitoring

Since site activities and weather conditions change, a continuous program to monitor atmospheric changes must be implemented using a combination of periodic area monitoring with direct-reading instruments, stationary sampling equipment, and personal monitoring devices.

Possible Conditions to Monitor.

- Combustible gases/vapors.
- Oxygen deficiency.
- Toxic atmospheres.
- Radioactive environment.

IMPORTANT CONSIDERATIONS OF DIRECT-READING INSTRUMENTS

Portability

The ability to withstand rough use is essential. Also, either an internal power supply or battery power is important for portability. The instrument shouldn't be so heavy or bulky that it is difficult for the worker to carry.

Reliable, Useful Results

Response time (the interval between the time the instrument senses a contaminant and when it generates data) is very important. Response times for direct reading instruments range from a few seconds to several minutes. Another important factor is operating range (difference between the lower detection limit—the lowest concentration the instrument will respond to—and the saturation concentration, the upper use limit). Also important is precision or reproducibility: the extent to which the instrument will show the same or near same reading under identical situations. A final element here is accuracy: the extent to which the instrument will accurately show the concentration of contaminant present.

SAFE USE OF FIELD EQUIPMENT 157

Selectivity/Sensitivity

Selectivity is the ability of an instrument to detect/measure a specific chemical or group of chemicals, while sensitivity is the ability of an instrument to accurately measure small concentrations of a particular contaminant.

Inherent Safety

Electrical devices, including instruments, must be constructed in such a way as to prevent ignition of a combustible atmosphere. The source of such ignition could be an arc generated by the power source itself or the associated electronics or a flame or heat source necessary for functioning of the instrument. The NFPA has created minimum standards in the National Electrical Code for electrical devices used in flammable or ignitable atmospheres. The standard establishes several classifications of hazardous atmospheres based on the type of material producing the flammable atmosphere and on how often the material generates the atmosphere. The standard provides for three methods of construction to prevent a potential source from igniting a flammable atmosphere: explosion-proof, intrinsically safe, and purged.

SOME COMMONLY USED DIRECT-READING INSTRUMENTS

Radioactive Atmospheres

Geiger Counter. The Dosimeter Corp. model 3700 detects gamma and beta radiation (see Fig. 8-1). With the probe window closed it detects gamma radiation and with the probe window open detects beta also. The printed meter scale goes to 0.5 mR/hr and 300 CPM (counts per minute). It has three ranges: $\times$ 1, $\times$ 10, and $\times$ 100, so it can actually detect up to 100 times the highest number on the meter, which is 0.5 mR/hr. Thus, it can actually detect up to 50 mR/hr.

After sufficient period of time to warm up, gamma instruments indicate a background exposure rate of about 0.01–0.02 mR/hr. This is due to natural background radiation from various kinds of radionuclides found in the soil and high energy cosmic radiation from outer space. Generally, there is no background for alpha and beta radiation.

Federal guidelines for radiation exposure are: 5 rem/year = 1.25 rem/calendar quarter = 100 millirem/week = 2.5 millirem/hr (based on a 40 hr week). If measured exposure increases to 3–5 times above gamma background, a qualified health physicist should be consulted. At no time should work continue with an exposure rate of 10 mR/hr or above without the advice of a health physicist. The absence of gamma readings above background should not be interpreted as the complete absence of radioactivity. Radioactive materials emitting low-energy gamma, alpha, or beta radiation may be present, but for a

158 WORKER PROTECTION DURING HAZARDOUS WASTE REMEDIATION

Fig. 8-1. Geiger counter to measure radiation emissions.

number of reasons may not cause a response on the instrument. Unless airborne, these radioactive materials should present minimal hazards, but more thorough surveys should be conducted as site operations continue to completely rule out the presence of any radioactive material.

Oxygen Deficient/Combustible Atmospheres

Oxygen Meter. Oxygen meters are used to evaluate an atmosphere for the following:

- Oxygen content for respiratory purposes. Most oxygen meters are set to alarm if oxygen falls below 19.5%.
- Increased risk of combustion. Generally, concentrations above 25% (23.5%) are considered oxygen-enriched and increase the risk of combustion.
- Use of other instruments. Some instruments require sufficient oxygen for

operation (some combustible gas meters do not give reliable results at oxygen concentrations below 10%).
- Presence of contaminants. A decrease in oxygen content can be due to the consumption (by combustion or a reaction such as rusting) of oxygen or the displacement of air by a chemical. If the oxygen decrease is due to consumption then the concern is the lack of oxygen. If it is due to displacement then there is something present that could be flammable or toxic.

Limitations in use of the oxygen meter include:

- Operation of oxygen meters depends on the absolute atmospheric pressure. Pressurized or low-pressure samples will give erroneous oxygen percent readings. For atmospheric sampling at higher or lower altitudes the oxygen meter should be calibrated at the elevation where sampling is to take place.
- High concentrations of carbon dioxide (CO_2) shorten the useful life of the oxygen sensor. As a general rule, the unit can be used in atmospheres greater than 0.5% CO_2 only with frequent replacing or rejuvenating of the sensor. Lifetime in a normal atmosphere (0.04% CO_2) can be from one week to one year depending on the manufacturer's design. In the case of the MSA 260 the sensor life will be reduced to 2 days in 100% CO_2, 50 days in 5%, and 100 days in 1%.
- Temperature can affect the response of oxygen indicators. The MSA 260 is temperature compensated in the range of 32–104°F (0–40°C). Use down to 0°F is possible when the sensor is calibrated at the temperature and if more sampling time is allowed for slow sensor response—approximately 3 minutes when a sample line is not used. Below 0°F the sensor may be damaged by the solution freezing.
- Strong oxidizing chemicals, like ozone and chlorine, can cause increased readings and indicate high or normal O_2 content when the actual content is normal or even low.

Combustible Gas Indicator. Combustible gas indicators (CGIs) measure the concentration of a flammable vapor or gas in air, indicating the results as a percentage of the lower explosive limit (LEL) of the calibration gas. The MSA 260 consists of two completely separate indicators—an oxygen meter and a CGI—in a single housing. The MSA 260 detects only combustible gases and vapors in air—it will not indicate the presence in the air of combustible airborne mists or dusts such as lubricating oils, coal dust, or grain dust.

CGIs use a combustion chamber containing a filament that combusts the flammable gas. To facilitate combustion the filament is heated or is coated with a catalyst (like platinum or palladium), or both.

Limitations in use of the CGI include;

160   WORKER PROTECTION DURING HAZARDOUS WASTE REMEDIATION

- The CGI is intended for use only in normal-oxygen atmospheres. Oxygen-deficient atmospheres will produce lowered readings. Also, the safety guard that prevents the combustion source from igniting a flammable atmosphere is not designed to operate in an oxygen-enriched atmosphere.
- Certain materials in the sampled atmosphere affect the catalytic material on the filament and may cause the indicator to respond incorrectly. These materials include organic lead compounds such as those used in leaded gasoline and silicon compounds in the form of silanes, silicones, and silicates (often found in hydraulic fluids).
- The MSA 260 CGI is calibrated to pentane (as being representative of the flammability characteristics of most commonly encountered combustible gases). Being calibrated to pentane, the 260 may give a different response to other chemicals; this factor is called *relative response*. Table 8-1 shows the relative response of selected chemicals for an MSA 260 CGI calibrated to pentane. To the extent that the relative response of the gas or vapor being detected is less than the relative response of the calibrant gas, the meter reading will understate the actual concentration of the gas being detected. Likewise, to the extent that the relative response of the gas/vapor being detected is greater than the relative response of the calibrant gas, the meter reading will overstate the actual concentration of the gas being detected. If the identity of the gas/vapor being detected is known, the actual concentration can be calculated; but this cannot be done if the identity of the detected gas is not known.

Factors to consider in using the MSA 260 O_2/CGI include:

- Accessory sampling lines are available. Lines over 100 feet in length are not recommended. A 50-foot line will increase the initial response time to approximately 30 seconds and the final response to approximately 3 minutes. Two 50-foot lines connected together will increase initial response to 60 seconds and final response to 6 minutes.

Table 8-1.   Relative Response of Selected Chemicals for the MSA 260 Combustible Gas Indicator Calibrated to Pentane.

Chemical	LEL	Concentration, % LEL*	Meter Response, % LEL	Relative Response
Methane	5.3%	50	85	170%
Acetylene	2.5%	50	60	120%
Pentane	1.5%	50	53	106%
1,4-Dioxane	2.0 ppm	50	37	74%
Xylene	1.1%	50	27	54%

*Actual concentration, as % LEL, of the detected gas/vapor.

- Upon initially opening and probing an enclosed area, move the probe into the area slowly while watching the meters to provide the earliest possible indication of a potentially dangerous condition.
- If readings approach or exceed 10% of the LEL, extreme caution should be exercised in continuing the investigation. If readings approach or exceed 25% of LEL, personnel should be withdrawn immediately. Before resuming any on-site activities, project personnel in consultation with experts in fire or explosion prevention must develop procedures for continuing operations. However, in view of the possibility of understated (or overstated) readings when unknown contaminants are detected, project personnel might wish to modify these percentage guidelines in any particular situation.
- An atmosphere that shows no flammability hazard can still be toxic to workers. Also, a tank or vessel which is safe before work is begun may be rendered unsafe by work activities which cause a temperature increase, or by stirring or handling bottom sludge in petroleum tanks.

Toxic Atmospheres

When the presence or types of *organic* vapors/gases are unknown, direct-reading instruments should be used to detect organic vapors. Until specific constituents can be identified, the readings from such instruments will be total airborne substances to which the instrument is responding. Identification of the individual vapor/gas constituents may permit the instruments to be calibrated to these substances and used for more specific and accurate analysis.

Sufficient data should be obtained during the initial entry to map or screen the site for various levels of organic vapors. These gross measurements can be used on a preliminary basis to: (1) determine levels of personnel protection, (2) establish site work zones, and (3) select candidate areas for more thorough study.

The number of direct reading instruments with the capability to detect and quantify nonspecific *inorganic* vapors and gases is extremely limited. Currently the photoionization process has very limited detection capability, while the flame ionization process has none. If specific inorganics are known or suspected to be present, measurements should be made with appropriate instruments, if available. Colorimetric tubes are practical *if* the substances present are known or can be narrowed to a few.

The direct-reading instruments to be discussed herein for the purpose of monitoring for toxic atmospheres are:

- Photoionization detector.
- Flame ionization detector.
- Colorimetric tubes.

The Photoionization Detector. This instrument provides a direct reading of the concentration of a variety of trace gases, particularly organics. It employs the principle of photoionization. This process involves the absorption of ultraviolet light (a photon) by a gas molecule, leading to ionization. The sensor consists of a sealed ultraviolet light source that emits photons with an energy level high enough to ionize many trace species, but not high enough to ionize the major components of air. Ions formed by the absorption of photons are collected on an electrode, producing a current which is measured and converted to a concentration reading on the meter (in ppm).

The analyzer consists of a probe, readout assembly, and battery charger. The probe contains the sensing and amplifying circuitry; the readout assembly contains the meter, controls, power supply and battery (see Fig. 8-2). The energy required to remove the outermost electron from the molecule is called the *ionization potential* (IP) and is specific for any compound or atomic species. IP is measured in electron volts (eV). The standard HNU probe uses a 10.2 eV lamp; 2 optional probes use 9.5 and 11.7 eV lamps. To change the lamp, the entire probe must be changed.

This unit is most useful in situations where the identity of the detected gas is known. In cases where the identity of the detected gas is unknown, the photoionizer can give only a very gross indication of what the detected contaminant(s) may be. This occurs by referring to tables showing the IP of various gases. Reference to these tables will show which gases have an IP equal to or less than the strength of the lamp used, but it will not tell which of the gases in that category is actually present.

Other limitations are:

- The factor of relative response. For instance, the unit is factory calibrated to benzene (although other optional calibrations are available). When the

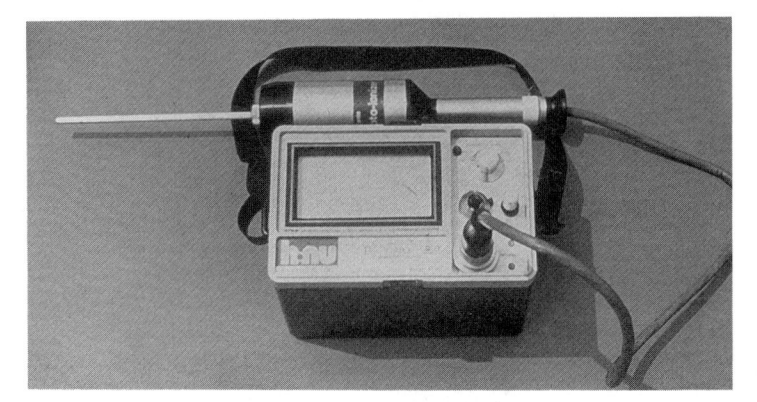

Fig. 8-2. HNU photoionization detector used to direct-read organic vapors.

gas being detected has a relative response less than the relative response of the calibrant gas, the reading shown on the HNU will understate the actual concentration of the gas being detected. The opposite is true for a detected gas with a relative response higher than that of the calibrant gas. If the identity of the detected gas is known the actual concentration can be calculated, but if the identity is unknown, this cannot be done. Table 8-2 shows the meter reading on an HNU calibrated to benzene when actually detecting 10 ppm of any of the gases/vapors shown in the table (when using a 10.2 eV probe).

- The response to a gas or vapor may radically change when the gas or vapor is mixed with other materials. This could result in a substantial over or understatement of the actual concentration of the contaminants.

Table 8-2. Photoionization Sensitivity.* Relative to Reference Standard Benzene.

Chemical	Photoionization Sensitivity ppm	Chemical	Photoionization Sensitivity ppm
p-Xylene	11.4	Dimethyl sulfide	4.3
m-Xylene	11.2	Allyl alcohol	4.2
Benzene	10.0	Propylene	4.0
(reference standard)		Mineral spirits	4.0
Toluene	10.0	2,3-Dichloropropene	4.0
Diethyl sulfide	10.0	Cyclohexene	3.4
Diethyl amine	9.9	Crotonaldehyde	3.1
Styrene	9.7	Acrolein	3.1
Trichloroethylene	8.9	Methyl methacrylate	3.0
Carbon disulfide	7.1	Pyridine	3.0
Isobutylene	7.0	Hydrogen sulfide	2.8
Acetone	6.3	Ethylene dibromide	2.7
Tetrahydrofuran	6.0	n-octane	2.5
Methyl ethyl ketone	5.7		
Methyl isobutyl ketone	5.7	Acetaldehyde oxime	2.3
Cyclohexanone	5.1	Hexane	2.2
Naptha (85‰ aromatics)	5.0	Phosphine	2.0
Vinyl chloride	5.0	Heptane	1.7
Methyl isocynate	4.5	Allyl chloride	1.5
Iodine	4.5	(3-Chloropropene)	
Methyl mercaptan	4.3	Ethylene	1.0
Epichlorohydrin	0.7	Isopropanol	1.0
Nitric oxide	0.6	Ethylene oxide	1.0
Beta pinene	0.5	Acetic anhydride	1.0
Citral	0.5	Alpha pinene	0.7
Ammonia	0.3	Debromochloropropane	0.7
Acetic acid	0.1	Methane	0.0
Nitrogen dioxide	0.02	Acetylene	0.0

*Reading in ppm when measuring 10.0 ppm of particular gas with monitor calibrated for benzene and connected to 10.2 eV probe.

- Dust in the atmosphere can collect on the lamp and block the transmission of UV light, causing a reduction in instrument reading.
- Humidity can cause two problems: When a cold instrument is taken into a warm, moist atmosphere, the moisture can condense on the lamp, reducing the available light. Also, moisture in the air reduces the ionization of chemicals and can cause a reduction in readings.
- Radio-frequency interference from pulsed DC or AC power lines, transformers, generators, and radio wave transmission may produce an error in response.
- In some cases, at high concentrations the instrument response can decrease. While the response may be linear (i.e., one to one) from 1 to 600 ppm, a concentration of 900 ppm may only give a meter response of 700.

The fan draws gas in through the probe at a rate of approximately 100 cc/minute. The HNU measures gases in the vicinity of the operator, a high reading, when measuring toxic or explosive gases, should be cause for action for operator safety.

Flame Ionization Detector. Flame ionization detectors are a means to detect toxic *organic* vapors (see Fig. 8-3). Flame ionization detectors (FIDs) use a

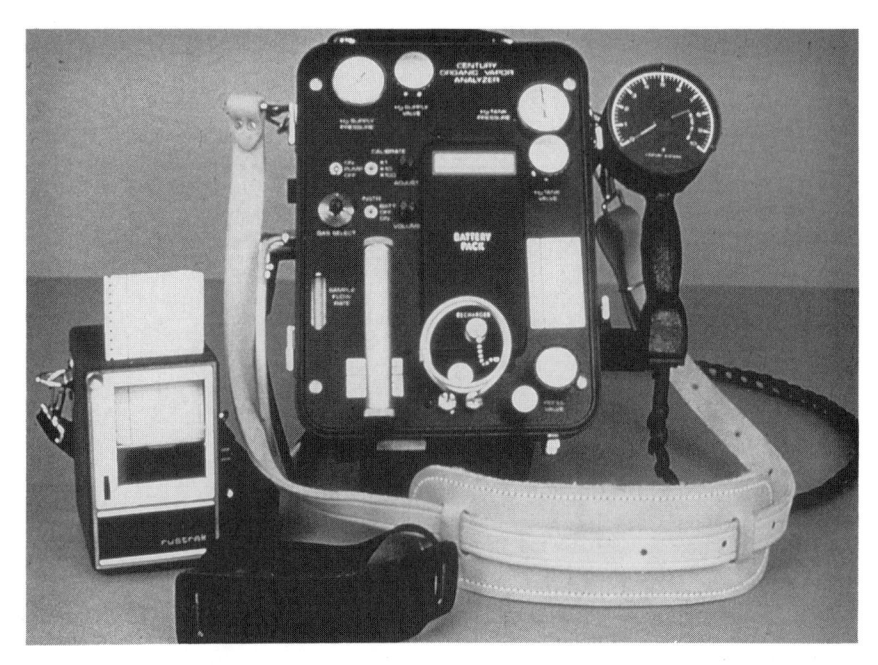

Fig. 8-3. Foxboro Organic Vapor Analyzer used to analyze vapors for organics.

hydrogen flame as the means to ionize the vapors. The FID responds to virtually all compounds containing carbon-hydrogen or carbon-carbon bonds. Inside the detector chamber, the sample is exposed to a hydrogen flame which ionizes the organic vapors. When the vapors are ionized, positively charged ions are collected on an electrode, producing a current which is converted to a concentration reading on the meter display. As noted above, FIDs respond only to organic compounds. Thus, they do not detect inorganic compounds like chlorine, hydrogen cyanide, or ammonia.

The Foxboro Organic Vapor Analyzer (OVA) can operate in two modes, survey and gas chromatography.

Survey Mode. In this mode the sample is ionized and the resulting current is translated on the meter for directly reading concentration as total organic vapors. As with the other direct-reading instruments previously discussed, the OVA responds differently to different chemicals. The OVA is factory calibrated to methane.

If the gas or vapor actually being detected is not methane, then the concentration shown on the meter will either understate or overstate the actual concentration of the detected chemical, depending on whether the relative response of the detected chemical is lower or higher than the relative response of the calibrant gas. If the identity of the detected chemical is known, the actual concentration of the detected chemical can be calculated. However, if the identity of the detected chemical is not known, the actual concentration cannot be calculated. Table 8-3 gives information on relative response of some materials.

Gas Chromatography Mode. If the OVA in use has a gas chromatography option available, it is possible to use the OVA to establish the identity of unknown detected organic vapors. The sample is separated into its various components and the identity and relative amounts of the components are established. This is done by injecting the sample of air into a column packed with an inert solid. As the carrier gas (for the OVA it is hydrogen) forces the sample throughout the column, the separate components of the sample are retained on the column for different periods of time. The amount of time a substance remains on the column is called its retention time.

As the components leave the column, they flow into the detector, which can be connected to a strip chart recorder which will record separate peaks for each component (see Fig. 8-3). This readout is called a *gas chromatogram* (see Fig. 8-4). The retention time is defined as the period of time that elapses between the injection of the compound into the column and the time the compound's components leave the column as represented by a peak. If the retention time of an unknown chemical agrees with the retention time of a known chemical recorded under the same set of analytical conditions, the unknown is tentatively identified.

The operating temperature range of the OVA is 50–104°F; minimum ambient

Table 8-3. Relative Response of OVA Calibrated to Methane.

Compound	Relative Response
Methane	100
Ethane	90
Propane	64
n-Butane	61
n-Pentane	100
Ethylene	85
Acetylene	200
Benzene	150
Toluene	120
Acetone	100
Methyl ethyl ketone	80
Methyl isobuytl ketone	100
Methanol	15
Ethanol	25
Isopropyl alcohol	65
Carbon tetrachloride	10
Chloroform	70
Trichloroethylene	72
Vinyl chloride	35

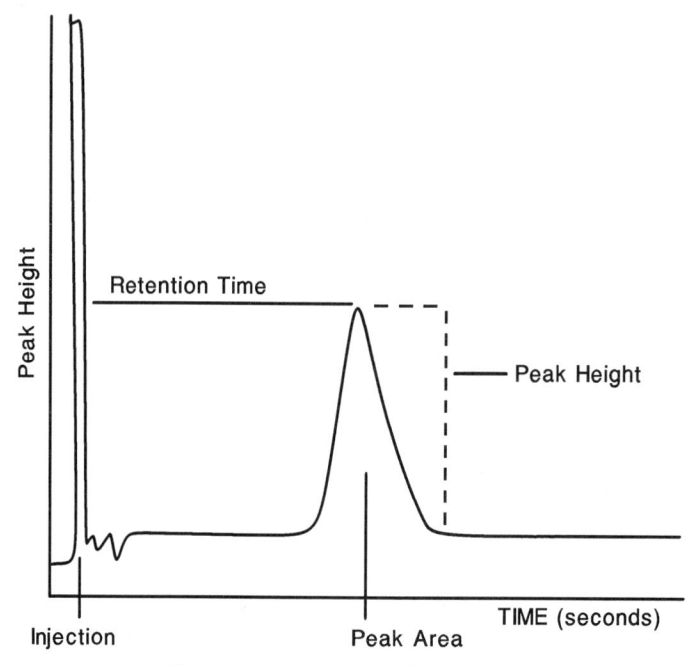

Fig. 8-4. Example of gas chromatogram.

temperature for flame ignition (cold start) is 50°F. To the extent that the relative response of the particular gas/vapor is lower than the relative response of the calibrant gas (methane), the OVA will understate the actual concentration of the gas/vapor. Gas/vapor with a higher relative response that the calibrant gas will be overstated.

Colorimetric Tubes. Colorimetric indicator tubes consist of a glass tube impregnated with an indicating chemical (see Fig. 8-5). The tube is connected to a piston or bellows-type pump and a known volume of contaminated air is pulled at a predetermined rate through the tube by the pump. The contaminant reacts with the indicator chemical in the tube, producing a change in color whose length is proportional to the contaminant concentration.

Detector tubes are normally chemical specific; however, some manufacturers do produce tubes for groups of gases such as aromatic hydrocarbons or alcohols. Concentration ranges may be in parts per million or percent. A preconditioning filter may precede the indicating chemical to remove contaminants other than the one in question or to remove humidity or to react with a contaminant to change it into a compound that reacts with the indicating chemical.

Limitations of the tubes are:

- Detector tubes have the disadvantage of poor accuracy and precision. In the past NIOSH tested and certified detector tubes but has since then discontinued this practice. Manufacturers report error factors of up to 50% for some tubes.
- The chemical reactions involved in the use of the tubes are affected by temperature. Cold weather slows the reactions and thus the response time. To reduce this problem it is recommended that the tubes be kept warm (for example, inside a coat pocket) until they are used if the measurement is done in cold weather. Hot temperatures increase the reaction and can cause a problem by discoloring the indicator when a contaminant is not present. This can happen even in unopened tubes; therefore, the tubes should be stored at a moderate temperature.
- Some tubes do not have a prefilter to remove humidity and may be affected by high humidity. The manufacturer's instructions should indicate if humidity is a problem.
- The chemical used in the tubes wears out over time. Thus, they are assigned a shelf life. This varies from 1 to 3 years.
- Although detector tubes are usually chemical specific, some tubes will respond to interfering compounds. The manufacturers provide information with the tubes on interfering gases and vapors.
- Interpretation of results can be a problem. Since the tube's length of color change indicates the concentration, the user must be able to see the end of

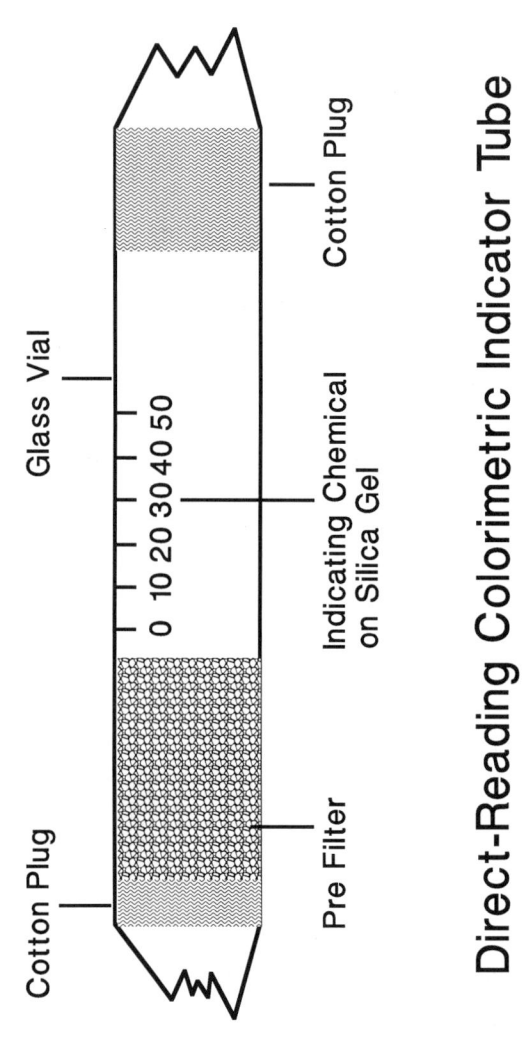

Fig. 8-5. Colorimetric tube used for direct-read applications.

the stain. Some stains are diffused and are not clear cut, or they may have an uneven endpoint. When in doubt use the highest value.
- Because of these considerations, it is very important to read the instructions that are provided with and are specific to a set of tubes.
- While there are many limitations and considerations for using detector tubes, they allow the versatility of being able to measure a wide range of chemicals with a single pump.

The Drager bellows pump can be operated with one hand and draws in 100 cm^3 per stroke. For inaccessible points of measurement, Dragger makes an extension hose which is 3 meters long. There are several manufacturers of colorimetric tubes and their accessories.

PART TWO. PERSONAL SAMPLING INSTRUMENTS

INTRODUCTION

Another means of monitoring a site is by sampling the immediate surroundings of persons working in a contaminated environment. Working on hazardous waste sites requires that individuals as well as the general environment be continually monitored for dangerous gases, vapors and particulates. One means of doing this is by sampling the immediate surroundings of persons working a contaminated environment. Placing sampling devices on the affected workers in their breathing zone areas can aid in determining and verifying contamination levels previously obtained by direct-reading equipment. This part of the chapter will focus on personal sampling equipment.

OBJECTIVES

- The reader will be able to identify the various types of atmospheric devices.
- The reader will know why and where these devices are worn on hazardous waste sites.
- The reader will know what airborne contaminants the devices collect.

CHOOSING SAMPLING METHODS

The atmosphere may be sampled during a hazardous waste cleanup operation to identify and quantify any gases, vapors, or particulates to which workers may be exposed. Such information may be obtained by two methods:

- Area sampling, which involves the placement of collection devices within designated areas and operating them over specific periods of time.
- Personal sampling, which involves the collection of samples from within the breathing zone of an individual, sometimes by the individual wearing a sampling device.

Once the sampling method has been selected, the type of sample desired must be determined. Prevailing conditions, the scope of site operations, and the intended use of the resulting information dictate the type collected.

- Instantaneous or grab-type samples are collected over brief time periods. They are useful in examining stable contaminant concentrations or peak levels of short duration. Instantaneous samples may require highly sensitive analytical methods due to the small sample volume collected.
- Integrated samples are more typical of on-site measurements. They are collected when the sensitivity of an analytical method requires minimum sample periods or volumes, or when comparison must be made to an 8-hour, time-weighted average threshold limit value (TLV) or the OSHA standard, permissible exposure limit (PEL).

Two types of sampling systems are used for the collection of integrated samples:

- Active samplers mechanically move contaminated air through a collection medium.
- Passive samplers rely on natural rather than mechanical forces to collect samples. Passive samplers are classified as either diffusion or permeation devices, according to their principle of operation.

The sampling instrument or system chosen depends on a number of factors, including:

- Instrument or system efficiency.
- Operational reliability.
- Ease of use and portability.
- Availability of the instrument and component parts.
- Information or analysis desired.
- Personal preference.

ACTIVE SAMPLERS

General Considerations

Active sampling systems mechanically collect samples on or into a selected medium. The medium is then analyzed in the laboratory to identify and quantify the contaminant(s) collected. Such a system typically consists of the following components:

- An electrically powered pump to move the contaminated air. Such a pump should contain a flow regulator to control the rate of movement and a flow monitor to indicate that rate.
- A sampler consisting of an appropriate sampling medium and a container designed for that medium. The sampler used largely depends upon the contaminant(s) to be sampled and the selected sample method.
- Flexible, nonporous, inert tubing to link the sampler to the pump. Integrated samples are commonly collected over known time periods and at known fixed flow rates. Thus, sample pump calibration and accurate time measurement are critical to the proper interpretation of data collected by active systems.

Sampling Pumps

Active sampling systems typically rely on electrically powered pumps to mechanically induce air movement. The most practical electrical sampling pumps are powered by rechargeable batteries and can operate continuously at constant flow rates for at least 8 hours. Typically, they are compact, portable, and quiet enough to be worn by individuals when monitoring personal exposures.

The type of portable pump selected is generally determined by such factors as the physical properties of the contaminant, the collection medium, and the collection flow rates specified by the analytical method used.

Sample Collection Devices

Gases/Vapors. Active samplers for gaseous and vapor contaminants make use of a variety of collection media, including solids, liquids, long-duration colorimetric tubes, and sampling bags. All require a pump to ensure proper contact between the collection media and the contaminants.

Solid sorbents are the class of media most widely used in hazardous materials sampling operations. These materials collect by adsorption and are often the media of choice for insoluble or nonreactive gases or vapors. Their popularity stems from a number of factors, including high collection efficiencies, indefinite shelf lives while unopened, ease of use compared to liquid adsorbers, improved tube design, and specific analytical procedures. The solid sorbent to be used is generally specified in standard sampling and analytical methods. One such group of standard methods is the *NIOSH Manual of Analytical Methods.*

Besides sorbent and tube configuration, these standards also specify requirements such as maximum sample volumes and collection rates, sample train configuration, and sample storage. The two most widely used solid sorbents are:

- *Activated charcoal*, which has perhaps the broadest range of collection efficiencies. The highest efficiencies are for organic vapors with boiling points above 0°C, and the lowest for organic gases with boiling points below −150°C. Activated charcoal exhibits nonpolar qualities and has a greater affinity for organic gases and vapors than for water, which is polar. This property results in far greater retention of adsorbed organic vapors than silica gel. Glass construction sampling tubes contain two volumes of activated charcoal, the larger being the primary sample stage and the smaller backup stage (Fig. 8-6).
- *Silica gel,* which is the next most widely used solid sorbent. It exhibits polar characteristics, preferentially adsorbing more polar or polarizable compounds. Thus, in order of decreasing collection efficiency, silica gel will adsorb water, alcohols, aldehydes, ketones, esters, aromatic compounds, olefins, and paraffins. Silica gel is also packaged in glass sampling tubes of several sizes, containing two or three stages of sorbent in varying proportions as specified by the analytical method.
- A number of other synthetic sorbents are available for specific gas or vapor contaminants or groups of contaminants (Table 8-4.)

Liquid absorbers are used with impinger or bubblers and powered pumps to collect soluble or reactive gases and vapors (Table 8-5). Only a relatively few analytical methods call for collection by impinger or bubbler. Further, most of the common absorbers tend to be contaminant-specific and they have limited shelf lives.

Four types of active samplers are used to collect gases and vapors by liquid absorption. These samplers ensure that contaminants in the sampled air are completely absorbed by the liquid sampling medium selected.

A Typical 150 mg Charcoal Tube For Low Flow Organic Vapor Sampling

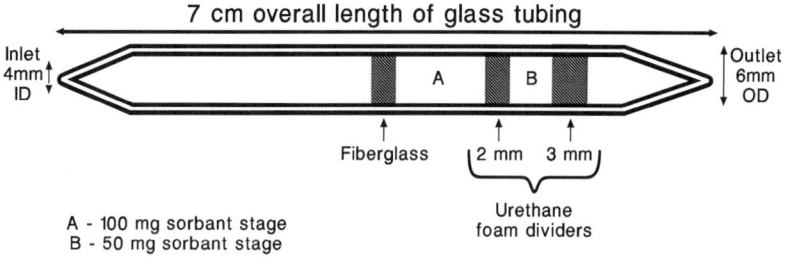

Fig. 8-6. Charcoal tube for personal sampling.

Table 8-4. Solid Sorbents Commonly Used in Gas and Vapor Sampling.

Solid Sorbent	Representative Gas or Vapor Adsorbed
Activated charcoal—coconut base	Organic solvents
Activated charcoal—petroleum base	1,2-Dibromo-3-chloropropane; ethylene; methyl bromide; n-propyl nitrate; 1,1,2,2-tetrachloroethane
Silica gel	Acetic acid; amines; amides
Molecular sieve 5A	Sulfur dioxide
Molecular sieve 13X	Acrolein
Tenax GC	Allyl glycidyl ether; diphenyl; ethylene glycol dinitrate; nitroglycerin; white phosphorus; trinitrotoluene
Floricil	Polychlorinated biphenyls
Chromosorb 101	*bis*-Chloromethyl ether
Chromosorb 104	Butyl mercaptan
Porpak Q	Furfuryl alcohol; methyl cyclohexanone
XAD-2	DDVP; Demeton; ethyl silicate; nitroethane; quinone; tetramethyl lead (as lead)

- The impinger is the most widely used gas absorber device (see Fig. 8-7). This device, usually made of glass, contains a measured volume of absorber liquid and is connected to the pump by flexible tubing. When the pump is turned on, the contaminated air is channeled down through the liquid. The popularity of impingers rests on such qualities as simple construction, ease of cleaning, the small quantity of liquid used (typically less than 25–30 milliliters), and size suitable for use as a personal monitor.
- Fritted bubblers are similar in use and appearance to impingers, but are

Table 8-5. Liquid Absorbers Commonly Used in Gas and Vapor Sampling.

Absorbing Liquid	Gas/Vapor Absorbed
0.1N H_2SO_4	Bases and amines
0.1N NaOH	Acids and phenol
0.1N HCl	Nickel carbonyl
Alkaline $CdSO_4$ ($CdSO_4$–NaOH)	Hydrogen sulfide
Methylene blue	Hydrogen sulfide
1‰ KI in 0.1N NaOH	Ozone
Nitro reagent (4-nitropyridyl propylamine in toluene)	Diisocyanates
0.3N H_2O_2	Sulfur dioxide
0.1‰ Aniline	Phosgene
1‰ $NaHSO_2$	Formaldehyde
Distilled water	Acids and bases

(A) Midget Impinger (B) Fritted Bubbler

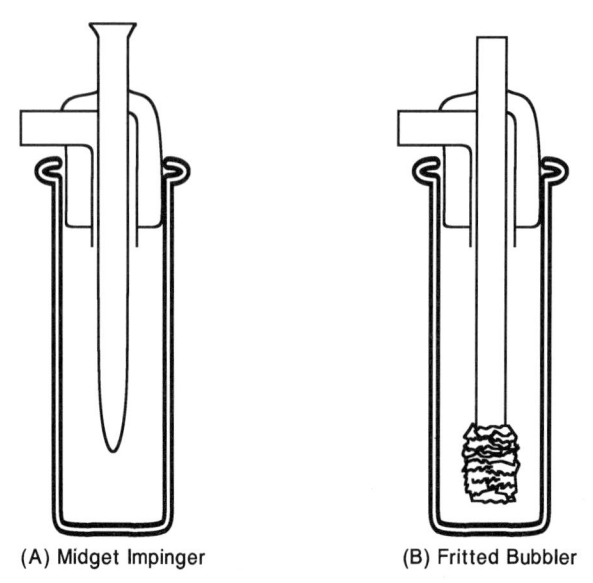

Fig. 8-7. Impinger/bubbler used for personal sampling.

generally used when a higher degree of air-liquid mixing is desired. With these devices, the contaminated air is forced through masses of porous glass, called frits, breaking the air stream into numerous small bubbles and increasing the surface area for air/liquid contact.
- The glass-bead column is used for special situations where the concentrated solution is required. Glass beads within the column are coated with the absorber liquid, thus providing greater surface area upon which the contaminant can be absorbed.
- Spiral and helical absorbers, somewhat similar in appearance to impingers and bubblers, used when slightly soluble or slow reacting contaminants must be sampled, and provide a prolonged period of contact with the absorber liquid.

Long duration, direct-reading colorimetric tubes are generally used in the 2–8 hour range, and may be used for time-weighted average sampling. Unlike their short-term counterparts, long-duration tubes require an electric rather than hand operated pump.

Sample bags offer alternatives in the collection of instantaneous and integrated samples of gases and vapors, and are particularly useful when representative samples are desired. Bag sampling begins by connecting the bag inlet valve with flexible tubing to the exhaust outlet of a typical sampling pump. The bag inlet valve is opened, the pump turned on, and the sample collected. Once sampling has been completed, the bag contents itself may be sampled, directly emptied into an analytical instrument, such as a gas chromatograph, or tested by colorimetric tube.

Particulates. Airborne particulates or aerosols include both dispersed liquids (mists and fogs) and solids (dusts, fumes, and smoke). The most common method of sampling particulates is to trap them on filters using active systems to collect integrated samples. The main difference between particulate sampling trains and other active systems is the type of sampler. Particulate collectors are generally used to selectively gather samples on the basis of particle size. Particulate samplers typically have an air inlet and a membrane or fiber filter on which the particulates are collected; a preselector may also be included ahead of the filter if the particulates are to be classified by size.

One group of preselectors are the centrifugal separators or cyclones, which are commonly used to separate and collect those particulates small enough to enter the respiratory system. Cyclones are conical or cylindrical in shape, with an opening through which particulate-laden air is drawn along a curved channel. Larger particles impact against the interior walls of the unit due to inertial forces and drop into a grit chamber in the base. The lighter particles continue on through and are drawn up through a tube at the center of the cyclone, where they are collected on a filter. Cyclones can be worn as personal respirable dust monitors.

Connected to most preselectors is a cassette containing the desired filter material. Typically, these cassettes are molded out of transparent polystyrene plastic to form a cylinder. They consist of two or three stacked sections, the number depending on the contaminant and the collection method. The sections of a cassette are molded to fit tightly when stacked and to tightly grip the outer edge of the filter. Each cassette has end plugs to seal the inlet and tubing connector port only when the desired collection period is completed (Fig. 8-8).

Two types of filter materials are used:

- *Fiber filters* are composed of irregular meshes of fibers forming openings or pores of 20 micrometers in diameter or less. A number of fiber filters are available (Table 8-6). The two with perhaps the greatest application to hazardous materials operations are *cellulose* and *glass*. Of the two, cellulose is the least expensive, and is available in a variety of sizes. Its greatest disadvantage is its tendency to absorb water, thus creating problems in weighing. For this reason, glass fiber filters are finding more applications.
- *Membrane filters* are microporous plastic films. This group of filters includes such materials as cellulose triacetate, polyvinylchloride, teflon, polypropylene, nylon, and silver (Table 8-6). These filters have an extremely low mass and ash content. Some are completely soluble in organic solvents

Assembly of a Three Piece Filter Cassette

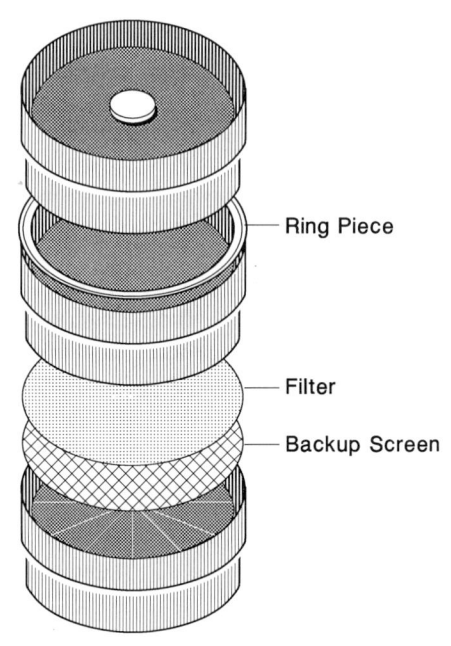

- Ring Piece
- Filter
- Backup Screen

Fig. 8-8. Filter cassette used for personal monitoring.

Table 8-6. Filter Media for Airborne Particulates.

Filter Medium	Representative Application
Cellulose ester, 0.45 micrometer pore	Metal fumes; acid mists
Cellulose ester, 0.8 micrometer pore	Asbestos; metal fumes; fibers; chlorodiphenyls (54‰ chlorine)
Fibrous glass	Total particulate; oil mists; pesticides; coal, tar, and pitch volatiles
Polyvinyl chloride, 5.0 micrometer pore	Weight analysis; hexavalent chromium
Polyvinyl chloride, 5.0 micrometer pore, in shielded cassette	Electrostatic dusts
Silver membrane	Total particulate; coal, tar, and pitch volatiles; free crystalline silica
Teflon	High temperature applications

enabling the concentration of collected particulates into small volumes for later analysis.

PASSIVE SAMPLERS

Quantitative passive samplers have become available only since the early 1970s. The key advantage of passive dosimeters is their simplicity (Fig. 8-9). These small, lightweight devices do not require a mechanical pump to move a contaminant through collection medium. Thus, calibration and maintenance are reduced or eliminated, although the sampling period must still be accurately measured. Despite this obvious advantage, such sources of error as observer interpretation, and the effects of temperature and humidity hold true for both active and passive systems.

The few passive samplers now available apply to gas and vapor contaminants only. These devices primarily function as personal exposure monitors, although they have some usefulness in area monitoring. Passive samplers are commonly divided into two groups, primarily on how they are designed and operated.

- *Diffusion samplers* function by the passive movement of contaminant molecules through a concentration gradient created within a stagnant layer of air between the contiminated atmosphere and the indicator material. Some diffusion samplers may be read directly, as are colorimetric length-of-stain tubes, while others require laboratory analysis similar to that performed on solid sorbents.

Diffusion-Type Passive Sampler

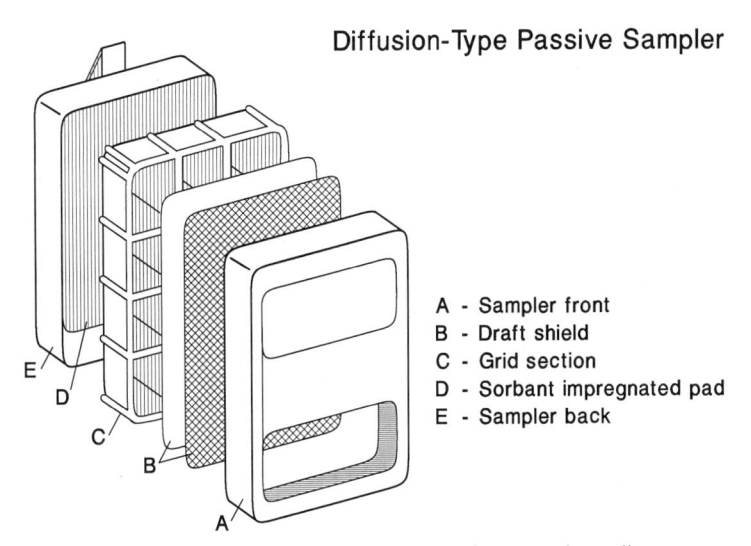

A - Sampler front
B - Draft shield
C - Grid section
D - Sorbant impregnated pad
E - Sampler back

Fig. 8-9. Passive sampler worn by workers for personal sampling.

- *Permeation samplers* rely on the natural permeation of a contaminant through a membrane. Permeation samplers are useful in picking out of a single contaminant from a mixture of airborne contaminants due to selective permeation of chemicals through the membrane. As with diffusion samplers, some passive samplers may be of the direct reading type while others may require laboratory analysis.

CALIBRATION

Atmospheric sampling systems must be accurately calibrated to a specific flow rate if the resultant data are to be correctly interpreted. Flow rate calibration of the electrically powered pump in active systems is important to achieving the constant flow rates often specified in standard analytical methods. Passive sampling systems, however, because of their simplicity in design and principles of operation, require no formal calibration. As a minimum, an active sampling system should be calibrated prior to use and following a prescribed sampling period. The overall frequency of calibration depends upon the general handling and use a sampling system receives. Pump mechanisms should be recalibrated after they have been repaired, when newly purchased, and following any suspected abuse.

PERSONAL SAMPLING PLAN

It is not easy to write a plan for sampling personnel on a hazardous waste cleanup site, for several reasons.

- It is often difficult to decide what to sample for.
- Many workers move around on the site, and some have many different tasks.
- Keeping track of the sampling data in an efficient manner designed to be informative requires attention and organization.

A monitoring standard for hazardous waste sites is being considered at this writing. Currently there are only a few substances for which a full standard exists. These include lead, asbestos, chromium, acrylonitrile, and others. No required sampling method exists for other substances and, if air concentrations do not exceed permissible exposure limits, personnel monitoring is not required. If PELs are exceeded on the site, the following guidelines may be used.

- Sample workers who are known to be in the most highly exposed locations.
- If the most hazardous locations have not been determined, sample all workers for a week.
- Group workers by exposure levels.

- Sample some members of each group every week.
- Change the plan each time conditions change on the site.

REFERENCES

ACGIH. 1978. *Air Sampling Instruments for Evaluation of Atmospheric Contaminants*. Cincinnati, OH: American Conference of Governmental Industrial Hygienists.

Dosimeter Corporation. Model 3700, 3708, and 3007 Instruction Manual. Cincinnati, OH: Dosimeter Corp.

Dragerwerk. 1985. Detector Tube Handbook. Lübeck, West Germany: Dragerwerk AG.

Foxboro. 1986. Instruction, Model OVA88 Century Organic Vapor Analyzer. Foxboro, MA: Foxboro Company.

HNU. 1986. Instruction Manual, Model PI 101, Portable Photoionization Analyzer. Newton, MA: HNU Systems, Inc.

Mine Safety Appliances Company. Instruction Manual, MSA 260 Combustible Gas and Oxygen Alarm. Pittsburgh, PA: Mine Safety Appliances Company.

NSC. 1983. *Fundamentals of Industrial Hygiene*. Chicago, IL: National Safety Council.

Patty, F. A. 1958. *Industrial Hygiene and Toxicology, Vol. 1*. New York: John Wiley & Sons, Inc.

U.S. EPA, Office of Emergency Remedial Response. *Air Surveillance for Hazardous Materials*. Washington, DC: U.S. Govt. Printing Office.

9
Personal Protective Equipment

INTRODUCTION

In order to maintain safety and health, workers involved in hazardous waste site activities must be protected from chemical contaminants and other hazards which may be present in the work area. Federal standards require that this protection be provided to the maximum extent practical through the use of engineering controls and work practices, with personal protective equipment (PPE) to be used only as a last resort. However, in most cleanup operations, engineering controls and work practices are impractical or insufficient to create a safe work environment. Thus, PPE is typically vital to worker protection on site. For this reason, it is important that all hazardous waste workers have sufficient knowledge and hands-on experience to utilize PPE effectively and safely on the job. This chapter is intended as an introduction to PPE.

OBJECTIVES

- Know the distinguishing characteristics of the major types of respiratory protective equipment.
- Know the selection considerations, advantages, and disadvantages of the different types of respirators.
- Understand the importance of respirator facepiece-to-face fit and know methods of fit testing.
- Appreciate the complexities involved in chemical protective clothing (CPC) selection.
- Know and understand the importance of using properly selected CPC.
- Be aware of available types of protective clothing and accessories.
- Be familiar with the EPA Levels of Protection for PPE ensembles.
- Be aware of various requirements for safe use of PPE.

PART 1. RESPIRATORS

RESPIRATORY PROTECTION REQUIREMENTS

The Need for Respiratory Protection

Respirators are designed to protect workers from the inhalation and ingestion of contaminants. These contaminants may be present in a given work area as dusts, fumes, mists, gases, and/or vapors. Respiratory protection is very critical to the hazardous waste worker because respiratory hazards are frequently encountered in the course of investigative and remedial actions on hazardous waste sites. For this reason, the information presented here is specific to those types of respirators which are most commonly used on hazardous waste sites.

OSHA Requirements

General Requirements. OSHA requires that workers be protected from exposure in excess of PELs for atmospheric contaminants as listed in 29 CFR 1910.1000. If engineering controls and/or work practices cannot feasibly be used to provide this protection, then respirators must be used.

Respiratory Protection Program. Under 29 CFR 1910.134, OSHA requires the employer to implement a written respiratory protection program covering all employees whose job assignments require the use of respirators. The respiratory protection program must meet the following requirements (OSHA, 1986):

- The program must be in writing.
- Respirator selection must be hazard-specific.
- Workers must be adequately trained in respirator use.
- Fit testing must be included.
- Only approved equipment may be used.
- Work area surveillance must be conducted to ensure that respiratory protection is appropriate and used as required by the program.
- Respirators must be decontaminated and sanitized after each use.
- Storage areas for respirators must be convenient, clean, and sanitary.
- Respiratory equipment must be regularly inspected, maintained, and repaired as needed.
- Workers must be medically determined to be physically fit to wear a respirator before being assigned to tasks requiring respiratory protection.
- The effectiveness of the program should be evaluated on a regular basis.

It should be noted that these requirements are generally applicable to all types of PPE. The requirements for a hazardous waste site PPE program, as required by 29 CFR 1910.120, are discussed in the fourth part of this chapter.

CLASSIFICATION OF RESPIRATORY PROTECTIVE EQUIPMENT

Respiratory protective equipment used in hazardous waste work can be generally categorized based on facepiece type and method of protection (U.S. NIOSH, 1985).

Facepiece Type

Various styles of facepieces are used with respiratory protective equipment. Based on the amount of facial coverage provided, facepieces suitable for use during cleanup operations may be divided into two major types:

- *Half masks* cover the face from below the chin to the bridge of the nose (see Fig. 9-1).
- *Full facepieces* cover the entire face, thus offering eye protection as well as a fit which is not easily disturbed (see Figs. 9-2 and 9-3).

Method of Protection

Respirators can also be classified as either air-purifying or atmosphere-supplying.

Air-Purifying Respirators (APRs). These devices use filters, neutralizing agents, and/or sorbent materials to purify the ambient atmosphere of the work area for breathing. The purifying materials are contained in disposable cartridge or canister-type purification elements (see Fig. 9-4) which must be replaced after a given duration of use. Canisters are appreciably larger than cartridges and are usually worn on a belt or harness and connected to the facepiece by a

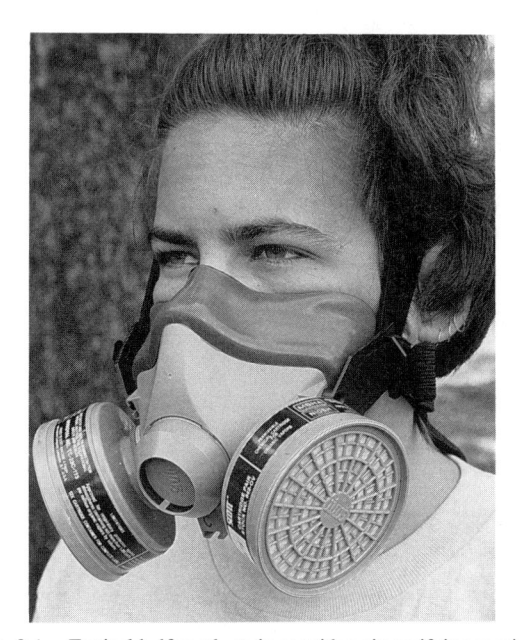

Fig. 9-1. Typical half-mask, twin-cartridge air-purifying respirator.

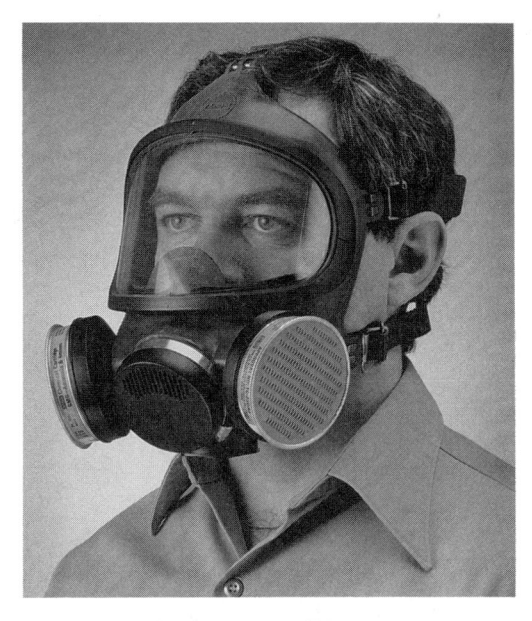

Fig. 9-2. Typical full-facepiece, twin-cartridge air-purifying respirator. (*Courtesy of Mine Safety Appliances Company*).

breathing tube, whereas cartridges typically mount directly to the respirator facepiece (compare Figs. 9-2 and 9-3).

Air-purifying respirators provide effective protection in atmospheres which contain relatively low concentrations of known contaminants and have near-normal oxygen levels. For atmospheres containing (1) high levels of contaminants, (2) extremely toxic contaminants, (3) unidentified contaminants, or (4) deficient oxygen levels, atmosphere-supplying respirators must be used. See Fig. 9-5.

Atmosphere-Supplying Respirators. These devices provide the user with breathing air from outside the contaminated work area. Two types of atmosphere-supplying equipment are commonly used:

- *Supplied-Air Respirators (SARs)*, or airline respirators, supply breathing air to the worker through an airline which is connected to a compressor and purification unit or a bank of air tanks located outside the contaminated area (see Fig. 9-6).
- *Self-Contained Breathing Apparatus (SCBA)* supplies breathing air from a tank worn on the user's back (see Fig. 9-7).

184 WORKER PROTECTION DURING HAZARDOUS WASTE REMEDIATION

Fig. 9-3. Typical full-facepiece, canister-type air-purifying respirator. (*Courtesy of Mine Safety Appliances Company*).

Based on facepiece type and method of protection utilized, respiratory protective equipment may be classified as follows:

- Half-mask, twin-cartridge APRs (Fig. 9-1).
- Full-facepiece, canister-type APRs (Fig. 9-3).
- Half-mask, supplied-air (or airline) respirators.
- Full-facepiece SCBAs, etc. (Fig. 9-7).

SELECTION OF RESPIRATORY PROTECTIVE EQUIPMENT

Each type of respiratory protective equipment has certain advantages and disadvantages. Thus, careful consideration is required to determine which type is appropriate for a specific work situation (NIOSH, 1985; NIOSH, 1987).

PERSONAL PROTECTIVE EQUIPMENT 185

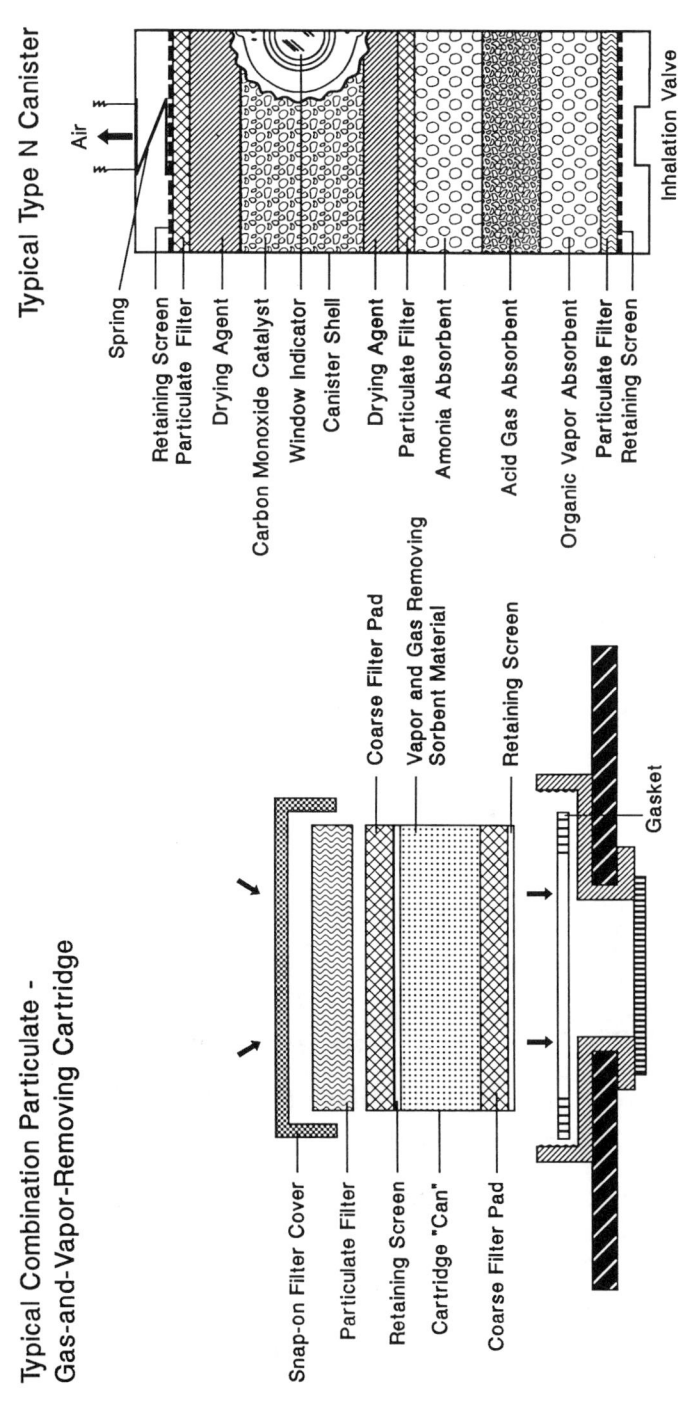

Fig. 9-4. Air-purifying cartridge and canister components.

CAN AN AIR-PURIFYING RESPIRATOR BE USED?

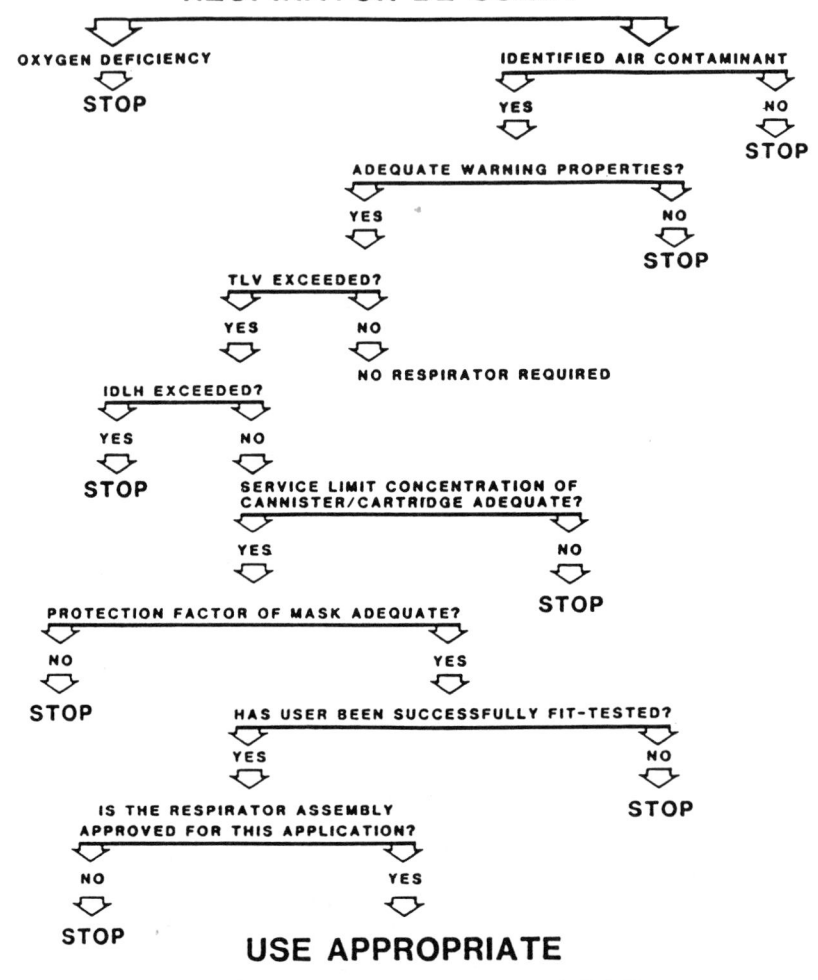

```
OXYGEN DEFICIENCY                    IDENTIFIED AIR CONTAMINANT
      ↓                                       ↓
    STOP                              YES          NO
                                       ↓           ↓
                                                 STOP
                           ADEQUATE WARNING PROPERTIES?
                              YES              NO
                               ↓               ↓
                                             STOP
              TLV EXCEEDED?
             YES         NO
              ↓          ↓
                      NO RESPIRATOR REQUIRED
   IDLH EXCEEDED?
   YES        NO
    ↓         ↓
  STOP    SERVICE LIMIT CONCENTRATION OF
          CANNISTER/CARTRIDGE ADEQUATE?
              YES              NO
               ↓               ↓
                             STOP
 PROTECTION FACTOR OF MASK ADEQUATE?
    NO              YES
     ↓               ↓
   STOP    HAS USER BEEN SUCCESSFULLY FIT-TESTED?
              YES              NO
               ↓               ↓
                             STOP
   IS THE RESPIRATOR ASSEMBLY
   APPROVED FOR THIS APPLICATION?
     NO              YES
      ↓               ↓
    STOP
```

USE APPROPRIATE AIR-PURIFYING RESPIRATOR

Fig. 9-5. Flow chart incorporating use limitations of APRs. (U.S. Army, 1988).

Selection Considerations for Air-Purifying Respirators

Advantages of APRs. APRs offer a number of advantages, in that they are light in weight, relatively inexpensive, relatively simple to maintain, and place minimal mobility restrictions on the user. Thus, as a general rule APRs are the

Fig. 9-6. Supplied-air respirator equipped with escape SCBA. (*Courtesy of Scott Aviation*).

preferred type of respirator for any situation in which they will provide adequate protection.

Disadvantages of APRs. A number of disadvantages of APRs must also be considered. For the most part, these disadvantages are related to work area conditions which preclude the use of APRs. These conditions are incorporated into Fig. 9-5, which can be utilized in determining if the use of APRs is safe in a specific work situation. Disadvantages of APRs are as follows:

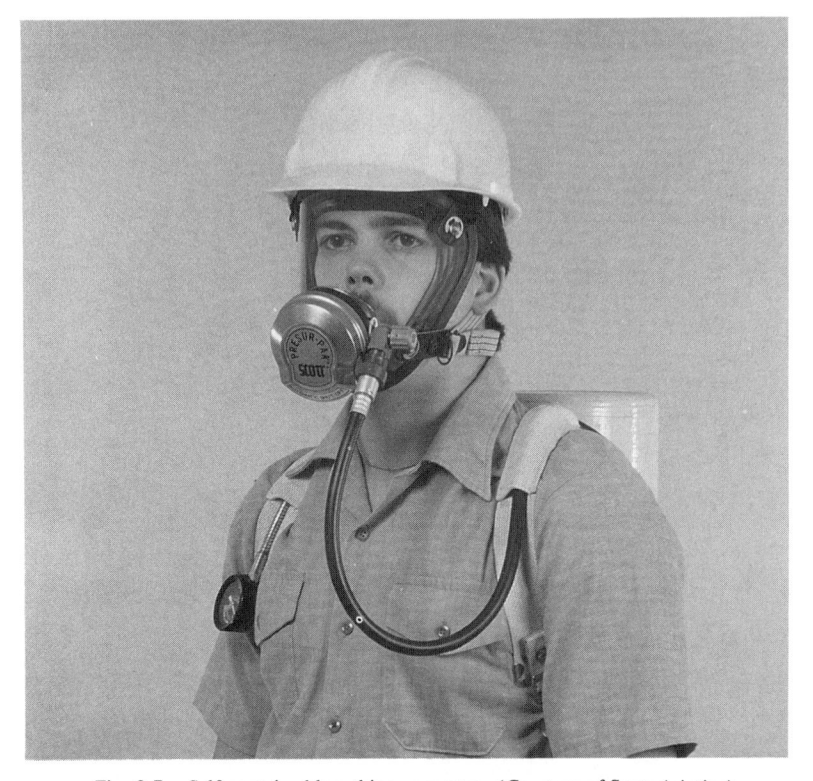

188 WORKER PROTECTION DURING HAZARDOUS WASTE REMEDIATION

Fig. 9-7. Self-contained breathing apparatus. (*Courtesy of Scott Aviation*).

- APRs cannot be used if contaminants which are highly toxic in small concentrations (such as hydrogen cyanide) are present.
- APRs cannot be used in oxygen-deficient atmospheres (that is, atmospheres containing less than 19.5% oxygen).
- All contaminants in the work area must be identified.
- APRs cannot be used if contaminant concentrations are excessively high.
- Selection of purification elements *must be hazard-specific*, since even "universal" elements are not effective against all potential contaminants. It should be noted that a system of color coding (as shown in Table 9-1) can be used to ensure that appropriate cartridges or canisters are used.
- Face-to-facepiece seal is critical, since most APRs work in the negative pressure mode (as discussed below). Exceptions are Powered Air Purifying Respirators (or PAPRs), which are fitted with a motor designed to feed a continuous flow of air to the facepiece, thus maintaining a positive pressure within the facepiece (as discussed below).
- Canisters and cartridges have a finite service life and must be discarded before the service life is exceeded and "breakthrough" occurs.

Table 9-1. Color Code for Cartridges and Gas Mask Canisters.

Atmospheric Contaminants to be Protected Against	Color Assigned
Acid gases	White
Organic vapors	Black
Ammonia gas	Green
Carbon monoxide gas	Blue
Acid gases and organic vapors	Yellow
Acid gases, ammonia, and organic vapors	Brown
Acid gases, ammonia, carbon monoxide, and organic vapors	Red
Other vapors and gases not listed above	Olive
Radioactive materials (except tritium and noble gases)	Purple
Dusts, fumes, and mists (other than radioactive materials)	Orange

1. A purple stripe shall be used to identify radioactive materials in combination with any vapor or gas.
2. An orange stripe shall be used to identify dusts, fumes, and mists in combination with any vapor or gas.
3. Where labels only are colored to conform with this table, the canister or cartridge body shall be a gray or metal canister, or cartridge body may be left in its natural metallic color.
4. The user shall refer to the wording of the label to determine the type and degree of protection the canister or cartridge will afford.

Source:
ANSI K13.1-1973, Identification of Air Purifying Respirator Canisters and Cartridges.

- Breathing through APRs requires greater than normal effort. However, PAPRs can sometimes be used to diminish this problem.
- High relative humidity may reduce the effectiveness of sorbent materials used in cartridges and canisters.

Other Considerations for Using APRs. The use of APRs can only be considered safe when the user has some way of knowing when the end of the canister or cartridge service life has been reached. Thus, APRs should only be used if contaminants involved have *adequate warning properties*. Warning properties allow the APR user to smell, taste, or experience irritation from concentrations of the contaminant *below appropriate exposure limits* in the event of breakthrough. In some restricted instances, OSHA will allow the use of APRs for specific contaminants with poor warning properties, provided one of the following conditions is met (NIOSH, 1985):

- Cartridges or canisters used have a *known service life* to which a liberal *safety factor* has been applied.
- Canisters used have an *end of service life indicator* designed to detect breakthrough of the specific chemicals involved.

In using APRs, the wearer should follow specific cartridge or canister replacement procedures. These purification elements should not be used if the manufacturer's expiration date has passed. They should be used immediately once the package seal is broken, and discarded after one work shift, at the end of service life, or when breakthrough begins to occur, *whichever comes first*.

APRs cannot be used if the concentration of air contaminants in the work area exceeds the *service limit concentration* of the cartridges or canisters used. Depending upon the specific contaminants involved, service limit concentrations vary from 10 ppm (0.001%) to 1000 ppm (0.1%) for cartridges and from 0.5% (5000 ppm) to 3% (30,000 ppm) for canisters. Applicable service limit concentrations should be clearly printed on cartridge or canister labels. However, in many instances, APRs will not provide adequate protection even though atmospheric contaminants are present in concentrations which are appreciably lower than the applicable service limit concentrations of the cartridges or canisters used. For further information pertaining to this, see the section below on adequacy of respiratory protection.

Selection Considerations for Supplied-Air (or Airline) Respirators

Advantages of SARs. The primary advantage of SARs is that they allow extended work periods (in comparison to the self-contained breathing apparatus or SCBAs) in atmospheres requiring atmosphere-supplying respiratory protection. Also, SARs are much less cumbersome than SCBAs.

Disadvantages of SARs. Disadvantages associated with SARs are primarily related to the airline. For example, worker mobility is restricted by the airline, in that workers must retrace their previous steps to exit a work area. Also, the maximum airline length allowable is 300 feet. The airline may be cut, torn, caught, or entangled, thus trapping the user and/or cutting off the air supply. The airline may also be contaminated or permeated by chemicals.

Other Considerations for Using SARs. Only those SARs which operate in the positive-pressure mode (as described below) should be used on hazardous waste sites. Work in IDLH conditions requires an escape air supply of at least 5 minutes duration when using SARs (see Fig. 9-6).

Air compressors and purification units used with SARs must be designed specifically to supply breathable air of at least grade D quality. Airlines must contain couplings located no further than 100 feet apart and all airline couplings must be incompatible with couplings on hoses containing any substance other than breathable air on site.

Selection Considerations for Self-Contained Breathing Apparatus

Advantages of SCBAs. SCBAs offer the primary advantage of providing atmosphere-supplying respiratory protection without the airline-related problems discussed above. When operated in the positive-pressure mode (as discussed below), SCBAs are considered to offer the highest level of respiratory protection currently available.

Disadvantages of SCBAs. Disadvantages of SCBAs are primarily related to the air tank. The presence of the tank makes SCBA units heavy and cumbersome. Passage through some small openings (such as manholes) while wearing SCBA may be impossible. Also, duration of air supply is limited. For example, when using an open-circuit SCBA system (as described as follows), air supply limits work duration to approximately 30 minutes *maximum* using a low-pressure air tank, or 60 minutes *maximum* using a high-pressure tank. SCBAs are expensive and require detailed maintenance procedures. Also, air tanks must be hydrostatically tested periodically (at 5-year intervals for steel tanks and 3-year intervals for aluminum-core, fiberglass-wound tanks).

Other Considerations for Using SCBAs. Only those SCBAs which operate in the positive-pressure mode (as discussed below) should be used on hazardous waste sites.

The most commonly used SCBAs operate as *open-circuit* systems, in which exhaled air exits the system. Extended work times may be achieved by utilizing *closed-circuit SCBAs*, in which exhaled air is purified, enriched with oxygen, and rebreathed. However, closed-circuit SCBAs tend to generate heat during the purification process. Thus, heat stress may be a greater problem when using the "rebreathers" as compared to the conventional, open-circuit SCBAs.

Hybrid SCBA/SAR combination systems are currently available which allow users to enter a work area using the SCBA system then plug into an airline for an extended work period using the SAR system. Air remaining in the tank is then saved for exiting the work area. When this type of system is used, no more than 20% of the SCBA air supply should be used during entry. An interesting variation on the hybrid SCBA/SAR concept is the "quick-fill" type system (Fig. 9-8). This system allows the user to periodically replenish the SCBA air supply by briefly plugging into an airline located within the immediate work area. Thus, an extended work period can be achieved while avoiding the airline-related problems discussed previously.

Some SCBA units are designed to be used only for escape purposes during emergency situations. These units are compact and can be donned quickly (Fig. 9-9). Escape-only SCBA units have a very limited air supply, which is fed to the wearer in a continuous flow after the unit is donned and activated. Therefore, these units are suitable only for use during emergency escapes.

THE IMPORTANCE OF RESPIRATOR FIT

Fit and Fit Testing

The face-to-facepiece fit of a respirator is of the utmost importance if the wearer is to be adequately protected. A poor fit will allow dangerously high volumes

Fig. 9-8. Hybrid system for quick fill of SCBA air tanks within a contaminated work area. (*Courtesy of Mine Safety Appliances Company*).

of contaminated air to enter the facepiece. Thus, fit tests should *always* be conducted as part of the donning procedure. Methods of fit testing may be described as follows (NIOSH, 1987):

- *Qualitative fit tests* are simple tests designed to determine whether or not an acceptable fit has been achieved. Qualitative fit tests may be conducted as follows:
    - *Negative-pressure tests* are conducted by blocking the inhalation pathway of the facepiece, inhaling gently, and holding the breath for 10 seconds while checking for leakage (Fig. 9-10). If fit is acceptable, the respirator should be pulled back toward the face by the vacuum generated through inhalation and remain there as the wearer holds his or her breath.
    - *Positive-pressure tests* can be performed by blocking the facepiece exhalation valve and gently exhaling (Fig. 9-11). Failure to generate a

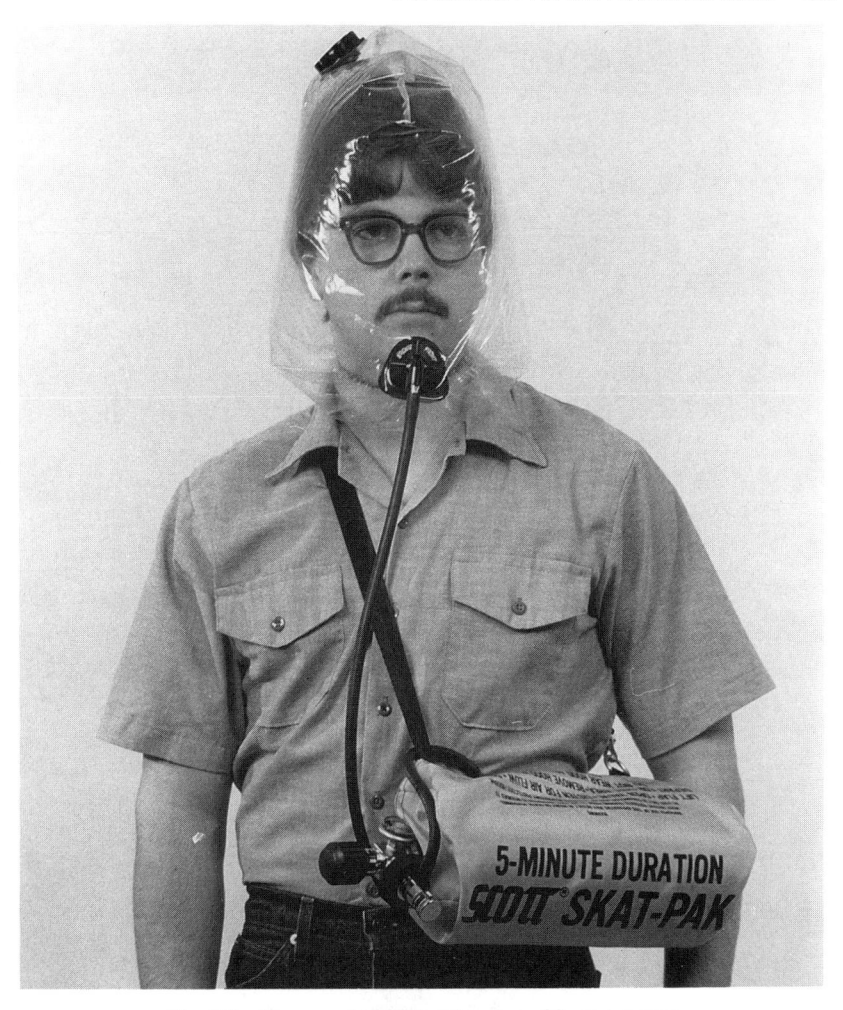

Fig. 9-9. Escape-only SCBA. (*Courtesy of Scott Aviation*).

positive pressure inside the facepiece indicates a poor fit. It should be noted that the exhalation port construction of some respirators makes it impossible to perform this test.

— *Irritant smoke, odorous vapor, and sweetener tests* are performed by exposing the wearer to irritants (such as stannic chloride) or substances which have distinctive odors or tastes (such as banana oil or saccharin mist). If the facepiece fit is good, the wearer should experience no reactions or sensations related to the substance used. Qualitative fit test apparatuses can be obtained in kit form from respirator manufacturers

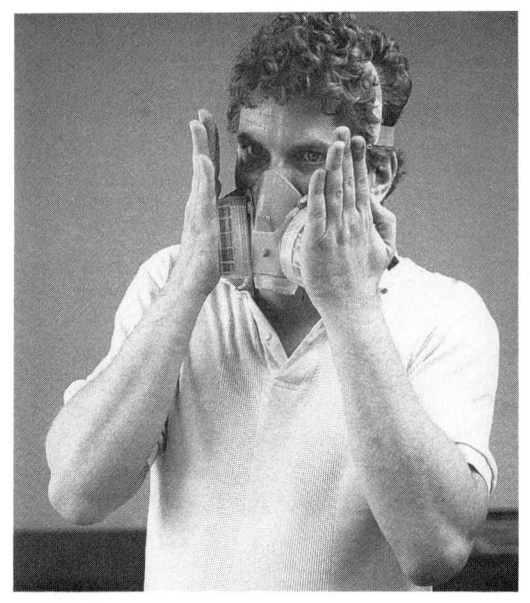

194 WORKER PROTECTION DURING HAZARDOUS WASTE REMEDIATION

Fig. 9-10. Negative-pressure fit check technique.

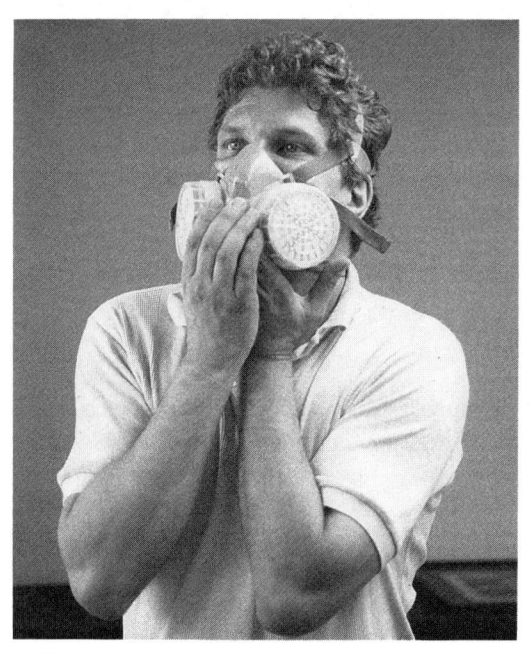

Fig. 9-11. Positive-pressure fit check technique.

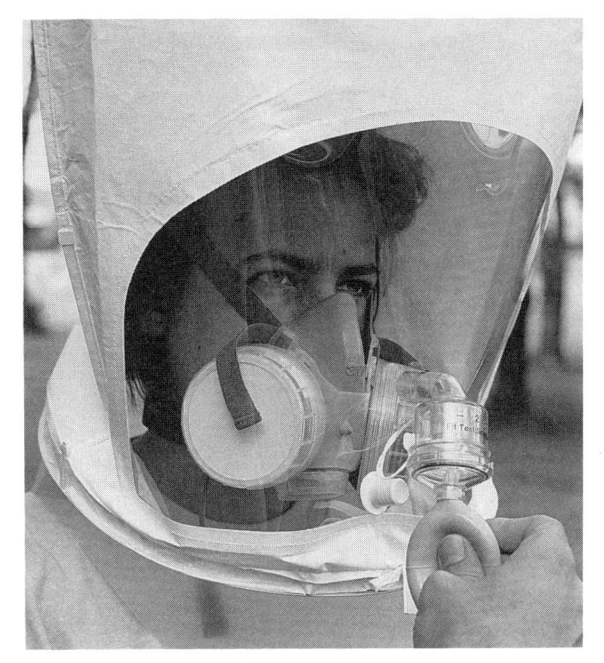

Fig. 9-12. Qualitative fit testing using saccharin mist.

(Fig. 9-12), or assembled from commonly available items. Specific fit testing protocols should be used (NIOSH, 1987).
- *Quantitative fit tests* are complicated tests designed to produce a numerical value (or fit factor) indicating the degree of fit. Quantitative testing is typically performed by placing the wearer in an enclosure containing a known concentration of a contaminant (Fig. 9-13). A sample is drawn from within the facepiece and analyzed to determine the concentration of the contaminant within the facepiece. The fit factor is then calculated as follows:

$$\text{Fit Factor} = \frac{\text{Airborne Concentration of Contaminant}}{\text{Concentration of Contaminant within Facepiece}}$$

Assigned Protection Factors

The use of quantitative fit testing to determine an actual fit factor for a given respirator is generally impractical outside of a laboratory setting. Thus, protection factors have been assigned to the various types of respirators by the American National Standards Institute (ANSI). Examples of assigned protection factors are shown in Table 9-2. The use of assigned protection factors in assessing

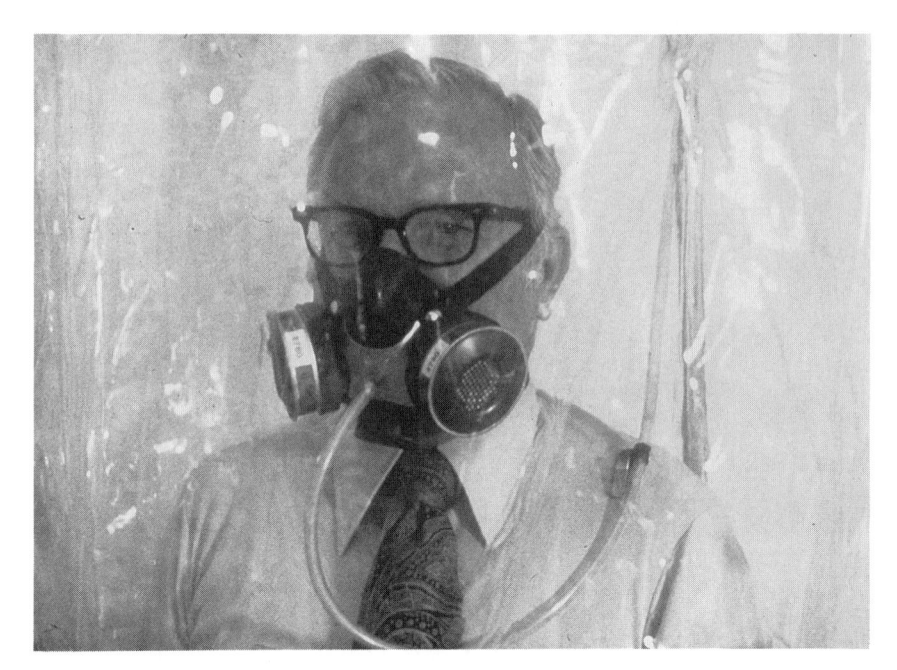

Fig. 9-13. Quantitative fit testing apparatus.

Table 9-2.

Respirator Type	Protection Factor
Half-mask APRs	10
Full-facepiece APRs	50
Supplied-air with full facepiece in positive-pressure mode	2,000
SCBA with full facepiece in positive pressure mode	10,000

the adequacy of respiratory protective equipment is discussed later in this chapter.

Positive-Pressure Versus Negative-Pressure Modes of Respirator Operation

Based on the pressure generated within the facepiece during use, respirators can be classified as follows (NIOSH, 1985):

- *Negative-pressure respirators* require the wearer to inhale and generate a negative pressure or vacuum within the facepiece in order to receive breathing air. Thus, a poor facial fit (or any other source of leakage) will allow large volumes of contaminated air to enter the facepiece during inhalation.

For this reason, a good fit is absolutely critical when using any respirator operating in the negative-pressure mode. Respirators which operate in this mode include all APRs (excluding PAPRs), and any SARs or SCBAs which operate in the demand mode.
- *Positive-pressure respirators* are designed to maintain a slight positive pressure within the facepiece at all times, so that any leaking air will theoretically move from the inside out so as to prevent contaminants from entering the facepiece. Thus respirators which operate in the positive-pressure mode have an appreciably higher protection factor than equivalent respirators operating in the negative-pressure mode. Only those SCBAs and SARs which operate in the positive-pressure mode should be used on hazardous waste sites. Two positive-pressure designs are currently used:
  — *Continuous-flow respirators* maintain positive pressure by feeding a continuous stream of breathing air to the facepiece. All PAPRs and some SARs are of the continuous-flow design.
  — *Pressure-demand respirators* are designed to maintain a slight positive pressure within the facepiece at all times. Breathing air flows into the facepiece only during inhalation, so that air consumption is much lower than with a continuous flow respirator.

It should be noted that during times of peak inhalation while performing strenuous tasks, workers can temporarily overcome the positive pressure within the facepiece of a respirator operating in the positive-pressure mode. If the face-to-facepiece seal is poor, contaminants may enter the facepiece during these intervals of negative pressure. Thus, a good facepiece fit is vital, even when using positive-pressure respirators.

WHAT CONSTITUTES EFFECTIVE RESPIRATORY PROTECTION?

The effectiveness of respiratory protective equipment can be assessed using the following equation:

$$\frac{CC_{wa}}{PF} = CC_{fp}$$

where:

CC_{wa} = Contaminant Concentration in work area
PF = Protection Factor assigned to the type of respirator under consideration
CC_{fp} = Contaminant Concentration within facepiece

Obviously, if contaminant concentration within the facepiece is in excess of exposure limits for the containment of concern, the protection provided the worker is inadequate and some type of respiratory protective equipment providing a higher protection factor should be used.

PART 2. CHEMICAL PROTECTIVE CLOTHING AND ACCESSORIES

Numerous substances likely to be encountered on hazardous waste sites pose a threat to the skin of workers involved in site activities. These substances may directly attack the skin, or pass through the skin to reach and attack other organs of the body. Thus, it is frequently necessary to place a barrier, in the form of chemical protective clothing and related accessories, between the worker and the hazards to which he or she is potentially exposed on site. This section of the text is intended to serve as an introduction to the topic of chemical protective clothing and accessories.

SELECTION OF CHEMICAL PROTECTIVE CLOTHING (CPC)

Selection Considerations

CPC is virtually worthless if not properly selected. Selection is a complex task which should be undertaken only by personnel with appropriate knowledge, training, and experience. Proper selection requires full consideration of the following factors (NIOSH, 1985):

- *Specific chemical contaminants present* in the work area must be identified, if possible, prior to selection, since no single protective material is effective against all potential chemical assaults.
- The *performance characteristics of available protective clothing* in resisting chemical attack and physical damage must be evaluated.
- *Site- and/or task-specific requirements and limitations* must also be considered. For example, the selection process must take into account:
  — The physical state of contaminants (for example, liquid versus vapor).
  — The "exposure time" required for a given task.
  — The likelihood of direct exposure, such as through chemical splashing, during a given procedure.
  — The degree of stress (particularly heat stress) placed on the wearer by a given article of protective clothing.

Attacks on CPC

In order to provide adequate protection for the wearer, an article of CPC must be sufficiently resistant to attacks by chemicals and physical agents present in the work area. These attacks can be classified as follows:

- *Permeation* refers to the process by which a chemical dissolves into or passes through a chemical protective material at the molecular level (that is, in the vapor phase).
- *Penetration* refers to the bulk movement of liquids through pores or small flaws in an article of CPC. Penetration may occur through imperfect seams, zippers, or pinholes in an article of CPC.
- *Degradation* refers to the loss of chemical resistance or physical competency of a protective material. This may occur due to chemical exposure or physical wear and abrasion within the work area.

Resistance to Chemical Attacks

Chemical resistance of CPC is typically reported in the following terms:

- *Breakthrough time* is the time elapsing between the introduction of a chemical to the outer surface of a protective material and the initial detection of that chemical on the inner surface of the material.
- *Permeation rate* is the rate at which a permeating chemical moves through a given material after breakthrough, as determined under a set of test conditions. Permeation rate is reported in terms of the mass of contaminant passing through a specific area of material during a specific length of time (for example micrograms per square centimeter per minute).

Availability of Information on Performance Characteristics of CPC

Data which may be used in selecting CPC are available from various sources. These sources may be roughly categorized as follows:

- *Governmental agency publications* are best represented by "Guidelines for the Selection of Chemical Protective Clothing" by the American Conference of Governmental Industrial Hygienists (Schwope et al., 1985). This publication contains research results as well as qualitative recommendations regarding the use of various CPC materials for several hundred chemicals.
- *Vendor Literature* (such as shown in Table 9-3) is published in various forms by manufacturers of CPC. Vendor literature typically contains one or both of the following types of information:

Table 9-3. Typical Vendor Data on Chemical Resistance of Chemical Protective Clothing.

Chemical Permeation Resistance Chart	VAUTEX 23 mils thick 19.1 oz/yd²						BETEX 19 mils thick 15.9 oz/yd²					
	Breakthrough time in minutes			Permeation Rate (micrograms per cm² per minute)			Breakthrough time in minutes			Permeation Rate in micrograms per cm² per minute		
Chemical	Test #						Test #					
	1	2	3	1	2	3	1	2	3	1	2	3
Acetone†	15	15	15	33.4	24.8	36.7	135	135	135	12.9	10.8	7.5
Acetonitrile†	20	20	20	4.0	3.3	3.8	165	165	165	.04	.05	.05
Acrylonitrile	<5	<5	<5	24	21	23	125	125	125	0.18	0.22	0.10
Adiponitrile	240+	240+	240+	NA	NA	NA	240+	240+	240+	NA	NA	NA
Allyl Alcohol	240+	240+	240+	NA	NA	NA	240+	240+	240+	NA	NA	NA
Ammonia Gas	240+	240+	240+	NA	NA	NA	240+	240+	240+	NA	NA	NA
Ammonia Liquid Splash	5	5	5	.121	.025	.096	5	5	5	.117	.076	.022
	60	60	60	.020	.010	.014	60	60	60	.004	.007	.002
	240	240	240	.014	.000	.005	240	240	240	.000	.001	.000
Aniline	240+	240+	240+	NA	NA	NA	240+	240+	240+	NA	NA	NA
Benzene	65	65	165	.05	.08	.09	<5	<5	<5	144	147	147
Carbon Disulfide†	240+	240+	240+	NA	NA	NA	<5	<5	<5	5.8	6.0	5.5
Chlorine Gas	240+	240+	240+	NA	NA	NA	240+	240+	240+	NA	NA	NA
Dichloromethane†	20	20	20	50.1	52.6	55.1	<5	<5	<5	193	310	259
Diethylamine	35	35	35	53.7	50.1	51.9	<5	<5	<5	78.8	102.1	55.5
Dimethyl Formamide†	30	30	30	6.8	7.7	8.4	240+	240+	240+	NA	NA	NA
Ethyl Acetate†	15	15	15	57.2	59.1	59.1	30	30	30	5.5	7.0	5.7
Ethyl Acrylate	<5	<5	<5	42.6	49.6	46.4	25	25	25	4.4	4.7	4.4
Ethylene Dibromide	240+	240+	240+	NA	NA	NA	15	15	15	24.4	23.3	21.8
Formaldehyde	240+	240+	240+	NA	NA	NA	240+	240+	240+	NA	NA	NA
Gasoline	240+	240+	240+	NA	NA	NA	25	25	25	300	212	287
n-Hexane†	240+	240+	240+	NA	NA	NA	<5	<5	<5	36.6	36.6	36.6
Hexamethylene Diamine	240+	240+	240+	NA	NA	NA	240+	240+	240+	NA	NA	NA
Hydrazine	240+	240+	240+	NA	NA	NA	240+	240+	240+	NA	NA	NA

PERSONAL PROTECTIVE EQUIPMENT 201

Table 9-3. (Continued)

Chemical Permeation Resistance Chart	VAUTEX 23 mils thick 19.1 oz/yd²						BETEX 19 mils thick 15.9 oz/yd²					
	Breakthrough time in minutes			Permeation Rate (micrograms per cm² per minute)			Breakthrough time in minutes			Permeation Rate in micrograms per cm² per minute		
Chemical	Test #						Test #					
	1	2	3	1	2	3	1	2	3	1	2	3
Hydrochloric Acid	240+	240+	240+	NA	NA	NA	240+	240+	240+	NA	NA	NA
Hydrofluoric Acid	240+	240+	240+	NA	NA	NA	240+	240+	240+	NA	NA	NA
Hydrogen Cyanide Gas	80	80	80	0.7	0.7	0.6	240+	240+	240+	NA	NA	NA
Isobutylamine	45	45	45	15.4	16.0	13.2	50	50	50	6.9	6.8	6.4
Methyl Alcohol† (Methanol)	240+	240+	240+	NA	NA	NA	240+	240+	240+	NA	NA	NA
Methyl Ethyl Ketone	<5	<5	<5	62.4	68.6	64.6	30	30	30	12.4	12.2	12.6
Nitric Acid	240+	240+	240+	NA	NA	NA	240+	240+	240+	NA	NA	NA
Nitrobenzene†	240+	240+	240+	NA	NA	NA	240+	240+	240+	NA	NA	NA
Nitrogen Tetroxide†	15	15	15	2*	<1*	<1*	15	15	15	0*	<1*	<1*
	30	30	30	7*	<1*	<1*	30	30	30	>40*	<1*	>40*
	45	45	45	>40*	>40*	<7*	45	45	45	>40*	>40*	

NA - Not Applicable if breakthrough time exceeds 240 minutes.
† Listed in ASTM F1001
* Concentrations observed expressed in parts per million. Permeation rates not available. Complete degradation occurred after reaching 40 ppm.

(Courtesy of Mine Safety Appliances Company)

- *Qualitative recommendations* rate protective materials using terms such as "excellent," "good," "fair," "poor," or "not recommended" for protection against specific chemicals.
- *Quantitative chemical resistance data* report actual permeation rates and/or breakthrough times for a given protective material under attack by specific chemicals.

Problems with Information Available on CPC

Information currently available on the performance characteristics of CPC has several deficiencies which may complicate the selection process. Each of the following limitations must be carefully considered when selecting CPC suitable for usage on a hazardous waste site (NIOSH, 1985):

- Permeation rate and breakthrough time are currently determined by testing small swatches of protective material while ignoring the performance of the suit or article of clothing as a whole. For example, chemical resistance of seams, zippers, and visors are not considered in test results of this type. Also, boots and gloves used with a suit may be constructed of a different material than the suit itself.
- For the most part, protective materials are currently tested using single-chemical assaults. However, most hazardous wastes are mixtures, which are sometimes much more aggressive in attacking chemical protective materials than a single component. Data on resistance to multicomponent chemical attacks are currently inadequate.
- Permeation rates and breakthrough times published for a given material are determined in a carefully controlled experimental setting. However, performance characteristics of CPC in actual use on a hazardous waste site may vary widely with variations in factors such as type and concentration of chemicals, ambient temperature, and humidity.
- Procedures for making qualitative recommendations are not standardized within the CPC industry and may vary appreciably from one manufacturer to another. Thus, one manufacturer may assign a rating of "excellent" to a material which is resistant to a given chemical for two hours while another manufacturer would require a material to be resistant to the same chemical for eight hours in order to receive the same rating (Rodgers, 1988).

Basic Principles of CPC Selection

In selecting and using CPC on a hazardous waste site, the following basic principles should be kept in mind at all times.

- No single chemical protective material offers protection against all chemicals. Selection *should* be chemical-specific, if the identity of chemicals present is known. Data presented in Table 9-3 offer excellent examples of the importance of chemical-specific selection.
- No protective material currently available is truly impermeable or able to provide an effective barrier to *prolonged* exposure. Breakthrough is simply a matter of time. The key question is whether or not breakthrough will occur before a given task can be completed.
- CPC should be selected which offers the widest range of protection available against the specific chemicals known or expected to be on site. Materials which offer the longest breakthrough times and lowest permeation rates for the chemicals of concern are most desirable.

CPC Materials and Technologies

In the past, CPC technologies have focused almost solely on reusable fabrics such as butyl rubber, Viton®, urethane, nitrile rubber, neoprene, polyvinyl chloride, and others (Langley, 1988). Each of these materials offers excellent protection against certain chemicals. In recent years, manufacturers have begun to offer reusable CPC constructed of two or more layers of different materials laminated together, thus offering a widened range of protection. Reusable CPC materials are generally quite durable and offer adequate protection if properly selected. However, the reusables are expensive to purchase, require complete decontamination after every use, and may present matrix permeation problems (as discussed in the section on use of PPE).

In recent years disposable items of CPC have come to be used commonly in certain situations. Tyvek® is probably the most popular of the disposable materials, offering adequate protection against particulate contaminants in many instances. Materials constructed of Tyvek® coated with polyethylene or saran offer protection against certain liquids, gases, and vapors, as well as particulate contaminants. The disposables offer advantages, in that they are relatively inexpensive to purchase and require no decontamination after use. However, the disposables typically have low durability relative to the reusables.

The most recent additions to the disposable CPC market have been constructed of laminated film-based materials. Examples include the Responder by Life Guard, Inc. and the Chemrel suit by Chemron, Inc. These new, high-tech materials offer protection against a wide range of chemicals for an extended period of time and are quite inexpensive relative to reusable materials. Because of this, it is anticipated that they will revolutionize the CPC industry within the next few years (Langley, 1988).

Responsibility of the Employer

Given the complexities discussed above, and the potential consequences of improper selection, it is obvious that selection of CPC is a task best left to experts. The information presented here is primarily intended to sensitize the worker to the importance of using properly selected equipment as instructed by those experts. For example, two gloves which appear to be "typical rubber gloves" to the casual observer may actually be composed of different materials which have completely different protective capabilities.

Under OSHA regulations contained in Title 29 CFR parts 1910.132 through 1910.137, the employer is required to ensure that PPE provides adequate employee protection and is appropriate for the work situation in which used. Thus, the *ultimate* responsibility for selection of CPC rests with the employer.

TYPES OF CPC AND ACCESSORIES

Numerous types of CPC and related accessories are available to protect workers involved in hazardous waste site activities. As in all cases involving PPE, chemical protective garments must be appropriate for the hazards and work place conditions in which they are used.

Chemical Protective Clothing

Table 9-4 presents a summary of general information on various types of protective clothing, some of which are discussed below.

Totally-Encapsulating Chemical Protective Suits. Totally-encapsulating chemical-protective (TECP) suits (Figs. 9-14 and 9-15) completely enclose the wearer. These suits are designed to seal out gases and vapors in addition to protecting against liquid splashes and particulate contaminants. TECP suits are also required to be capable of maintaining a positive internal pressure, so that any leakage will theoretically be from the inside out.

TECP suits have traditionally been constructed of reusable materials, but have more recently been produced using disposable materials. It should be noted that problems in maintaining adequate positive pressure have been encountered with some reusable TECP suits.

Non-Encapsulating Suits. Non-encapsulating chemical protective (NECP) suits cover most of the wearer's skin, thus providing good protection from splashes and/or particulate contaminants (Figs. 9-16, 9-17, 9-18, and 9-19). Protection can be enhanced by tape-sealing the cuffs to boots and gloves. However, gases and vapors present in the workplace are able to reach the wearer's skin. NECP suits are available in both reusable and disposable models.

Gloves and Boots. Gloves and boots, as shown in various figures within this section, are very important items of protective equipment because they cover the parts of the body most likely to routinely contact hazardous waste materials. They are also important for protection against conventional safety hazards (for example, the use of steel-toed boots to prevent physical damage to the feet).

Items for Additional Protection. Additional protection can be provided to the worker by the use of items such as aprons and sleeve protectors (see Table 9-4). Items such as these can be worn over a chemical protective suit in order to provide additional coverage to those areas most likely to directly contact waste materials (as in the case of a chemical splash.)

Disposable Overgarments. Disposable overgloves and overboots or boot covers are recommended to minimize exposure of the underlying chemical-protective garments and enhance worker protection. For example, a cheap glove can be worn over an expensive glove, so that the cheaper outer glove will bear the brunt of wear and gross contamination. Thus, the expensive inner glove is protected. At the time of decontamination, the outer glove can simply be dis-

PERSONAL PROTECTIVE EQUIPMENT   205

Table 9-4. Protective Clothing and Accessories.

Type of Clothing or Accessory	Description	Type of Protection	Use Considerations
Totally encapsulating chemical protective suit.	One-piece garment. Boots and gloves may be integral, attached and replaceable, or separate.	Protects against splashes, dust, gases, and vapors.	Does not allow body heat to escape. May contribute to heat stress in wearer, particularly if worn in conjunction with a closed-circuit SCBA; a cooling garment may be needed. Impairs worker mobility, vision, and communication.
Non-encapsulating chemical protective suit.	Jacket, hood, pants, or bib overalls and one-piece coveralls.	Protects against splashes, dust, and other materials but not against gases and vapors. *Does not* protect parts of head or neck.	Do not use where gas-tight or pervasive splashing protection is required. May contribute to heat stress in wearer. Tape-seal connections between pant cuffs and boots and between gloves and sleeves.
Aprons, leggings, and sleeve protectors.	Fully sleeved and gloved apron. Separate coverings for arms and legs. Commonly worn over non-encapsulating suit.	Provides additional splash protection of chest, forearms, and legs.	Whenever possible, should be used over a non-encapsulating suit (instead of using a fully encapsulating suit) to minimize potential for heat stress. Useful for sampling, labeling, and analysis operations. Should be used only when there is a low probability of total body contact with contaminants.

Table 9-4. (Continued)

Type of Clothing or Accessory	Description	Type of Protection	Use Considerations
Gloves and sleeves	May be integral, attached, or separate from other protective clothing.	Protect hands and arms from chemical contact.	Wear jacket cuffs over glove cuffs to prevent liquid from entering the glove. Tape-seal gloves to sleeves to provide additional protection.
	Overgloves.	Provide supplemental protection to the wearer and protect more expensive undergarments from abrasions, tears, and contamination.	
	Disposable gloves.	Should be used whenever possible to reduce decontamination needs.	
Safety boots	Boots constructed of chemical-resistant material.	Protect feet from contact with chemicals.	
	Boots constructed with some steel materials (e.g., toes, shanks, insoles).	Protect feet from compression, crushing, or puncture by falling, moving, or sharp objects.	All boots must at least meet the specifications required under OSHA 29 CFR Part 1910.136 and should provide good traction.
	Boots constructed from nonconductive, spark-resistant materials or coatings.	Protect the wearer against electrical hazards and prevent ignition of combustible gases or vapors.	
Disposable shoe or boot covers	Made of a variety of materials. Slip over the shoe or boot.	Protect safety boots from contamination. Protect feet from contact with chemicals.	Covers may be disposed of after use, facilitating decontamination.

Adapted from NIOSH, 1985.

posed of, so that problems related to decontamination are minimized. Likewise, cheap disposable chemical protective suits can be used as oversuits to be worn over expensive reusable chemical protective suits and then discarded at the time of decontamination.

Protective Clothing for Unique Hazards

Situations which pose unique hazards to workers may arise during work on a hazardous waste site. Examples of such situations are fires or potential explosions onsite. Types of protective clothing which may be required in responding to such situations are described in Table 9-5. It should be noted that activities requiring these types of protective items fall within the realm of emergency response or extremely hazardous operations and are thus outside the scope of this text. Table 9-5 is included only for reference.

Accessories

In addition to the various articles of protective clothing previously described, numerous accessory items may be used to enhance protection and allow for safer work on a hazardous waste site. Some of these accessories are described in Table 9-6 and shown in Fig. 9-22.

PART 3. LEVELS OF PROTECTION

Individual components of respiratory protective equipment, chemical protective clothing, and various accessories may be assembled into personal protective ensembles providing protection as demanded by site-specific hazards. EPA has created recommendations for levels of protection for PPE ensembles (NIOSH, 1985). These recommendations are shown in Tables 9-7 through 9-10 and briefly discussed below.

LEVEL A PROTECTION

Level A protection requires the highest possible degree of both respiratory protection and skin and eye protection. Thus, the TECP suit can be considered the definitive protective item of the Level A ensemble (see Fig. 9-14).

Information on Level A protection is included in Table 9-7. Level A protection is required whenever gases or vapors which pose a high degree of hazard to the skin are present in the work area (see Fig. 9-15).

Table 9-5. Articles of Protective Clothing for Unique Hazards.

Type of Clothing or Accessory	Description	Type of Protection	Use Considerations
Firefighters' protective clothing	Gloves, helmet, running or bunker coat, running or bunker pants (NFPA No. 1971, 1972, 1973), and boots.	Protects against heat, hot water, and some particles. Does not protect against gases and vapors, or chemical permeation or degradation. NFPA Standard No. 1971 specifies that a garment consist of an outer shell, an inner liner, and a vapor barrier with a minimum water penetration of 25 lbs/in^2 (1.8 kg/cm^2) to prevent the passage of hot water.	Decontamination is difficult. Should not be worn in areas where protection against gases, vapors, chemical splashes, or permeation is required.
Proximity garment (approach suit)	One- or two-piece overgarment with boot covers, gloves, and hood of aluminized nylon or cotton fabric. Normally worn over other protective clothing, such as chemical-protective clothing, firefighters' bunker gear, or flame-retardant coveralls.	Protects against brief exposure to radiant heat. Does not protect against chemical permeation or degradation. Can be custom-manufactured to protect against some chemical contaminants.	Auxiliary cooling and an SCBA should be used if the wearer may be exposed to a toxic atmosphere or needs more than 2 or 3 minutes of protection.

Flame/fire retardant coveralls	Normally worn as an undergarment.	Provides protection from flash fires.	Adds bulk and may exacerbate heat stress problems and impair mobility.
Blast and fragmentation suit	Blast and fragmentation vests and clothing, bomb blankets, and bomb carriers.	Provides some protection against very small detonations. Bomb blankets and baskets can help redirect a blast.	Does not provide hearing protection.
Radiation-contamination protective suit	Various types of protective clothing designed to prevent contamination of the body by radioactive particles.	Protects against alpha and beta particles. Does NOT protect against gamma radiation.	Designed to prevent skin contamination. If radiation is detected on site, consult an experienced radiation expert and evacuate personnel until the radiation hazard has been evaluated.

Source: NIOSH, 1985.

Table 9-6. Accessory Items for Protective Ensembles.

Type of Clothing or Accessory	Description	Type of Protection	Use Considerations
Flotation gear	Life jackets or work vests. (Commonly worn underneath chemical protective clothing to prevent flotation gear degradation by chemicals.)	Adds 15.5 to 25 lbs (7 to 11.3 kg) of buoyancy to personnel working in or around water.	Adds bulk and restricts mobility. Must meet USCG standards (46 CFR Part 160).
Cooling garment	One of three methods: (1) A pump circulates cool dry air throughout the suit or portions of it via an air line. Cooling may be enhanced by use of a vortex cooler, refrigeration coils, or a heat exchanger. (2) A jacket or vest having pockets into which packets of ice are inserted. (3) A pump circulates chilled water from a water/ice reservoir and through circulating tubes, which cover part of the body (generally the upper torso only).	Removes excess heat generated by worker activity, the equipment, or the environment.	(1) Pumps circulating cool air require 10 to 20 ft^3 (0.3 to 0.6 m^3) of respirable air per minute, so they are often uneconomical for use at a waste site. (2) Jackets or vests pose ice storage and recharge problems. (3) Pumps circulating chilled water pose ice storage problems. The pump and battery add bulk and weight.
Safety helmet (hard hat)	For example, a hard plastic or rubber helmet.	Protects the head from blows.	Helmet shall meet OSHA standard 29 CFR Part 1910.135.

PERSONAL PROTECTIVE EQUIPMENT 211

Helmet liner	Insulates against cold. Does not protect against chemical splashes.	
Hood	Commonly worn with a helmet.	Protects against chemical splashes, particulates, and rain.
Face shield	Full-face coverage, eight-inch minimum.	Protects against chemical splashes. Does not protect adequately against projectiles.	Face shields and splash hoods must be suitably supported to prevent them from shifting and exposing portions of the face or obscuring vision. Provides limited eye protection.
Splash hood		Protects against chemical splashes. Does not protect adequately against projectiles.
Safety glasses		Protect eyes against large particles and projectiles.	If lasers are used to survey a site, workers should wear special protective lenses.
Goggles		Depending on their construction, goggles can protect against vaporized chemicals, splashes, large particles, and projectiles (if constructed with impact-resistant lenses).
Ear plugs and muffs		Protect against physiological damage and psychological disturbance.	Must comply with OSHA regulation 29 CFR Part 1910.95. Can interfere with communication. Use of ear plugs should be carefully reviewed by a health and safety professional because chemical contaminants could be introduced into the ear.

Table 9-6. (Continued)

Type of Clothing or Accessory	Description	Type of Protection	Use Considerations
Headphones	Radio headset with throat microphone.	Provide some hearing protection while enabling communication.	
Personal dosimeter		Measures worker exposure to ionizing radiation and to certain chemicals.	Highly desirable, particularly if emergency conditions arise. To estimate actual body exposure, the dosimeter should be placed inside the fully-encapsulating suit.
Personal locator beacon	Operated by sound, radio, or light.	Enables emergency personnel to locate victim.	
Two-way radio		Enables field workers to communicate with personnel in the Support Zone.	
Knife		Allows a person in a fully-encapsulating suit to cut his or her way out of the suit in the event of an emergency or equipment failure.	Should be carried and used with caution to avoid puncturing the suit.

Source: NIOSH, 1985.

Fig. 9-14. Totally encapsulating chemical protective suit, SCBA and other components of a typical level A ensemble of PPE.

LEVEL B PROTECTION

Level B protection requires the same degree of respiratory protection as Level A, but requires a lesser degree of skin protection. Thus, the Level B ensemble is quite similar to the Level A ensemble, except for the use of a non-encapsulating suit, such as a two-piece splash suit (Fig. 9-16). Information on Level B protection is included in Table 9-8. Level B is used whenever the conditions for the use of APRs are not met and threats to the skin are non-IDLH and in the form of liquid splashes or particulate contaminants (Fig. 9-17).

LEVEL C PROTECTION

Level C Protection involves the same degree of skin protection as level B, but a lesser degree of respiratory protection. The Level C ensemble is quite similar

Fig. 9-15. Use of level A protection.

to the level B ensemble, except that respiratory protection is provided by an APR rather than an SCBA (Fig. 9-18). Information on Level C equipment is shown in Table 9-9. Level C should only be used in work areas where all criteria for the use of APRs are met (Fig. 9-19).

LEVEL D PROTECTION

Level D personal protective equipment provides protection only against "normal" workplace safety hazards. Hard hats, steel-toed boots, safety glasses, and cotton work clothes are basic components of the Level D ensemble (see Fig. 9-20 and Table 9-10). Level D should be used only in those work areas in which both respiratory hazards and skin hazards are absent (Fig. 9-21).

An important aspect of PPE is protection from "normal" workplace safety hazards, as well as the chemical-related hazards typical of the hazardous waste site. Thus, all levels of protection should incorporate the basic safety equipment (as represented by the Level D ensemble) as needed to ensure worker safety from *all* site hazards.

MODIFIED LEVELS OF PROTECTION

In considering levels of protection, it should be noted that the four levels as presented here are highly generalized and should be fine-tuned to provide the

PERSONAL PROTECTIVE EQUIPMENT 215

Table 9-7. Equipment and Use Considerations for Level A Protection.

Level of Protection	Equipment	Protection Provided	Should be Used When:	Limiting Criteria
A	Recommended: • Pressure-demand, full-facepiece SCBA or pressure-demand supplied-air respirator with escape SCBA. • Fully-encapsulating, chemical-resistant suit. • Inner chemical-resistant gloves. • Chemical-resistant safety boots/shoes. • Two-way radio communications. Optional: • Cooling unit. • Coveralls. • Long cotton underwear. • Hard hat. • Disposable gloves and boot covers.	The highest available level of respiratory, skin, and eye protection.	• The chemical substance has been identified and requires the highest level of protection for skin, eyes, and the respiratory system based on either: — measured (or potential for) high concentration of atmospheric vapors, gases, or particulates or — site operations and work functions involving a high potential for splash, immersion, or exposure to unexpected vapors, gases, or particulates of materials that are harmful to skin or capable of being absorbed through the intact skin.	• Fully-encapsulating suit material must be compatible with the substances involved.

Table 9-7. (Continued)

Level of Protection	Equipment	Protection Provided	Should be Used When:	Limiting Criteria
			• Substances with a high degree of hazard to the skin are known or suspected to be present, and skin contact is possible. • Operations must be conducted in confined, poorly ventilated areas until the absence of conditions requiring Level A protection is determined.	

Source: NIOSH, 1985.

PERSONAL PROTECTIVE EQUIPMENT   217

Table 9-8. Equipment and Use Considerations for Level B Protection.

Level of Protection	Equipment	Protection Provided	Should be Used When:	Limiting Criteria
B	Recommended: • Pressure-demand, full-facepiece SCBA or pressure-demand supplied-air respirator with escape SCBA. • Chemical-resistant clothing (overalls and long-sleeved jacket; hooded, one- or two-piece chemical splash suit; disposable chemical-resistant one-piece suit). • Inner and outer chemical-resistant gloves. • Chemical-resistant safety boots/shoes. • Hard hat. • Two-way radio communications. Optional: • Coveralls. • Disposable boot covers. • Face shield. • Long cotton underwear.	The same level of respiratory protection but less skin protection than Level A. It is the minimum level recommended for initial site entries until the hazards have been further identified.	• The type and atmospheric concentration of substances have been identified and require a high level of respiratory protection, but less skin protection. This involves atmospheres: — with IDLH concentrations of specific substances that do not represent a severe skin hazard; or — that do not meet the criteria for use of air-purifying respirators. • Atmosphere contains less than 19.5 percent oxygen. • Presence of incompletely identified vapors or gases is indicated by direct-reading organic vapor detection instrument, but vapors and gases are not suspected of containing high levels of chemicals harmful to skin or capable of being absorbed through the intact skin.	• Use only when the vapor or gases present are not suspected of containing high concentrations of chemicals that are harmful to skin or capable of being absorbed through the intact skin. • Use only when it is highly unlikely that the work being done will generate either high concentrations of vapors, gases, or particulates or splashes of material that will affect exposed skin.

Source: NIOSH, 1985.

218 WORKER PROTECTION DURING HAZARDOUS WASTE REMEDIATION

Fig. 9-16. Non-encapsulating chemical protective suit, SCBA, and other components of a typical level B ensemble of PPE.

specific degree of protection required for a specific task. Thus, any number of "modified" levels of protection may be used on site. Accessory items, as described in Table 9-6 and shown in Fig. 9-22, can be used to enhance the protection provided by various personal protective ensembles.

Reportedly, the National Fire Protection Association (NFPA) has proposed a more specific classification of PPE which takes into consideration additional factors, such as the degree of flammability of various chemical protective materials.

PART 4. USE OF PPE

Properly selected PPE can provide adequate protection only if properly used. Proper use of respirators, CPC, and various accessories requires attention to a number of specific considerations (OSHA, 1985). This section of the text is

Fig. 9-17. Use of level B protection.

intended to make the reader aware of various practical concerns for the proper use of PPE.

WRITTEN PPE PROGRAM

A Written PPE program must be included in the site-specific safety and health plan (29 CFR 1910.120). While using PPE onsite, all employees must adhere strictly to the provisions of the program. OSHA requires that the PPE program address each of the following topics:

- Selection.
- Use and limitations.
- Work mission duration.
- Maintenance.

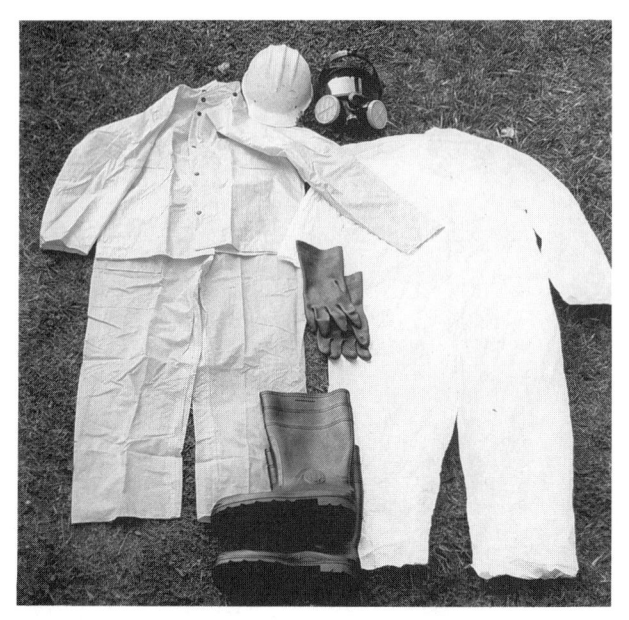

Fig. 9-18. Non-encapsulating chemical protective clothing, APR, and other components of a typical level C ensemble of PPE.

Fig. 9-19. Use of level C protection.

Table 9-9. Equipment and Use Considerations for Level C Protection.

Level of Protection	Equipment	Protection Provided	Should be Used When:	Limiting Criteria
C	Recommended: • Full-facepiece, air-purifying, canister-equipped respirator. • Chemical-resistant clothing (overalls and long-sleeved jacket; hooded, one- or two-piece chemical splash suit; disposable chemical-resistant one-piece suit). • Inner and outer chemical-resistant gloves. • Chemical-resistant safety boots/shoes. • Hard hat. • Two-way radio communications. Optional: • Coveralls. • Disposable boot covers. • Face shield. • Escape mask. • Long cotton underwear.	The same level of skin protection as Level B, but a lower level of respiratory protection.	• The atmospheric contaminants, liquid splashes, or other direct contact will not adversely affect any exposed skin. • The types of air contaminants have been identified, concentrations measured, and a canister is available that can remove the contaminant. • All criteria for the use of air-purifying respirators are met.	• Atmospheric concentration of chemicals must not exceed IDLH levels. • The atmosphere must contain at least 19.5 percent oxygen.

Source: NIOSH, 1985.

Table 9-10. Equipment and Use Considerations for Level D Protection.

Level of Protection	Equipment	Protection Provided	Should be Used When:	Limiting Criteria
D	Recommended: • Coveralls. • Safety boots/shoes. • Safety glasses or chemical splash goggles. • Hard hat. Optional: • Gloves. • Escape mask. • Face shield.	No respiratory protection. Minimal skin protection.	• The atmosphere contains no known hazard. • Work functions preclude splashes, immersion, or the potential for unexpected inhalation of or contact with hazardous levels of any chemicals.	• This level should not be worn in the Exclusion Zone. • The atmosphere must contain at least 19.5 percent oxygen.

Source: NIOSH, 1985.

Fig. 9-20. Typical components of a level D personal protective ensemble.

- Storage.
- Decontamination and disposal.
- Training and proper fitting.
- Donning and doffing.
- Inspection procedures.
- Limitations during temperature extremes.
- Program evaluation.

TRAINING IN PPE USE

Training in PPE use is requried by OSHA in 29 CFR Parts 1910.120 and 1910 subparts I and Z.

Before entry into an area requiring PPE, all workers should be trained sufficiently to:

- Establish user familiarity and confidence.
- Make the user aware of capabilities and limitations of the equipment used.

Fig. 9-21. Use of level D protection.

- Maximize the protective efficiency of the equipment.
- Maximize the ability to work efficiently in PPE.

An adequate PPE training program should cover all major points presented below. However, the PPE program must be largely site-specific, since both equipment and hazard-specific considerations of use will vary widely.

WORK MISSION DURATION

Work mission duration must be estimated before work in PPE actually begins. Factors limiting mission duration are described next.

Air Supply Consumption

Air supply consumption with an SCBA unit may be significantly increased (thus reducing time on task) by factors such as strenuous work rate, lack of fitness of user, and/or large body size of user. Shallow, rapid, irregular breathing patterns, or hyperventilation, can also lead to rapid air consumption. These conditions may result from heat-stress, anxiety, lack of acclimitization, or lack of familiarization with the SCBA.

PERSONAL PROTECTIVE EQUIPMENT   225

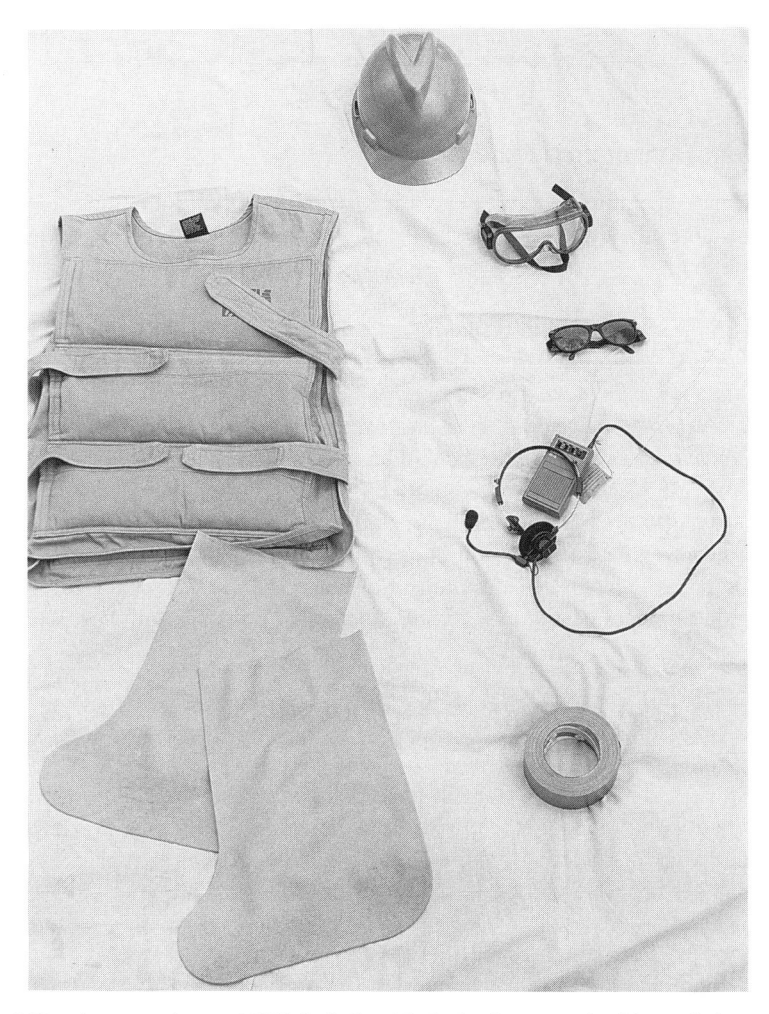

Fig. 9-22. Accessory items of PPE, including (clockwise from top): hard hat, splash goggles, safety glasses, voice-actuated two-way radio, duct tape, disposable boot covers, and cooling vest.

Permeation and Penetration of Protective Clothing or Equipment

Work mission duration cannot exceed the length of time during which items of CPC used can be expected to provide adequate protection. Thus, penetration and permeation of CPC (as described above) are of major concern.

Penetration may occur due to leakage of fasteners or valves on PPE, particularly under extreme temperature conditions (as discussed below). *Permeation* may occur due to improper selection of material, or prolonged use of equipment in a given atmosphere.

Ambient Temperature Extremes

Ambient temperature extremes can affect work duration in a number of ways. For example, the effectiveness of PPE may be reduced as hot or cold temperatures affect:

- *Valve operation* on suits and/or respirators.
- *Durability and flexibility* of CPC materials.
- *Integrity of fasteners* on suits.
- *Breakthrough time and permeation rate* of chemicals.
- *Concentration* of airborne contaminants.

In many instances, *heat stress* is the most immediate hazard to the wearer of PPE, and the greatest limitation upon work mission duration. *Coolant supply* will directly affect mission duration in instances in which cooling devices are required to prevent heat stress. Specific methods of dealing with heat stress are discussed below.

HEAT STRESS AND OTHER PHYSIOLOGICAL FACTORS

Heat stress and other physiological factors directly affect the ability of workers to operate safely and effectively while wearing PPE and thus are important considerations in selecting equipment and making work assignments. Factors that may predispose a worker to reduced work tolerance are:

- Poor physical condition or obesity.
- Alcohol and drug use (including prescription drugs).
- Dehydration or sunburn.
- Old age.
- Infection, illness, or disease.
- Lack of acclimitization.
- Work environments with elevated temperatures.
- Work environments requiring burdensome amounts of PPE.
- High workloads.

Monitoring for the Effects of Heat Stress

Monitoring of the physiological condition of workers using Levels A, B, or C PPE should begin whenever the temperature exceeds 70°F in the work area

(NIOSH, 1985). Monitoring for heat strain can incorporate the following factors and procedures:

Heart Rate. Radial pulse should be measured during a 30-second interval at the beginning of a rest period. If the heart rate exceeds 110 beats per minute, shorten the next work cycle by one-third, while keeping the rest period the same length. If the heart rate exceeds 110 beats per minute at the beginning of the next rest period, shorten the following work cycle by one-third. Continue monitoring and shortening work cycles until the heart rate is less than 110 beats per minute.

Oral Temperature. Oral temperature should be measured at the end of a work period by placing a clinical thermometer under the tongue for 3 minutes *before* drinking. If the oral temperature exceeds 99.6°F, shorten the next work cycle by one-third while keeping the rest period the same length. If the oral temperature still exceeds 99.6°F at the end of the next work period, shorten the following work period by one-third. Continue to monitor and shorten work periods until the worker's temperature is less than 99.6°F. No worker should be permitted to wear semipermeable or impermeable clothing if oral temperature exceeds 100.6°F.

Body Water Loss. Body water loss can be monitored by weighing a worker both before and after a work shift. If body water loss during a single shift exceeds 1.5% of the workers' total body weight, then actions should be taken to increase the worker's fluid intake and/or decrease the work load.

Frequency of physiological monitoring (and, therefore, the length of work cycles) depends initially on ambient air temperature as shown in Table 9-11.

Table 9-11. Suggested Frequency of Physiological Monitoring for Fit and Acclimatized Workers.[a]

Adjusted Temperature[b]	Normal Work Ensemble[c]	Impermeable Ensemble
90°F or above	After each 45 minutes of work	After each 15 minutes of work
87.5–90°F	After each 60 minutes of work	After each 30 minutes of work
82.5–87.5°F	After each 90 minutes of work	After each 60 minutes of work
77.5–82.5°F	After each 120 minutes of work	After each 90 minutes of work
72.5–77.5°F	After each 150 minutes of work	After each 120 minutes of work

[a] For work levels of 250 kilocalories/hour.
[b] Calculate the adjusted air temperature (t_{aadj}) by using this equation: $t_{aadj}°F = t_a°F + (13 \times \%\ sunshine)$. Measure air temperature (t_a) with a standard mercury-in-glass thermometer, with the bulb shielded from radiant heat. Estimate % sunshine by judging what percentage of time the sun is not covered by clouds that are thick enough to produce a shadow. (100% sunshine = no cloud cover and a sharp, distinct shadow; 0% sunshine = no shadows.)
[c] A normal work ensemble consists of cotton coveralls or other cotton clothing with long sleeves and pants.
Source: NIOSH, 1985.

Heat Injury Prevention

Prevention of heat injury may be accomplished through a number of managerial actions (NIOSH, 1985). For example, *work schedules* should be adjusted to coincide with work/rest schedules as described in Table 9-11. Work slowdowns should be mandated as needed. Personnel on work teams should be rotated as required to prevent heat injury. Also, it is advisable to work during cool hours of the day, or at night, during periods of hot weather.

Heat injury can also be prevented by providing cool rest areas. It is also vital to encourage the drinking of large quantities of fluids during periods of heavy sweating. This may be accomplished by providing water at 60°F with small disposable cups which hold about 4 ounces. In some instances cooling devices, as described above, are vital for preventing heat injury.

Workers should be encouraged to remain physically fit and avoid obesity. Several days should be allowed for acclimatization, during which time new employees gradually work up to a full work load. Most importantly, all workers should be trained to recognize symptoms of heat stress and treat heat injuries.

PERSONAL FACTORS AFFECTING RESPIRATOR USE

A number of personal use factors of workers may diminish the effectiveness of respirators. For example, any facial hair or long hair which comes between the respirator facial seal gasket and the wearer's skin will prevent a good respirator fit. Thus, beards are not allowable for workers required to use respirators. Individual facial features, such as scars, hollow cheeks, deep skin creases, missing teeth, etc. may also prevent a good respirator fit. Chewing gum and tobacco should be prohibited during respirator use.

Likewise, the temple pieces on conventional eyeglasses interfere with respirator fit. However, spectacle kits are available which can be used to mount the corrective lenses within the facepiece. Contact lenses may not currently be worn with a full-facepiece respirator under OSHA regulations (29 CFR 1910.134). However, the American National Standards Institute (ANSI 1988) has proposed that the use of contact lenses be allowed with full-facepiece respirators, provided the wearer can demonstrate the ability to do so without problems.

DONNING PPE

In donning an ensemble of PPE, an established routine should be followed. All equipment should be fully inspected as a part of the donning procedure. Donning and doffing of PPE should always be done with the aid of an assistant. Field tests for respirator fit should *always* be performed as part of the donning procedure. After donning, all ensemble components should be checked for proper fit, proper functioning, and relative comfort before entering a hazardous area.

INSPECTION OF PPE

PPE should be fully inspected before each use. PPE inspection checklists, such as the following, may be used:

Inspecting CPC

General Inspection Procedure (applicable to all items of CPC).

- Determine that the clothing material is correct for the specified task at hand.
- Visually inspect for imperfect seams, nonuniform coatings, tears, and malfunctioning closures.
- Hold up to a light and check for pinholes.
- Flex the product and observe for cracks and observe for other signs of shelf deterioration.
- If the product has been used previously, inspect inside and out for signs of chemical attack, such as discoloration, swelling, and stiffness.

Inspecting Fully Encapsulating Suits.

- Check the operation of pressure relief valves.
- Inspect the fitting of wrists, ankles, and neck.
- Check faceshield, if so equipped, for cracks, crazing and/or fogginess.
- TECPs may require pressure testing or whole suit in-use testing (as described in appendix A of 29 CFR 1910.120)

Inspecting Gloves.

- Before use, check for pinholes. Blow into glove, then roll gauntlet towards fingers and hold under water. No air should escape.

Inspecting Respirators

General Procedures (applicable to all types of respirators).

- Check material condition of harness, facial seal, and breathing tube (if so equipped), for pliability, signs of deterioration, signs of discoloration, and damage.
- Check faceshields and lenses for cracks, crazing, and fogginess.
- Check inhalation and exhaust valves for proper operation.

Inspecting Air-Purifying Respirators.

- Inspect air-purifying respirators:
  - Before each use to be sure they have been adequately cleaned.
  - After each use, during cleaning.
  - At least monthly if in storage for emergency use.
- Examine cartridges or canisters to ensure that:
  - They are the proper type for the intended use.
  - The expiration date has not passed.
  - They have not been opened or used previously.

Inspecting SCBAs.

- Inspect SCBAs:
  - Before and after each use.
  - At least monthly when in storage.
  - Every time they are cleaned.
- Check air supply.
- Check all connections for tightness.
- Check for proper setting and operation of regulators and valves (according to manufacturers' recommendations).
- Check operation of alarms.

Inspecting Supplied-Air Respirators.

- Inspect SARs:
  - Daily when in use.
  - At least monthly when in storage.
  - Every time they are cleaned.
- Inspect airline prior to each use, checking for cracks, kinks, cuts, frays, and weak areas.
- Check for proper setting and operation of regulators and valves (according to manufacturers' recommendations).
- Check escape air supply (if applicable).
- Check all connections for tightness.

IN-USE MONITORING OF PPE

While working in PPE, workers should constantly monitor equipment performance. If indications of possible in-use equipment failure are noted, the worker should exit the contaminated area immediately and investigate.

Degradation of ensemble components during use may be indicated by discoloration, swelling, stiffening, softening, etc. of materials. Likewise, any tears, punctures, or splits at seams or zippers should be noted, as should any unusual residues on items of PPE.

Perception of odors; irritation of skin, eyes, and/or respiratory tract; and general discomfort may be indications of equipment failure. Also, symptoms commonly associated with chemical exposure and oxygen deficiency may be the end result of equipment failure. These symptoms include rapid pulse, nausea, chest pain, difficulty in breathing, or undue fatigue. Restrictions of movement, vision, or communication may also result from equipment failure.

DOFFING PPE

Like donning, doffing of PPE should be done according to an established routine. Furthermore, doffing routines should be well integrated with decontamination and disposal procedures for used PPE.

STORAGE OF PPE

Storage is an important aspect of PPE use. Improper storage may lead to damage due to contact with dust, moisture, sunlight, damaging chemicals, extreme temperatures and physical abrasion.

The following considerations should be observed in storing CPC:

- Potentially contaminated clothing should be stored in an area separate from street clothing.
- Potentially contaminated clothing should be stored in a well ventilated area, with good air flow around each item, if possible.
- Different types and materials of clothing and gloves should be stored separately to prevent issuing the wrong material by mistake.
- Protective clothing should be folded or hung in accordance with manufacturers' recommendations.

The following considerations should be observed in storing respirators:

- SCBAs, supplied-air respirators, and air-purifying respirators should be dismantled, washed, and disinfected after each use.
- SCBAs should be stored in storage containers supplied by the manufacturer.
- Air-purifying respirators should be stored individually in their original cartons or carrying cases.
- All respirator facepieces should be sealed inside a plastic bag for storage.

REUSE OF CPC

Items of CPC must be completely decontaminated prior to reuse. Otherwise, the items cannot be considered safe to use.

In some instances, contaminants may permeate the matrix of CPC and be difficult or impossible to remove. Such contaminants may continue to diffuse through the CPC material toward the inner surface during storage, thus posing the threat of direct skin contact to the next wearer. If matrix permeation is possible, the article of CPC should be hung in a warm, well-ventilated place to allow the item to release the permeated contaminant.

Extreme care should be taken to ensure that permeation and degradation have not rendered CPC unsafe for reuse. It should also be noted that permeation and degradation may occur without any visible indications.

MAINTENANCE OF PPE

Proper maintenance is vital to the proper functioning of PPE. Thus, the site PPE program should include specific maintenance schedules and procedures in accordance with manufacturers' recommendations for all reusable items of PPE.

Maintenance can generally be divided into three levels as follows:

- Level 1. User or wearer maintenance, requiring a few common tools or no tools at all.
- Level 2. Shop maintenance, which can be performed by the employer's maintenance shop.
- Level 3. Specialized maintenance that can be performed only by the factory or an authorized repair person.

In some instances, it may be advisable for an employer to send selected employees through a manufacturer's training course in order to establish an in-house PPE maintenance program.

REFERENCES

ANSI. 1973. *Identification of Air Purifying Respirator Canisters and Cartridges*, K13.1. New York, NY: American National Standards Institute.

ANSI. 1980. *Practices for Respiratory Protection*, Z88.2. New York, NY: American National Standards Institute.

ANSI. 1988. *Proposed Revisions to ANSI Z88.2-1980*. New York, NY: American National Standards Institute.

Langley, J. 1988. "Choosing Adequate Personal Protective Clothing for the Haz Mat Team." *Hazardous Waste Management Magazine*, June, 10–11.

Rodgers, S. J. 1988. "How to Select and Use Personal Protective Garments." *Industrial Hygiene News*, Sept., 44–55.

Schwope, A. D.; Costas, P. P.; Jackson, J. O.; and Weitzman, D. J. 1985. *Guidelines for the Selection of Chemical-Protective Clothing*, 2nd Edition. Cincinnati, OH: American Conference of Governmental Industrial Hygienists, Inc.

NIOSH/OSHA/USCG/EPA 1985. *Occupational Safety and Health Guidance Manual for Hazardous Waste Site Activities.* NIOSH Publication No. 85-115. Washington, DC: U.S. Govt. Printing Office.

NIOSH, 1987. *NIOSH Guide to Industrial Respiratory Protection.* NIOSH Publication No. 87-116. Washington, DC: U.S. Govt. Printing Office.

OSHA, 1986. *Respiratory Protection.* OSHA 3079. Washington, DC: U.S. Govt. Printing Office.

U.S. Army, 1988. *Air Purifying Respirators.* Washington, DC: U.S. Govt. Printing Office.

U.S. Dept. of Labor. Title 29 Part 1910. Washington, DC: U.S. Govt. Printing Office.

10
Safe Sampling Techniques

INTRODUCTION

One of the most common tasks on a Superfund site is the taking of both environmental and hazardous waste samples. These samples, once taken, are the basis for many decisions pertaining to worker safety and disposal techniques used. Improperly taken samples can produce erroneous results that could lead to (1) injury to the worker, and/or (2) carrying out activities that are inappropriate for the material on site. It is, therefore, important for persons working on site to know the correct and safe techniques for taking samples.

OBJECTIVES

The intent of this section is to describe a collection of methods and materials sufficient to address most sampling situations that arise during routine waste site and hazardous spill investigations. It includes a compilation of methods, the purpose being to supply detailed, practical information directed at providing field investigators with a set of functional operating procedures. These procedures should be set forth in a sampling plan. Thus, one important objective of this section is to become familiar with the various elements of a sampling plan, including (but not limited to) the devices available for sampling use. Understanding the importance of the components of the sampling plan will help ensure that the plan is fully implemented, thus ensuring worker safety.

PURPOSE OF SAMPLING

The purpose of sampling is to determine the characteristics of a larger source of material by examining a selected amount (a sample) of that larger source. The intent of the whole process involves attributing characteristics of the sample to the source from which it was taken. In view of this, it is essential that the sample be as representative as possible of the source from which it is taken.

A sample may be either an environmental or a hazardous waste sample.

- Environmental samples are taken in an area surrounding a spill or hazardous waste site. They are off-site samples from soils, rivers, lakes, etc. Such samples usually contain relatively dilute pollutant concentrations, and do

not usually require the special handling procedures used for concentrated wastes.
- Hazardous waste samples, on the other hand, are collected from on-site sources of hazardous waste and require special handling due to hazardous characteristics.

Use of samples:

- Samples may be used to determine the extent of remedial action needed. This involves determining the boundaries of the site requiring remedial action. It also involves determining what sources of material within a hazardous waste site require treatment as a hazardous waste, because not every source of material within a site may display hazardous characteristics.
- Once the site boundaries are determined and it is known what specific waste sources must be disposed of as hazardous wastes, additional samples may be needed to submit to the proposed hazardous waste disposal site. Such disposal sites generally require a sample of the waste along with a written description (waste profile) of it before they will accept it for disposal.
- Samples of hazardous wastes may also be required in litigation to determine who is responsible for the presence of hazardous waste at a particular site.

WHY SAMPLING IS DANGEROUS

Sampling can be one of the most hazardous activities to workers' health and safety because it often involves direct contact with unidentified wastes. Contact with unidentified waste puts the worker in an extremely vulnerable position in regard to whether he or she is wearing adequate PPE and in regard to what types of danger a particular waste poses. Fig. 10-1 shows the potential for sampling personnel to be exposed to the waste being sampled. For example, consider a situation in which an unmarked or improperly marked container is opened for sampling and is found to contain a liquid which, unbeknownst to the sampler, is spent hydrofluoric acid. Opening the container, taking a sample, and transferring it to a sample container are activities during which the sampler could easily breathe in vapors, be splashed, or otherwise come in direct contact with the liquid. Hydrofluoric acid liquid or vapor in contact with the eyes can cause prolonged or permanent visual defects or total destruction of the eyes. Skin contact can result in severe burns.

The remainder of this section deals with how to minimize these dangers through development and implementation of the sampling plan.

DEVELOPMENT OF A SAMPLING PLAN

A detailed, written sampling plan should be prepared prior to occurrence of any sampling activity. Elements of such a plan are described here.

Fig. 10-1. Drum sampling to illustrate potential for worker exposure. (EPA, 1987).

Background Information

Research of available background information regarding the type of wastes which may be present on site. Such information may include: paperwork filed with EPA and state environmental authorities regarding hazardous waste generated, treated, or disposed of on-site; files of the local health department regarding site visits or citations; interviews with past or present employees of the firm or firms involved with the site; interviews with persons living or working near the site; etc.

Sample Location

Decide at what points inside and outside the site boundaries samples should be taken. If wastes are containerized, decide which containers will be sampled.

How Many Samples Per Sampling Point

Decide how many samples will be taken at each sampling point. This decision should take into consideration the purpose(s) of the proposed sample, such as determination of hazardousness, submission to a disposal facility, or litigation.

Volume Per Sample

Decide on the volume needed of each individual sample. This decision should be made in consultation with the persons or laboratory to whom the sample is to be submitted. If it is to be submitted for laboratory analysis, the lab should be consulted to determine how large a sample it needs to run the particular test in question. Likewise, if the sample is to be submitted to a disposal facility, the facility in question should be contacted to determine what size sample it requires.

Sample Containers

Decide what type of sample containers to use. Again, consult with the proposed recipient of the sample regarding any special requirements or recommendations. Some laboratories may have special requirements regarding containers for certain types of samples. Some points to consider are discussed next.

Compatibility with the Chemical the Container Is to Hold. Plastic or glass containers are generally used for collection and storage of hazardous waste samples. Plastic is used only when the constituents of the material are known not to react with the plastic. Among the materials not generally compatible with plastic are petroleum distillates, chlorinated hydrocarbons, pesticides, and solvents.

Glass is relatively inert to most chemicals and can be used to collect and store almost all hazardous waste samples except strong alkali solutions and hydrofluoric acid. The technicians shown in Fig. 10-2 are preparing an assortment of glass containers for taking samples.

Containers should have tight, screw-type lids. Lids are sometimes required to have liners. For instance caps with Teflon® liners are recommended with hydrocarbons, pesticides, and petroleum residues.

Resistance to Breakage. Depending on how much handling of the sample is anticipated before it reaches its ultimate destination, the strength of the sample containers may be of greater or lesser concern.

Volume. The container must be large enough to hold the required volume. It also may need to be large enough to hold the entire discharge from the sampling instrument. Further, depending on the type of sampling instrument used, the mouth of the container may need to be large enough to receive the discharge valve of the sampling device.

Fig. 10-2. Drum sampling personnel shown preparing sample containers (EPA, 1987).

Selection of Sampling Equipment

Because of difficulties in decontaminating equipment it is recommended to use disposable equipment where feasible. If disposable equipment cannot be used, equipment should be used which can easily be decontaminated.

Hazardous wastes are usually complex mixes of liquids, solids, and semi-solids. Liquids and semi-solids may vary greatly in viscosity, corrosivity, volatility, explosiveness, and flammability. Solid wastes can range from granular to big lumps. The wastes may be in drums, barrels, sacks, bins, tanker or vacuum trucks, ponds, lagoons, piles, or other forms or containers. In view of this, there is no one type of sampling device which can collect representative samples of all types of waste. A number of different samples are commonly used; most, however, have certain limitations which the user should be aware of.

Some of the more commonly used devices are discussed below, according to the type of material being sampled.

Soil Sampling. Soil types can vary considerably within a hazardous waste site. These variations, along with variations in vegetation, can affect the rate of contaminant migration through the soil. Because of this, it is important to keep a detailed record of sample location, depth, and soil grain size, color, and odor.

Subsurface conditions such as temperature, available oxygen, and light penetration are often stable at a given location. Because of this, soil samples should be kept at their in-ground temperature or lower, should be protected from direct light, sealed tightly in glass containers, and analyzed as soon as possible.

The depth from which samples can be collected and the sampling method to be used depend on the physical properties of the soil, soil grain size, soil cohesiveness, soil moisture, depth to bedrock, and the depth to the water table.

Shallow Depths. *Pick Axe/Shovel/Scoop.* These can be used to collect samples up to 2 or 3 feet below the surface. In most cases these samples would be disturbed, that is, the sequence of soil layering would not be maintained. However, soil can be shoveled out to a certain depth and an undisturbed sample troweled or hand-cored out at that depth.

Hand Auger. This is difficult to use in hard or rocky soil. Unless the soil is very damp and the auger screw tightly compressed, the hand auger will not yield an undisturbed sample.

Post Hole Digger. This is difficult to use in hard, rocky soil or to get below 2 feet in depth. It can yield an undisturbed sample.

Depths Between 3 and 16 Feet. *Viemeyer Sampler.* This is used for core samples of most soil types except stony or very wet soils.

Greater Depths. *Split Spoon.* This is used for undisturbed soil samples from a wide variety of soil types and from depths greater than with other types of equipment. It is generally used with a power operated drill rig.

Sludge/Sediment Sampling. *Sludge* is defined as semi-dry material ranging from dewatered solids to high viscosity liquids. However, sludges often form as a result of settling of the higher-density components of a liquid; in such a case the sludge may often have a liquid layer above it.

Sediments are the deposited material underlying a body of water. Sediments can be collected in much the same manner as sludges, with some qualifications: because there is likely to be substantial variation in sediment composition at different locations in the water body, it is important to record the exact sampling location. Also, the presence of rocks, debris, and organic matter may preclude the use of some samplers. Sediments may sometimes be exposed with no water overlying them.

The sludge/sediment sampling equipment listed below depends on the extent to which there is overlying water.

Sludge/sediment with no overlying water:

- *Scoop:* Yields a disturbed sample from shallow depth.
- *Sampling trier:* Yields an undisturbed sample, but must be inserted at an angle.

- *Hand corer:* Yields an undisturbed sample from depths of one to two feet.

Sludge/sediment with shallow overlying water layer.

- *Scoop:* Yields a disturbed sample. (*Note:* The person taking the sample shown in Fig. 10-3 is wearing no protective clothing other than a glove to keep his hand from coming in contact with the liquid and this is *not a recommended practice*).
- *Pond sampler:* Yields a fairly disturbed sample from shallow depth.
- *Hand corer:* Can be fitted with extension handle to retrieve an undisturbed sample.
- *Glass tube:* If the sludge is not too hardened, this can be used to sample sludge in 55-gallon drum.

Sludge/sediment with deep overlying water layer.

- *Gravity corer:* Yields an essentially undisturbed sample. Depending on

Fig. 10-3. Unprotected worker sampling water (OSHA, 1985).

sludge density and corer weight, it can penetrate up to 30 inches into sludge. If sludge is in a container, the corer could damage container bottom.
- *Ponar grab:* Capable of sampling most types of sludges/sediments. Collects a disturbed sample from the top 1 to 3 inches.

Bulk Materials Sampling. Bulk materials are generally homogeneous collections of single identifiable products. They are usually contained in bags, drums, or hoppers, or occasionally piled on the ground. Any surfaces that are exposed to the atmosphere may undergo some chemical change and should be avoided during sample collection. Also, as the process producing the bulk materials may show some variation over time, it is advisable to collect a series of samples as one composite.

The use of the sampling equipment listed below depends on whether the material is consolidated or unconsolidated.

Bulk materials in unconsolidated state:

- *Scoop:* To be used when a small amount of sample is needed.
- *Grain thief:* Used to take composites from a large amount of material. Most useful when solids are no larger than 1/4" in diameter.

Bulk materials which are moist/sticky:

- *Scoop:* To be used when a small amount of sample is needed.
- *Sampling trier:* Yields undisturbed sample from depths of one to two feet.

Liquids Sampling. By their nature liquids are relatively easy to collect; however, it is more difficult to obtain a representative sample. This is because density, solubility, temperature, currents, and other factors can change the composition of the liquid material over time and distance. Liquids include both aqueous and nonaqueous solutions, subdivided into surface waters, containerized liquids, and ground waters.

Surface waters at shallow depths near the edge of the water body:

- *Submerge sample container:* The major shortcoming is that the outside of the container becomes contaminated and must be decontaminated.
- *Submerge stainless steel beaker or scoop:* Contents can then be transferred into sample container.

Surface waters at shallow depths some distance from the edge of the waterbody:

- *Pond sampler:* Used to take samples up to 10 feet from the edge of ponds, pits, lagoons, etc.
- *Small peristaltic pump:* By using stiff-walled tubing extended laterally a sample can be taken some distance out from the edge of the water. Tubing must be compatible with the material being pumped.

Surface waters at greater depths:

- *Extended bottle sampler:* Useful for sampling depths up to 5 feet. The sampling bottle becomes exposed to contaminated material and must be decontaminated before reuse or shipping.
- *Peristaltic pump:* Can lift liquids from depths up to 18–24 feet (lift capacity declines with viscosity of the liquid).
- *Kemmerer bottle:* Currently the most practical way of collecting discrete samples from liquids at sampling depths exceeding the lift capacity of pumps.
- *Weighted bottle:* Difficult to use in very viscous liquids. The outside of the bottle is exposed to contaminated material.

Containerized Liquids. Sampling tanks, containers, and drums presents unique problems not associated with natural water bodies. Such containers are usually closed except for small access ports, manways, or hatches (on larger vessels) or taps or bungs (on smaller vessels). The size, shape, construction material, and location of access will limit the types of equipment used for sampling.

An important consideration with regard to sealed containers is that such vessels may have a vapor pressure buildup. Vapor buildup creates the potential for explosive reactions or release of toxic gases. In view of this, vessels should be opened with great caution. Preliminary sampling of head space gases may be needed.

Another consideration is that layering is very common in vessels which have been undisturbed for some time. In such cases it is important to have a sample representing the entire depth of the vessel. No attempt should be made to agitate the vessel and rehomogenize the contents.

Use of the sampling devices listed below depends on the depth of the vessel. Vessels less than $3-3\frac{1}{2}$ feet deep:

- *Glass tubes (drum thieves).* The advantage is that they are inexpensive and so disposable. However, the tubes should not be broken and left in the drum unless it is certain that this will not interfere with ultimate disposal of the waste. Fig. 10-4 shows a technician closing off the top of the sampling tube with his glove finger; another method of closing off the tube is

Fig. 10-4. Sampling with a drum thief (EPA, 1987).

to use a rubber stopper. If the liquid is of low viscosity it is hard to maintain a vacuum in the tube, and the liquid tends to dribble out as the tube is withdrawn from the drum. Glass tubing should not be used with materials containing hydrofluoric acid or strong alkali solutions.

- *PVC tubing.* Such tubing is sometimes used in the same manner as a glass drum thief. However, as noted earlier, equipment made of such material should only be used when it is certain that the liquid being sampled is compatible with it. PVC is not for materials containing ketones, nitrobenzene, dimethylforamide, mesityloxide, tetrahydrofuran, or many common solvents.
- *COLIWASA.* The major use is where a true representation of a multiphase waste is needed. The major drawback is that is very difficult to decontaminate in the field. Comes in PVC or glass.

Vessels greater than $3-3\frac{1}{2}$ feet deep:

- *Perisaltic pump.* Can lift liquids from depths up to 18–24 feet (lift capacity declines with viscosity of the liquid).

- *Submersible pump*. May be used if the depth of the vessel exceeds lift the capacity of a peristaltic pump. It must be compatible with the liquid it is to be submerged in.
- *Kemmerer sample*. See earlier discussion.
- *Bacon bomb*. A heavy sampler best suited for viscous materials in large storage tanks or in lagoons.

Groundwater sampling is covered in Chapter 11.

Personal Protective Equipment

Based on the available information on site conditions and wastes present, a trained health and safety professional should determine the appropriate personal protection needed by sampling personnel.

Standard Operating Procedure. Develop a standard operating procedure (SOP) for sampling. This sampling SOP should tie in to the drum handling SOP, as a certain amount of drum handling (in the form of drum moving and opening) is often necessary before sampling can occur.

One purpose of the sampling SOP is to develop an expeditious sampling procedure. An expeditious sampling procedure has several advantages: minimization of sampling personnel contact with the chemicals; reduction of fatigue from PPE; and maximization of sampling team productivity.

Elements which should definitely be included in the sampling SOP are:

- Designation of personnel to be used and division of labor—the sampling team should be well organized.
- Keep sampling personnel at a safe distance while containers are being opened.
- Do not lean over other containers to reach the container being sampled.
- Never stand on drums—this is very dangerous. Use mobile steps or other platform to achieve the needed height.
- Seal up drums after sampling.

Sample Integrity

The sampling plan should set forth what steps will be taken to maintain sample integrity between the time the sample is collected and the time it is delivered for analysis. Most samples have a limited holding time before analysis begins—that is, after a certain specific period of time sample characteristics may undergo alteration. If analysis is not performed by the end of that time the results of the analysis may not accurately reflect the material sampled.

All samples should be refrigerated to near 39°F (4°C) until received by the laboratory. Fig. 10-5 shows sample containers being placed in a styrofoam

Fig. 10-5. Collecting samples in a cooler; note the confined space (EPA, 1987).

cooler. The cooler should be partially filled with sample containers and then overfilled with ice. In some cases, if the identity of the material is known, preservatives may be added to the sample to maximize holding time. However, the lab doing the analysis should be consulted before any preservatives are used.

Decontamination of Sampling Equipment

Sampling equipment should be decontaminated between uses in order to prevent crosscontamination of samples. Typically this is done with a soapy wash, tap water rinse, followed by a mild solvent rinse (such as isopropyl alcohol) and a final rinse with triple deionized water. Personnel performing decontamination should wear appropriate protective equipment. Rinse solutions should be properly disposed of. In some cases it may be possible to have the solution accepted at a wastewater treatment plant; in other cases, the material may have to go to a hazardous waste disposal facility.

In some cases the outside of the sample container may need to be decontaminated. This procedure could begin by simply wiping off the container, then rinsing it with an appropriate solution. For example, containers with acid waste on the outside could be rinsed in a solution of water and baking soda, while containers of caustic wastes could be rinsed with mild acetic acid (vinegar). One way to minimize contamination of the outside of the sample container is to place a small plastic bag around the container as far up as the neck of the container, fastening it with a rubber band. After the sample is placed in the container, the bag can be removed and disposed of as a hazardous waste.

Recordkeeping

Samples and sample analysis are worthless unless an accurate record is maintained identifying the source of the sample and other pertinent information. As the purpose of sampling is to attribute to a larger source the characteristics of the sample taken from it, it is essential that the sample source can be pinpointed exactly. Sample containers should bear a label identifying: the container number or specific location from which the sample came, sample ID number, the name of the person sampling, date/time of sampling, and analysis requested.

A log should be kept with information entered into it as it is being done. Information should be entered in the log for each source sampled to include the following: exact location of sample source including container number (if it is a containerized waste); condition of container (does waste need recontainerizing?); how full does container appear to be?; color and appearance of waste; how many samples taken and what size taken; sampling date/time; and name of sampler.

In addition, a chain of custody record must accompany each sample from the time it is collected showing who has custody of the sample at any particular time. Fig. 10-6 shows a sample chain of custody sheet.

IMPLEMENTATION OF SAMPLING PLAN

Once the above information has been developed, implementation of the plan may proceed. In order for the plan to have any effectiveness, it is essential that everyone involved with implementation of the plan be familiar with it. This means everyone concerned—from supervisors to laborers. Not only must they be familiar with the plan's provisions but they must understand why the steps set out in the plan are important to their health and safety—otherwise they will have no incentive to follow it.

PACKAGING, MARKING, LABELING, AND SHIPPING OF HAZARDOUS MATERIAL SAMPLES

Introduction

Samples collected during a response to a hazardous material incident may have to be transported elsewhere for analysis. The Environmental Protection Agency

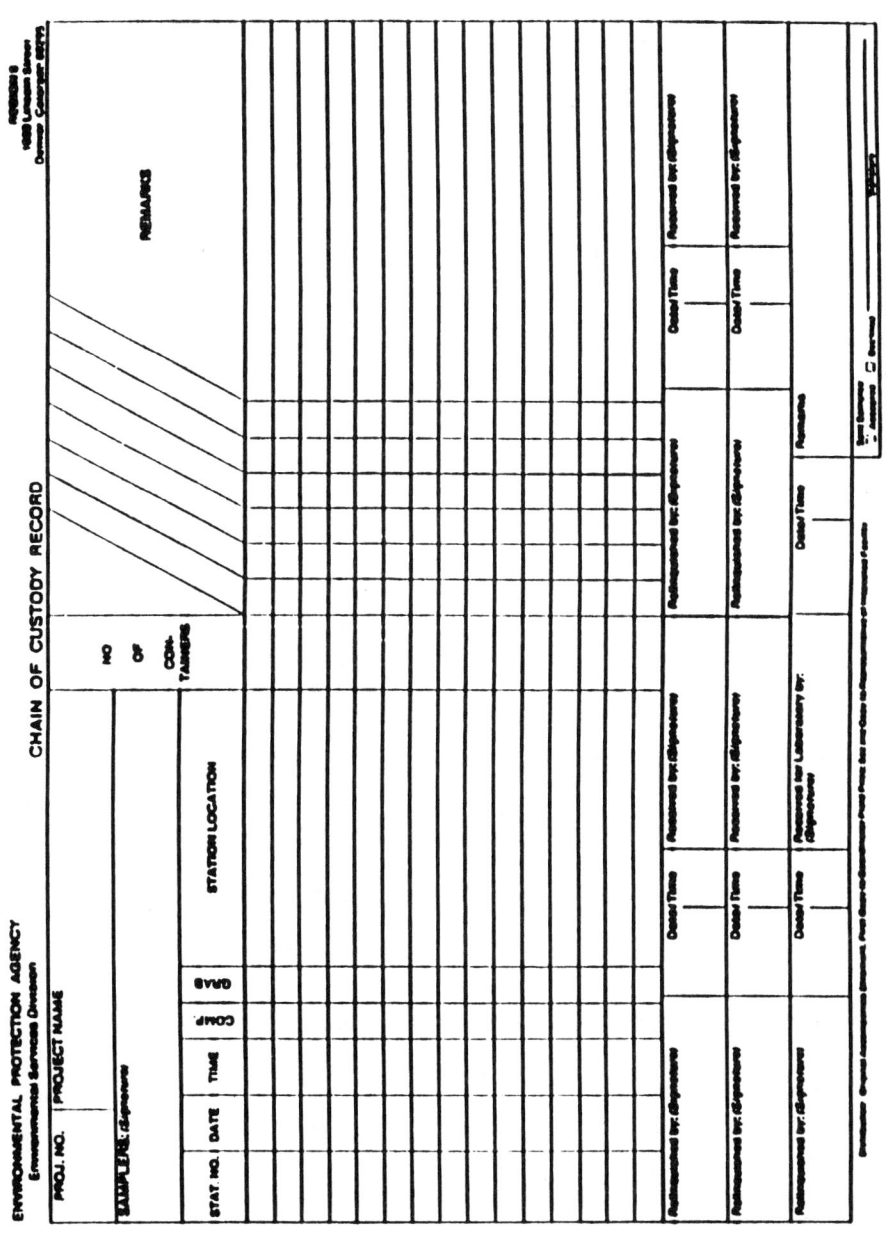

Fig. 10-6. Chain-of-custody sheet.

(EPA) encourages compliance with Department of Transportation (DOT) regulations governing the shipment of hazardous materials. These regulations (49 CFR parts 171 through 179) describe proper marking, labeling, packaging and shipment of hazardous materials, substances and wastes. In particular, part 172.402(h) of 49 CFR is intended to cover shipment of samples of unknown materials destined for laboratory analysis.

Environmental Samples Versus Hazardous Materials Samples

Samples collected at an incident should be classified as either *environmental* or *hazardous* material (or waste) samples. In general, environmental samples are collected off site (for example from streams, ponds, or wells) and are not expected to be grossly contaminated with high levels of hazardous materials. On-site samples (for example, soil, water, and materials from drums or bulk storage tanks, obviously contaminated ponds, lagoons, pools, and leachates from hazardous waste sites) are considered hazardous. A distinction must be made between the two types of samples in order to:

- Determine appropriate *procedures for transportation* of samples. If there is any doubt, a sample should be considered hazardous and shipped accordingly.
- *Protect the health and safety* of laboratory personnel receiving the samples. Special precautions are used at laboratories when samples other than environmental samples are received.

Environmental Samples

Environmental samples must be packaged and shipped according to the following procedures.

Packaging. Environmental samples may be packaged following the procedures outlined later in this section for samples classified as "flammable liquids" or "flammable solids." Requirements for marking, labeling, and shipping papers do not apply.

Environmental samples may also be packaged without being placed inside metal cans as required for flammable liquids or solids, see Fig. 10-7.

- Place sample container, properly identified and with a sealed lid, in a polyethylene bag, and seal bag.
- Place sample in a fiberboard container or metal picnic cooler which has been lined with a large polyethylene bag.
- Pack with enough noncombustible, absorbent, cushioning material to minimize the possibility of the container breaking.

SAFE SAMPLING TECHNIQUES

PACKAGING ENVIRONMENTAL SAMPLES

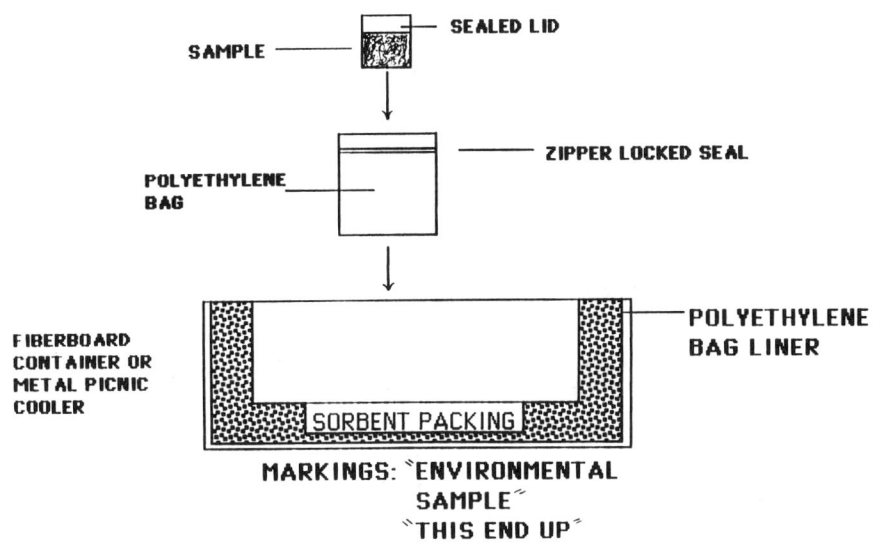

Fig. 10-7. Packaging environmental samples.

- Seal large bag.
- Seal or close outside container.

Marking/Labeling. Sample containers must have a completed sample identification tag and the outside container must be marked "Environmental Sample." The appropriate side of the container must be marked "This End Up" and arrows placed accordingly. No DOT marking or labeling is required.

Shipping Papers. No DOT shipping papers are required.

Transportation. There are no DOT restrictions on mode of transportation.

Rationale: Hazardous Material Samples

Samples not determined to be environmental samples or samples known or expected to contain hazardous materials must be considered hazardous substance samples and transported according to the following requirements:

- If the substance in the sample is known or can be identified, package, mark, label, and ship according to the specific instructions for the material (if it is listed) in the DOT Hazardous Materials Table, 49 CFR 172.101.
- For samples of hazardous materials of unknown content, part 172.402 of

49 CFR allows the designation of hazard class based on the shipper's knowledge of the material and selection of the appropriate hazard class from part 173.2 (see Table 10-1).

The correct shipping classification for an *unknown sample* is selected through a process of elimination, utilizing the DOT classification system (Table 10-1). Unless known or demonstrated otherwise (through the use of radiation survey instruments), the sample is considered radioactive and appropriate shipping regulations for "radioactive material" are followed. If radioactive material is eliminated, the sample is considered to contain "Poison A" materials, the next classification on the list. DOT defines "Poison A" as extremely dangerous poisonous gases or liquids of such a nature that a very small amount of gas, or vapor of the liquid, mixed with air is dangerous to life.

Most "Poison A" materials are gases or compressed gases and would not be found in drum-type containers. Liquid "Poison A"s would be found only in closed containers. All samples taken from closed drums do not have to be shipped as "Poison A"s, which provides for a "worst case" situation. Based upon information available, a judgment must be made whether a sample from a closed container is a "Poison A".

If "Poison A" is eliminated as a shipment category, the next two classifications ("flammable" or "nonflammable" gases) should be considered. In practice, few gas samples are collected. The next applicable category to be considered would be "flammable liquid". With the elimination of "radioactive material", "Poison A", "flammable gas", and "nonflammable gas," the sample can be classified as "flammable liquid" (or solid) and shipped accordingly. These procedures would also suffice for shipping any other samples classified below "flammable liquids" in the DOT Classification Table.

Table 10-1. DOT Hazard Classes.

Radioactive Material
Poison A
Flammable Gas
Nonflammable Gas
Flammable Liquid
Oxidizer
Flammable Solid
Corrosive Material (Liquid)
Poison B
Corrosive Material (Solid)
Irritating Materials
Combustible Liquid (containers greater than 110 gallons)
ORM-B
ORM-A
Combustible Liquid (containers less than 110 gallons)

For samples containing *unknown material*, other categories listed below flammable liquids/solids on the table are generally not considered because eliminating other substances as flammable liquids requires flashpoint testing, which may be impractical and possibly dangerous at a site. Thus, unless the sample is known to consist of material listed below flammable liquid on the table, it is considered to be a flammable liquid (or solid) and shipped accordingly.

Procedures: Samples Classified as Flammable Liquid (or Solid)

The following procedure is designed to meet the requirements for a "limited quantity" exclusion for shipment of flammable liquids and solids, as set forth in parts 173.118 and 173.153 of 49 CFR. By meeting these requirements, the DOT constraints on packaging are greatly reduced. Packaging according to the limited quantity exclusion requires notification on the shipping papers.

Packaging. Collect sample in a glass container (16 ounces or less) with nonmetallic, Teflon®-lined screw cap. To prevent leakage, fill container no more than 90% full at 130°F. If an air space in the sample container would affect sample integrity, place that container with a second container to meet 90% requirement.

Complete sample identification tag and attach securely to sample container.

Seal the container and place in 2-mil thick (or thicker) polyethylene bag, one sample per bag. Position identification tag so it can be read through the bag. Seal the bag.

Place sealed bag inside metal can and cushion it with enough noncombustible, absorbent material (for example, vermiculite or diatomaceous earth) between the bottom and sides of the can and bag to prevent breakage and absorb leakage. Pack one bag per can. Use clips, tape, or other positive means to hold can lid securely, tightly, and permanently, see Fig. 10-8.

Place one or more metal cans into a strong outside container, such as a metal picnic cooler or a DOT approved fiberboard box. Surround cans with noncombustible, absorbent, cushioning material for stability during transport.

Limited quantities of flammable liquids, for the purpose of the exclusion, are defined as one pint or less (49 CFR part 173.118(a)(2).

Limited quantities of flammable solids, for the purpose of this exclusion, are defined as one pound net weight in inner containers and no greater than *25 pounds net weight* in the outer container (49 CFR part 173.153.(a)(1).

Marking/Labeling.

- Use abbreviations only where specified.
- Place following information, either hand printed or in label form, on the metal can:

PACKAGING HAZARDOUS WASTE SAMPLES

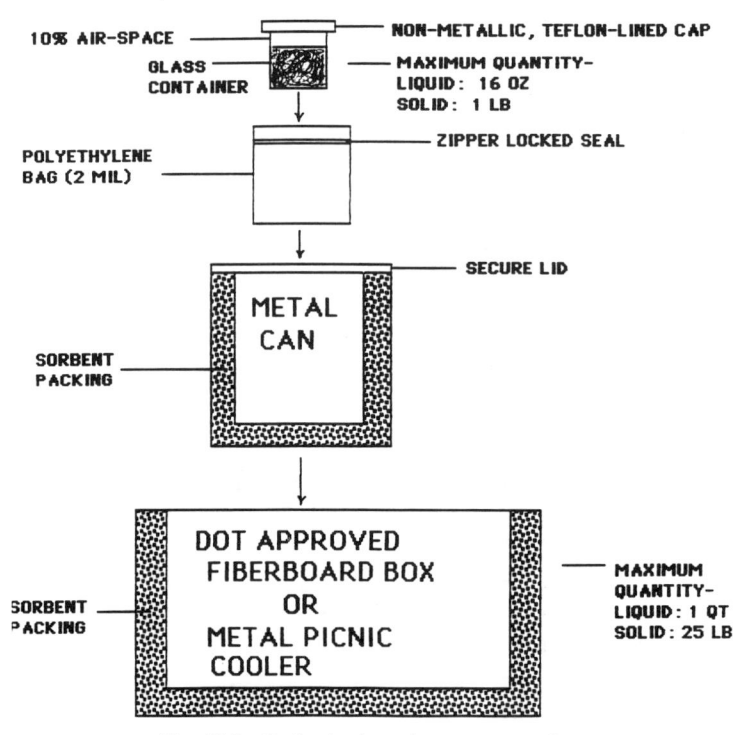

Fig. 10-8. Packaging hazardous waste samples.

Laboratory Name and Address
"Flammable Liquid, n.o.s. UN1993," or
"Flammable Solid, n.o.s. UN1325."

Not otherwise specified (n.o.s.) is not used if the flammable liquid (or solid) is identified. Then the name of the specific material is listed before the category (for example, "Acetone, Flammable Liquid") followed by its appropriate UN number found in the DOT hazardous materials table (172.101).

Place the following DOT labels (if applicable) on outside of can (or bottle):

- *"Flammable Liquid"* or *"Flammable Solid"*.
- *"Dangerous When Wet."* Must be used with "Flammable Solid" label if material meets the definition of a *water-reactive material.*
- *"Cargo Aircraft Only"*. Must be used if net quantity of sample in each

outer container is greater than 1 quart (for "Flammable Liquid, n.o.s.") or 25 pounds (for "Flammable Solid, n.o.s.").

Place all information on outside shipping container as on can (or bottle), specifically:

- Proper shipping name.
- UN or NA number.
- Proper label(s).
- Addressee and addressor.

Note that the previous two steps are EPA *recommendations* and is a DOT *requirement*.

Print *"Laboratory Samples"* and *"This End Up"* or *"This Side Up"* clearly on top of shipping container. Put upward pointing arrows on all four sides of container.

Shipping Papers. Use abbreviations only where specified.

Complete carrier-provided bill of lading and sign certification statement (if carrier does not provide, use the standard industry form). Provide the following information in the order listed (one form may be used for more than one exterior container):

- *"Flammable Liquid, n.o.s. UN1993"* or *"Flammable Solid, n.o.s. UN1325"*
- *"Limited Quantity"* (or *"Ltd. Qty."*)
- Net weight or net volume ("weight" or "volume" may be abbreviated) just before or just after "Flammable Liquid, n.o.s. UN1993" or "Flammable Solid, n.o.s. UN1325."
- Further descriptions such as "Laboratory Samples" or "Cargo Aircraft Only" (if applicable) are allowed if they do not contradict required information.

Include *chain-of-custody record*, properly executed, in outside container if legal use of samples is required or anticipated.

Transportation. Transport unknown hazardous substance samples classified as flammable liquids by rented or common carrier truck, railroad, or express overnight package services.

Do not transport by any passenger-carrying air transport system, even if they have cargo only aircraft. DOT regulations permit regular airline cargo only aircraft, but difficulties with most suggest avoiding them. Instead, ship by airlines that only carry cargo.

Transport by government-owned vehicle, including aircraft. DOT regulations do not apply, but EPA personnel will still use procedures described except for execution of the bill of lading with certification.

Other Considerations.

- Check with analytical laboratory for size of sample to be collected and if sample should be preserved or packed in ice.
- For EPA employees, accompany shipping containers to carrier and, if required, open outside container(s) for inspection.

Procedures: Samples Classified as Poison "A"

Packaging. Collect samples in a polyethylene or glass container with an outer diameter narrower than the valve hole on a DOT specification #3A1800 or #3AA1800 metal cylinder. To prevent leakage, fill container no more than 90% full (at 130°F). Seal sample container. Complete sample identification tag and attach securely to sample container.

Attach string or flexible wire to neck of the sample container; lower it into metal cylinder partially filled with noncombustible, absorbent cushioning material (for example, diatomaceous earth or vermiculite). Place only one container in a metal cylinder. Pack with enough absorbing material between the bottom and sides of the sample container and the metal cylinder to prevent breakage and absorb leakage. After the cushioning material is in place, drop the end of the string or wire into the cylinder valve hole.

Replace valve, torque to 250 foot-lbs (for 1-inch opening), and replace valve protector on metal cylinder, using Teflon tape. Place one or more cylinders in a sturdy outside container.

Marking/Labeling. Use abbreviations only where specified. Place the following information, either hand printed or in label form, on the side of the cylinder or on a tag wired to the cylinder valve protector:

- "*Poisonous Liquid, n.o.s. NA1955,*" or "*Poisonous Gas, n.o.s. NA1955.*"
- Laboratory name and address.
- DOT label "*Poisonous Gas*" (even if sample is liquid) on cylinder. Put all information on metal cylinder on outside container.

Print "*Laboratory Sample*" and "*Inside Packages Comply With Prescribed Specifications*" on top and/or front of outside container. Mark "*This Side Up*" on top of container and upward-pointing arrows on all four sides.

Shipping Papers. Use abbreviations only as specified. Complete carrier-provided bill of lading and sign certification statement (if carrier does not provide, use standard industry form). Provide the following information in the order listed (one form may be used for more than one exterior container):

- *"Poisonous Liquid, n.o.s. NA1955"*
- Net weight or net volume ("weight" or "volume" may be abbreviated), just before or just after "Poisonous Liquid, n.o.s. NA1955"

Include a *chain-of-custody record*, properly executed, in container or with cylinder if legal use of samples is required or anticipated.

For EPA employees, accompany shipping container to carrier and, if required, open outside container(s) for inspection.

Transportation. Transport unknown hazardous substance samples classified as Poison A only by ground transport or by Government-owned aircraft. Do not use air cargo, other common carrier aircraft, or rented aircraft.

Sample Identification

The sample tag is the means for identifying and recording the sample and the pertinent information about it. The sample tag should be legibly written and completed with an indelible pencil or waterproof ink. The information should also be recorded in a log book. The tag should be firmly affixed to the sample container. As a minimum, it should include:

- Exact location of sample.
- Time and date sample was collected.
- Name of sampler and witnesses (if necessary).
- Project codes, sample station number, and identifying code (if applicable).
- Type of sample (if known).
- Hazardous substance or environmental sample.
- Tag number (if sequential tag system is used).
- Laboratory number (if applicable).
- Any other pertinent information.

REFERENCES

Windholz, M.; Budavori, S.; Blumetti, R.; and Otterbein, E. *The Merck Index.* 1983. Rahway, NJ: Merck & Company.

U.S. EPA, Office of Emergency and Remedial Response. *Sampling For Hazardous Materials.* Washington, DC: U.S. Govt. Printing Office.

U.S. DOT. CFR 49 Section 173. Washington, DC. U.S. Govt. Printing Office.

11

Groundwater Principles and Monitoring Considerations

INTRODUCTION

[In terms of continued existence, environmental quality, and economic development, a plentiful supply of relatively pure water is of the utmost importance to man. For this reason, groundwater quality is given much attention prior to, during, and after remedial activities on hazardous waste sites. It is vital that groundwater monitoring be conducted prior to the beginning of cleanup operations in order to delineate the degree and extent of contamination, so that effective remedial actions can be planned. Monitoring is also important during and after cleanup operations so that the effectiveness of those operations can be assessed. Thus, it is important that workers involved in activities such as well installation and sampling have an understanding of the basic principles of groundwater and groundwater monitoring.] This chapter is designed to fill that need. However, the reader should realize that groundwater is a very complex subject. The information presented here is intentionally abbreviated and generalized, so as to be appropriate to the scope of this text.

OBJECTIVES

- Know and understand various factors affecting the occurrence, distribution, movement, properties, and environmental characteristics of subsurface waters.
- Be aware of potential sources of groundwater contamination and complexities involved in predicting contaminant migration.
- Be aware of general considerations for well installation and operation in conjunction with monitoring operations on hazardous wastes sites.
- Appreciate the complexities involved in effective groundwater monitoring and the importance of following proper procedures.
- Become familiar with specific procedures for purging and collecting samples from monitoring wells.

CHARACTERISTICS OF GROUNDWATER SYSTEMS

Groundwater Hydrology

Hydrology is the study of the occurrence, distribution, movement, and properties of the waters of the earth. The science of hydrology can be generally subdivided into *surface water hydrology* which involves the waters of the earth's surface, such as streams, lakes, and oceans, and *groundwater hydrology* which involves the waters beneath the earth's surface (Bates and Jackson, 1980; Fetter, 1980). While groundwater hydrology is the primary focus of this chapter, it should be noted that surface water and groundwater are closely related, as shown in Fig. 11-1.

The Hydrologic Cycle

The hydrologic cycle is a convenient concept for describing the various changes which the waters of the earth may undergo. Although the cycle has no beginning or end, it is convenient to assume that the cycle begins with precipitation and progresses through a number of steps, as shown in Fig. 11-1.

After *precipitation* falls to the earth's surface, some of the water seeps into the ground through the process of *infiltration*. If the rate of precipitation exceeds the rate of infiltration, the excess water flows as *runoff* across the earth's surface and into streams or other surface water bodies. Some infiltrating water clings

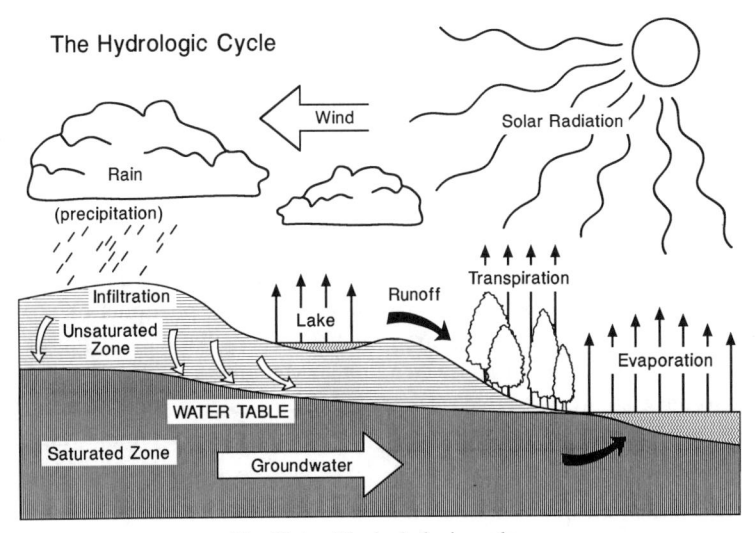

Fig. 11-1. The hydrologic cycle.

to soil particles. This water may be drawn into the rootlets of plants, used by the plants, and emitted as vapor into the atmosphere through the process of *transpiration*. Excess soil moisture is pulled downward by gravity. Upon reaching the *water table* (see Fig. 11-1), the water is considered part of the *groundwater system*. (Viessman et al., 1977; Fetter, 1980; Barfield et al., 1981).

Groundwater flows through rock and soil units of the earth to discharge at springs or as seepage into streams, lakes, or oceans, thus becoming *surface water*. The cycle is completed as *evaporation* from surface water bodies sets the stage for renewed precipitation.

Water Beneath the Earth's Surface

With regard to hydrology, the subsurface can be divided into two zones, the saturated zone and the unsaturated zone (see Fig. 11-1). *The unsaturated zone* is located above the water table. Pores or void spaces within this zone contain air and water vapor in addition to water, so that pore water pressure is less than atmospheric pressure. *The saturated zone* is located beneath the water table. All pores within this zone are filled with water, and pore water pressure is greater than atmospheric pressure. Water located in the saturated zone is referred to as *groundwater*. *The water table* is the surface along which pore water pressure is equal to atmospheric pressure. (Bates and Jackson, 1980).

Groundwater Flow

With regard to humid regions, some observations generally applicable to groundwater flow in unconfined aquifers (as defined below) can be made. As shown in Fig. 11-1, the water table generally slopes in the direction of groundwater flow, and has the same general shape as the earth's surface or topography, but tends to be flat in the absence of groundwater flow. Groundwater tends to flow away from topographic high areas (where recharge occurs) and toward topographic low areas (where discharge occurs).

Factors Affecting Groundwater Systems

The characteristics of groundwater systems are determined by a number of factors. "Natural" factors include; the amount and type of precipitation providing recharge, the topography of the area, and the characteristics of the geologic units through which the groundwater flows (Manning, 1988). The actions of man also have significant effects on groundwater systems. All of these factors must be considered when developing and implementing a groundwater monitoring program.

Properties of Geologic Units

The basic characteristics of a groundwater system are largely controlled by the properties of the geologic unit (or units) containing the groundwater. Major controlling properties are porosity and permeability. (Houston and Kasim, 1982; Manning, 1988).

Porosity refers to the percentage of pore space or void space within a geologic unit. Two types of porosity (as shown in Fig. 11-2) are *primary porosity*, which existed at the time of formation of the unit, and *secondary porosity*, which came into being after the unit was formed.

Permeability refers to the capacity of a geologic unit to transmit a fluid. This is a function of the degree of connection between pores or voids within the unit.

Hydraulic conductivity refers to the ease with which a material will conduct water as measured in gallons per day per square foot ($gal/day/ft^2$). This property is directly related to the porosity and permeability of the material involved.

As shown by Fig. 11-3, careful consideration of these types of properties is important when planning and executing a monitoring program to detect and evaluate subsurface contamination.

Aquifers

A geologic unit which is saturated with water and sufficiently porous and permeable to yield economically significant amounts of water is called an *aquifer*. Aquifers can be generally classified as either unconfined, confined, or perched aquifers.

Unconfined aquifers are relatively close to the surface, with continuous layers of high hydraulic conductivity extending from the land surface to the base of the aquifer (Fig. 11-4). Thus, the aquifer can receive recharge from the downward seepage of water through the overlying unsaturated zone. For this reason,

Types of Geologic Units

Primary Openings

Well-sorted Sand   Poorly-sorted Sand

Secondary Openings

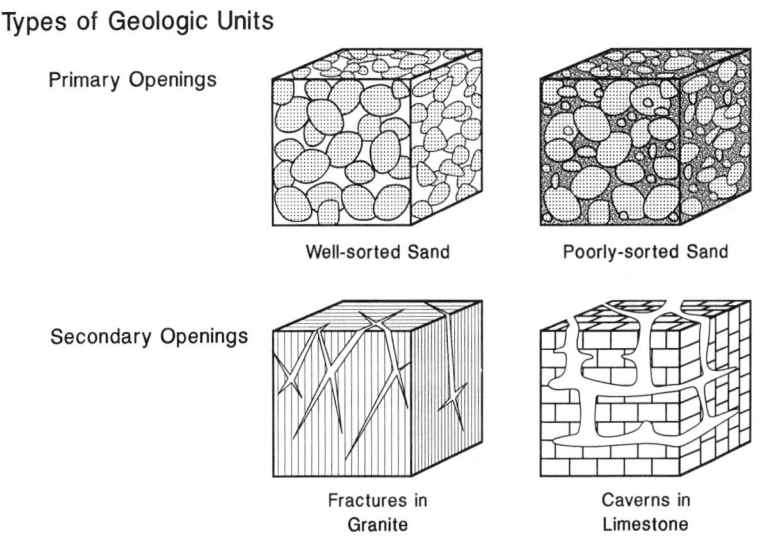

Fractures in Granite   Caverns in Limestone

Fig. 11-2. Types of geologic units.

Contaminant Migration through Fractured Porous Limestone

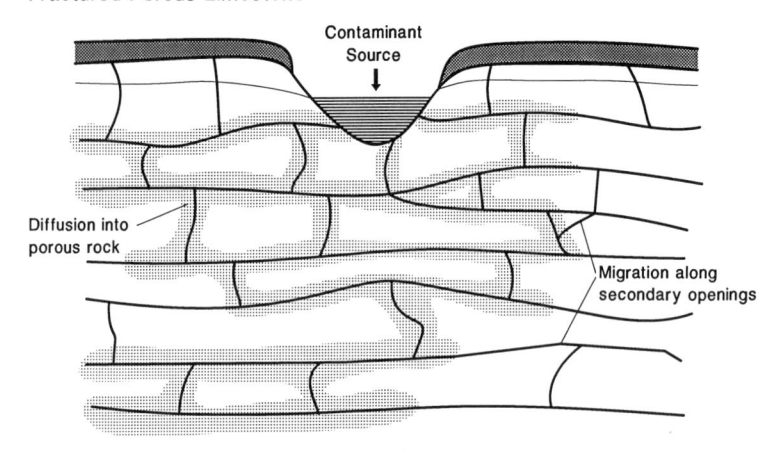

Fig. 11-3. Contaminant migration through fractured porous limestone.

confined aquifers can easily be locally contaminated by improper waste disposal. Note that if a tightly cased well is installed in an unconfined aquifer, the elevation of the water surface inside the well (i.e., the piezometric or potentiometric surface) will be that of the water table (Fig. 11-4).

Confined aquifers are overlain by a confining layer having little or no hydraulic conductivity. These aquifers tend to be recharged primarily at the area

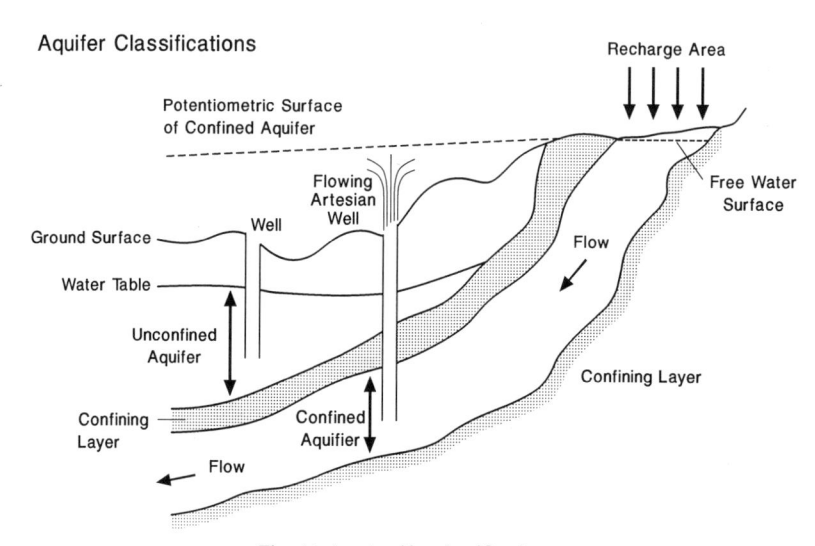

Fig. 11-4. Aquifer classifications.

of outcrop of the aquifer (Fig. 11-4). If a tightly cased well is installed in a confined aquifer, the water level in the well may rise under pressure to an elevation far above the top of the aquifer (i.e., to the elevation of the potentiometric or piezometric surface). This is referred to as an *artesian well*. If the potentiometric surface is above the land surface, water will flow from the well without the aid of a pump (see Fig. 11-4). Contaminants may enter confined aquifers at the area of recharge or through a leaky confining layer.

Perched aquifers form when water collects above a laterally discontinuous layer of low permeability at some distance above the main water table (Fig. 11-5) (Fetter, 1980; U.S. EPA, 1983).

Effects of Man's Activities on Hydrologic Systems

Modern technology allows man to dramatically change the characteristics of groundwater systems. With regard to this, two important factors, groundwater extraction and groundwater contamination, should be considered.

Extraction of groundwater from a well may alter the direction, as well as the amount, of groundwater flow. As shown by Fig. 11-6, pumping from a well tends to create a cone of depression. Water within the cone of depression flows toward the drawdown point at the well location. (Fetter, 1980; Viessman et al., 1977). Note that removal of groundwater may be directly related to groundwater contamination and therefore must be carefully considered with regard to groundwater monitoring.

Contamination of groundwater may result from any number of man's activities. All factors which may affect groundwater system characteristics (as discussed above) should be considered in establishing a groundwater sampling program. However, groundwater contamination, whether known or suspected, is the reason monitoring is required, and is also an important consideration for successful monitoring.

A Perched Aquifer

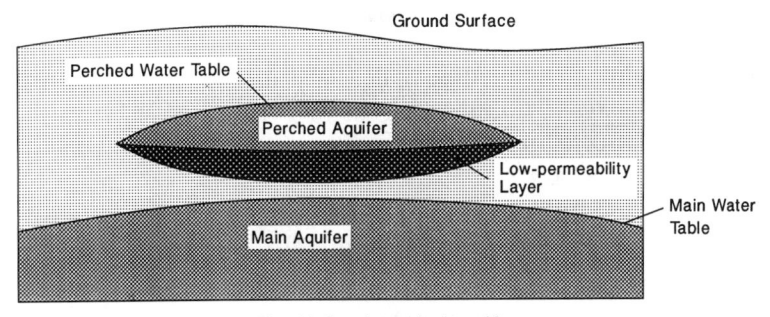

Fig. 11-5. A perched aquifer.

Contamination of Groundwater due to Induced Infiltration from a Contaminated Stream

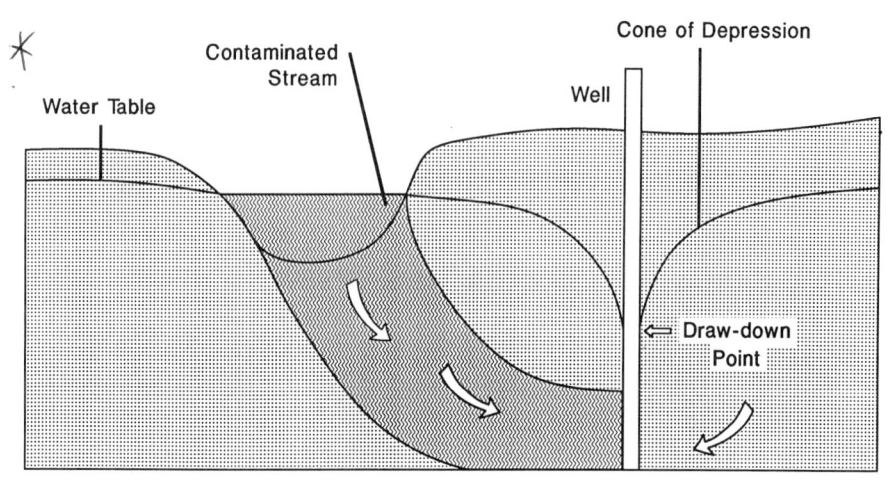

Fig. 11-6. Contamination of groundwater due to induced infiltration from a contaminated stream.

CONTAMINATION OF GROUNDWATER

Sources of Contamination

The rise of modern technology and related economic development has produced numerous potential sources of groundwater contamination. Identification of these sources is very important for monitoring purposes. Some contaminant sources are listed in Table 11-1. Note the predominance of hazardous waste-related sources of groundwater contamination.

Contamination due to improper waste disposal is illustrated in Fig. 11-7. At this site, infiltration has been enhanced by the excavation of the landfill, thus creating a recharge mound at the landfill site. Note that this has enhanced the leaching of contaminants out of the landfill and into the groundwater system. Current construction techniques for hazardous waste landfills are intended to prevent, or at least delay, this type of contamination.

Contamination due to leaking storage tanks is illustrated in Fig. 11-8. This is a common source of groundwater contamination both on and off hazardous waste sites.

Movement of Contaminants in Groundwater

Upon entering a groundwater system, the contaminant or solute forms a *contaminant plume* which begins to migrate within the system. The chemical com-

GROUNDWATER PRINCIPLES AND MONITORING CONSIDERATIONS 263

Table 11-1. Sources of Groundwater Contaminants (Manning, 1988).

Source	Potential Contaminants
Accidental spills	Inorganic and organic chemicals
Deep-well injection of wastes	Inorganic and organic compounds, radioactive materials, and radionuclides.
Hazardous waste disposal sites	Inorganic compounds (especially heavy metals) and organic compounds (such as pesticides and priority pollutants)
Industrial liquid-waste storage ponds and lagoons	Heavy metals, cleaning solvents, and degreasing compounds
Landfills, industrial	Inorganic and organic compounds
Land disposal of liquid and semi-solid industrial wastes	Organic compounds, heavy metals, and cleaning solvents and degreasers
Land disposal of municipal wastewater and waste sludges	Organic compounds, inorganic compounds, heavy metals, and microbiological contaminants
Municipal landfills	Heavy metals, gases, organic compounds, and inorganic compounds (such as calcium, chlorides, and sodium)
Underground storage tanks	Organic cleaning and degreasing compounds, petroleum products, and various other hazardous wastes

Water Table Recharge Mound and Groundwater Contamination at an Unlined Landfill Site

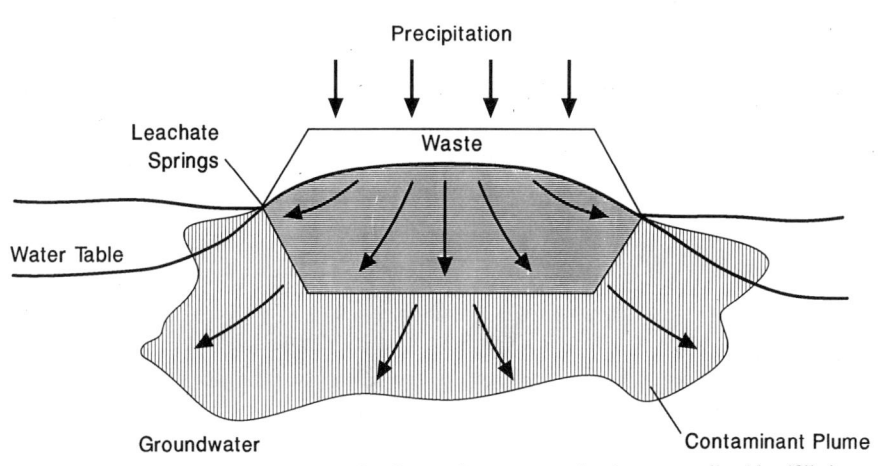

Fig. 11-7. Water table recharge mound and groundwater contamination at an unlined landfill site.

Groundwater Contamination

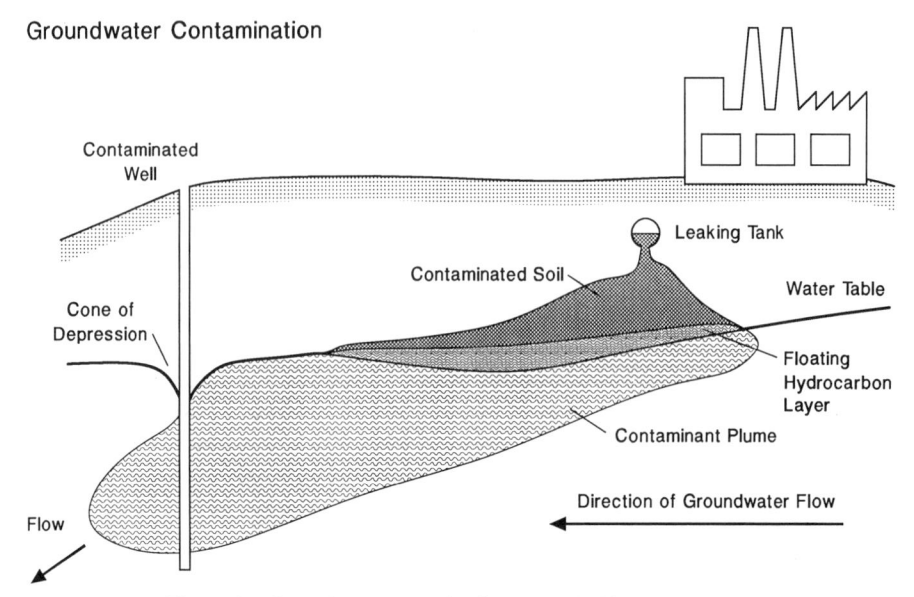

Fig. 11-8. Groundwater contamination due to leaking storage tanks.

position, shape, and rate of travel of a plume may be affected by a number of factors.

Factors Affecting Contaminant Migration. *Advection* is the process by which a chemical contaminant is transported by the bulk motion of groundwater. Thus the contaminant moves at a rate and in a direction equal to that of the groundwater.

Hydrodynamic dispersion is the tendency for a contaminant to deviate from the rate and path which would be expected based on advection alone. Hydrodynamic dispersion may occur due to factors such as:

- Complex mechanical or molecular forces operating within the system.
- The viscosity of the solute.
- Differing hydraulic conductivities of various rock layers in the aquifer.

Plume shapes may vary widely, as factors such as aquifer characteristics and dispersivities of contaminants vary both longitudinally (Fig. 11-9) and transverse (compare Figs. 11-8 and 11-10) to the direction of groundwater flow.

Sorption is the tendency for a solute to adhere to solid particles composing a geologic unit through which it passes. Differential sorption characteristics may act to separate the components of a multicontaminant plume (see Fig. 11-11).

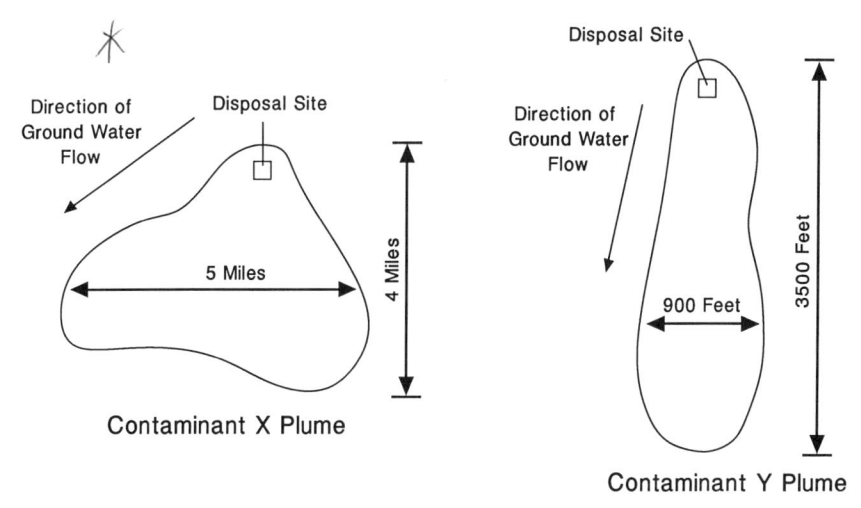

Fig. 11-9. Plume shape variations in plan view.

Biological decay of a contaminant may operate within a groundwater system to render the contaminant harmless. However, many contaminants are *persistent* and will not biodegrade. Also, biological decay may produce a contaminant which is more toxic than the original (as when trichloroethylene biodegrades into vinyl chloride) (Manning, 1988; Gillham and Cherry, 1982).

Migration of a Dense Contaminant Slug

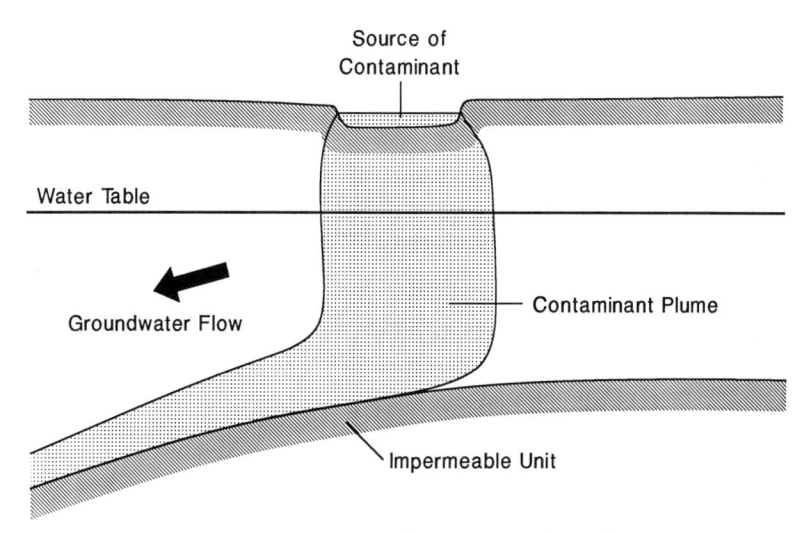

Fig. 11-10. Migration of a dense contaminant slug.

Effect of Differing Sorption Characteristics on a Multi-Contaminant Plume

A. Initial Plume Composition

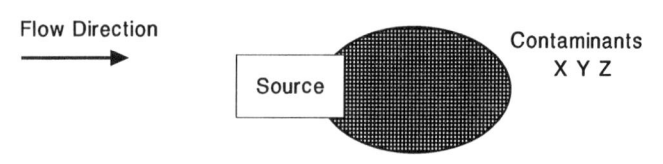

B. Plume Composition after Down-Gradient Migration

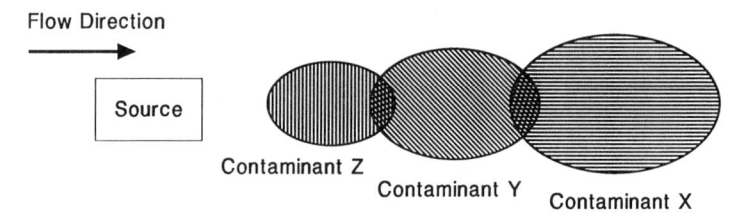

Fig. 11-11. Effect of differing sorption characteristics on a multicontaminant plume.

Predicting Contaminant Migration

Actual contaminant migration is the result of a complex interaction of all factors discussed above. The solute may move faster than, slower than, or at the same speed as the groundwater, and the resulting plume may assume any given shape and reach any given size, depending upon the specifics of the situation involved. Therefore, prediction of migration is a complex process requiring specific information pertaining to the characteristics of both the solute and the aquifer involved. Predictive modeling techniques are currently used to aid in the process, but are reportedly inaccurate for plumes in excess of 2,000 meters. Thus, extensive drilling and sampling are typically required to delineate or verify the extent of a contaminant plume.

SPECIFIC CONSIDERATIONS FOR WASTE SITE MONITORING WELLS

Remedial action on hazardous waste sites requires the installation and sampling of numerous wells, both onsite and offsite. The wells are required for monitoring and, in some cases, for treatment of groundwater contamination. As should be apparent from the information contained in this chapter, the installation and operation of these wells requires careful consideration of many specific and complex factors if desired objectives are to be achieved. Due to this, well installation and operation should be planned and supervised by personnel having extensive knowledge and experience in hydrology and related subjects. Furthermore, all employees involved in well installation and/or operation should be aware of considerations such as the following:

- Any material solid, liquid, or gas, which comes out of a drill hole or well may pose the danger of contamination of personnel. Thus, provisions of the site safety plan applicable to drilling and sample collection should be carefully followed, including the use of any required PPE. Due to the potential for contaminant migration, it is not safe to assume that samples collected off site are nonhazardous.
- Equipment decontamination procedures must be carefully followed between wells when drilling and sampling. Otherwise, crosscontamination of wells or samples may occur.
- Great care is required in executing well installation plans, since factors such as correct well location and depth may have a critical effect on whether or not a target plume is tapped.
- Great care should be exercised when it is possible that drilling may puncture an impermeable layer separating contaminated and uncontaminated aquifers, as crosscontamination may result.
- Overpumping of wells may unfavorably alter factors such as the direction of groundwater flow or contaminant migration (EPA, 1983).

The *Procedures Manual for Groundwater Monitoring at Solid Waste Disposal Facilities"* (EPA, 1977) provides detailed information on installation of wells and other information needed for establishing a monitoring program.

GROUNDWATER SAMPLING PROCEDURES

As in all sampling activities on hazardous waste sites, sampling for the purpose of groundwater monitoring requires extreme care if samples obtained are to be truly representative of actual site conditions (EPA, 1977; EPA, 1983). Given the inherent complexities of hydrology and contaminant migration as previously discussed, it should be obvious that very specific procedures must be used if valid samples are to be collected. For this reason, the site sampling plan must include specific procedures for groundwater sample collection, and all personnel involved in groundwater monitoring must follow those procedures to the letter.

A number of factors related to the monitoring well itself must be considered for monitoring purposes. These factors include well depth, casing diameter, casing material, type and size of screen (if used), vertical position of well screen or slotted section of casing, and type of backfill or packing used around casing (Fig. 11-12). All of this information must be obtained (for example, from driller logs) before monitoring begins.

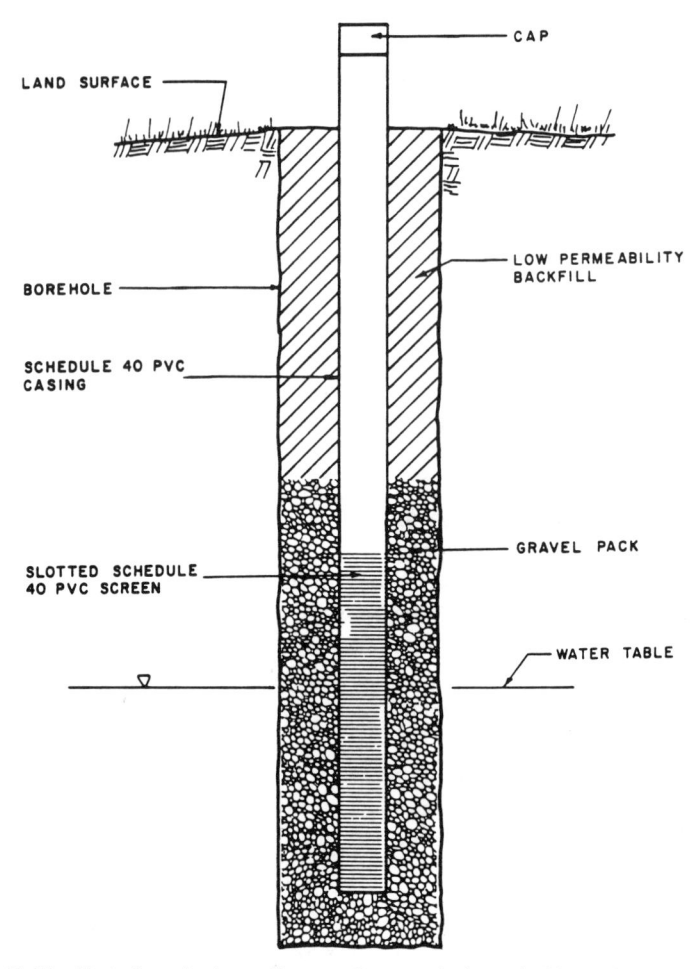

268   WORKER PROTECTION DURING HAZARDOUS WASTE REMEDIATION

Fig. 11-12. Typical monitoring well screened over a single vertical interval (EPA, 1977).

Purging Monitoring Wells Prior to Sampling

It is important for personnel responsible for sample collection to realize that the composition of water standing in and immediately surrounding a well is generally not representative of overall groundwater quality at the sampling site. This is because of the presence of stagnant water, and possibly drilling wastes or other extraneous contaminants, in and around the casing. Thus, it is critical to purge the well prior to sampling by pumping or bailing to flush standing water out of the casing and pull a sufficient amount of "fresh" water in from the aquifer (EPA, 1977; EPA, 1983).

The amount of water which is removed from the well during purging is important. The number of bore-volumes to be removed prior to sampling should be clearly stated in the sampling plan. Recommendations vary from two to ten bore-volumes. The bore volume can be calculated from the formula;

$$V = W.D. \times (C.R.)^2 \times 0.163$$

where

V = volume of water in well (gallons)
$W.D.$ = depth of water in well (feet)
$C.R.$ = radius of well casing (inches) (EPA, 1989).

Note that water depth can be determined readily by measuring the distance from the top of the well to the water surface and subtracting that distance from the total depth of the well.

An alternative to the bore-volume removal method for well purging is to monitor the water level in the well, and the temperature, conductivity, and pH of the water removed from the well during purging. These factors should change progressively as the well is purged. When these factors stabilize, the well can be considered sufficiently purged.

The depth at which water is removed from the well is an important aspect of the purging process. The most complete purging reportedly occurs if water is pumped from just below the water surface, rather than from some greater depth.

The rate at which wells are purged is also significant. If pumps are used to purge monitoring wells, it is important to avoid overpumping and excessive draw-down. Otherwise, nonrepresentative samples may result from changes in groundwater flow.

The disposal of purge water may present problems in some instances. If contamination is suspected, it is recommended that the purge water be containerized and stored until samples are analyzed. The purge water can then be properly treated or disposed of, once the identity and concentration of any contaminants present are determined.

Equipment Used for Removing Groundwater

Various types of equipment have been used to extract groundwater from monitoring wells, both for purging and sample collection. Three of these devices—the peristaltic pump, the gas pressure displacement system, and the bucket-type bailer—are described in detail in the section on specific purging and sampling procedures, below. It should be noted that submersible pumps are also used. Submersibles do an excellent job of removing large volumes of water at depth. However, they are frequently expensive, awkward, too large to fit small diameter wells, and difficult to decontaminate in the field (EPA, 1977).

Collecting Groundwater Samples

Many of the same rules applicable to collecting samples from other sources on the hazardous waste site also apply when groundwater samples are collected. For example, procedures for use of appropriate sample containers, avoidance of crosscontamination, proper labeling, and chain-of-custody documentation are important despite the type of substance being sampled. These types of general considerations for sample collection are covered in Chapter 10, and will not be repeated here. In many instances, the same equipment which is used to purge a monitoring well can also be used for collecting samples. Procedures for using this equipment in sample collection are given in the following section.

Specific Purging and Sampling Procedures

Purging with a Gas Pressure Displacement System. A pressure displacement system consists of a chamber equipped with a gas inlet line, a water discharge line and two check valves (see Fig. 11-13). When the chamber is lowered into the casing, water floods it from the bottom through the check valve. Once full, a gas (for example nitrogen or air) is forced into the top of the chamber under sufficient pressure to result in the upward displacement of the water out the discharge tube. The check valve in the bottom prevents water from being forced back into the casing and the upper check valve prevents water from flowing back into the chamber when the gas pressure is released. This cycle can be repeated as necessary until purging is complete (EPA, 1977).

The pressure lift system is particularly useful when the well depth is beyond the capability of a peristaltic pump. The water is displaced up the discharge tube by the increased gas pressure above the water level. The potential for increased gas diffusion into the water makes this system unsuitable for sampling for volatile organic or most pH-critical parameters.

When purging with the gas pressure displacement system, the following procedures should be used (EPA, 1977):

- Using clean uncontaminated equipment (such as an electronic level indicator) determine the water level in the well, then calculate the fluid volume in the casing.
- Determine depth to midpoint of screen or well section open to aquifer (consult driller's log).
- Lower displacement chamber until top is just below water level.
- Attach gas supply line to pressure adjustment valve on cap.
- Gradually increase gas pressure to maintain discharge flow rate.
- Measure rate of discharge frequently. A gallon bucket and stopwatch are usually sufficient.

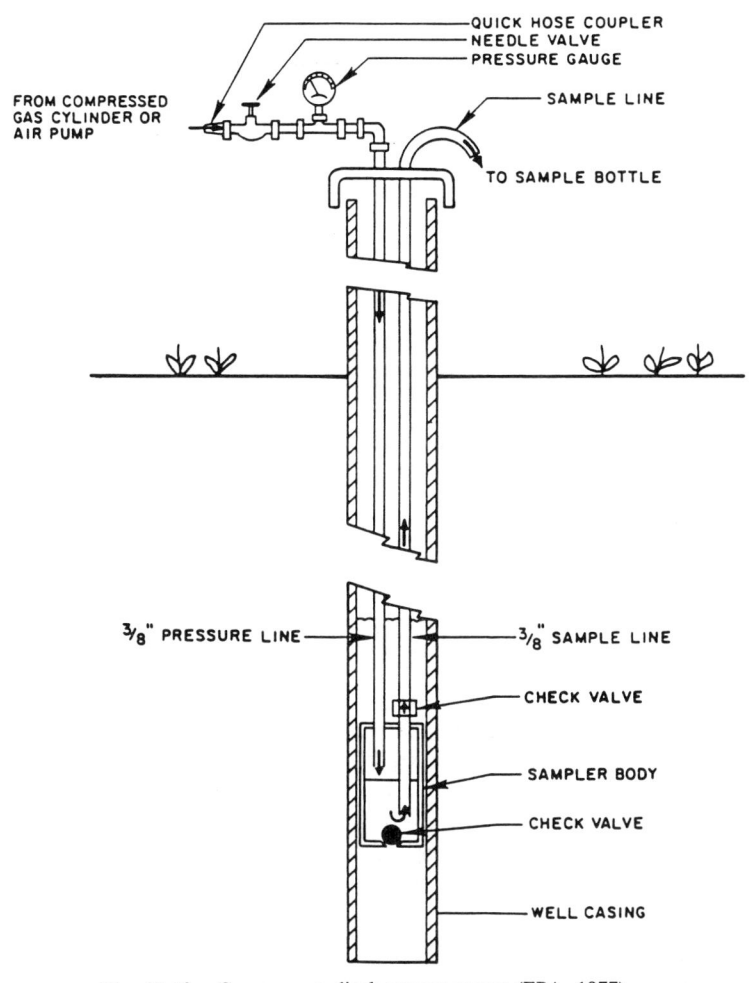

Fig. 11-13. Gas pressure displacement system (EPA, 1977).

- Purge a minimum of two casing volumes or until discharge characteristics stabilize (see discussion on well purging).
- After pumping, monitor water level recovery. Recovery rate may be useful in determining sample rate.

Purging with a Peristaltic Pump. This system consists of a peristaltic pump (Fig. 11-14) capable of achieving a pump rate of 1–3 liters per minute, and an assortment of Teflon® tubing for extending the suction intake (EPA, 1977). A battery operated pump is preferable as it eliminates the need for DC generators or AC inverters.

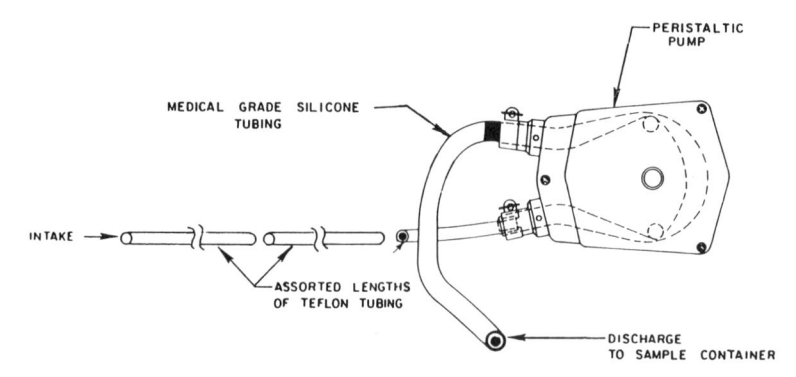

Fig. 11-14. Peristaltic pump for liquid sampling (EPA, 1977).

The use of a peristaltic pump for well purging is particularly advantageous, since the same system can later be utilized for sample collection. The application, however, is limited to wells with a depth of less than approximately 8 meters, due to the limited lift capabilities of peristaltic action.

When purging with the peristaltic pump, the following procedures should be used (EPA, 1977):

- Using clean equipment, sound well for total depth and water level, then calculate the fluid volume in the casing ("casing volume").
- Determine depth from casing top to midpoint of screen or well section open to aquifer. (Consult driller's log or sound for bottom.)
- If depth to midpoint of screen is in excess of 8 meters, choose alternate system.
- Lower intake into the well to a short distance below the water level and begin water removal. Collect or dispose of purged water in an acceptable manner. Lower suction intake, as required, to maintain submergence.
- Measure rate of discharge frequently. A gallon bucket and stopwatch are most commonly used.
- Purge a minimum of two casing volumes or until discharge, *p*H, temperature, or conductivity stabilize. See discussion on well purging.
- After pumping, monitor water level recovery. Recovery rate may be useful in determining sample rate.

Sampling Monitor Wells with a Peristaltic Pump.

A pump system is considerably advantageous when analytical requirements demand sample volumes in excess of several liters. The major drawback of a pump system is the potential for increased volatile component stripping as a result of the required lift vacuum. Samples for volatile organic analysis should be collected with a bailer as described below and should precede any sample collection which may further disturb the well bore content (Dunlap et al., 1977).

The peristaltic pump system can be used for monitor well sampling whenever the lift requirements do not exceed 8 meters. It becomes particularly important to use a heavy wall tubing in this application in order to prevent tubing collapse under the high vacuums needed for lifting from depth.

When sampling with the peristaltic pump, the following procedures should be used (Dunlap et al., 1977):

- Using clean, uncontaminated equipment, i.e., an electronic level indicator (avoid indicating paste), determine the water level in the well, then calculate the fluid volume in the casing.
- Purge well (as previously discussed).
- If soundings show suffcient level of recovery, prepare pump system. If insufficient recovery is noted, allow additional time to collect samples on a periodic schedule which will allow for recovery between sampling.
- Collect volatile organic analysis samples (if required) with bucket-type bailer.
- Install clean medical-grade silicon tubing in peristaltic pump head.
- Attach pump to required length of Teflon® suction line and lower to midpoint of well screen (if known), or slightly below existing water level.
- Consider the first liter of liquid collected as system purge/rinse. *Note:* If well yield is insufficient for required analysis this purge volume may be suitable for some less critical analysis.
- Fill necessary sample bottles by allowing pump discharge to flow gently down the side of bottle with minimal entry turbulence. Cap each bottle as filled.
- Preserve the sample if necessary as per guidelines (see Chapter 10).
- Check that a Teflon® liner is present in cap if required. Secure the cap tightly.
- Label the sample bottle with an appropriate tag. Be sure to complete the tag with all necessary information. Complete chain-of-custody documents and field log book entries.
- Place the properly labeled sample bottle in an appropriate carrying container maintained at 4°C throughout the sampling and transportation period.
- Allow system to drain, then disassemble. Return tubing to lab for decontamination.

Sampling Monitor Wells with a Bucket-Type Bailer.

Bucket-type bailers are tall, narrow buckets equipped with a check valve on the bottom. This valve allows water to enter from the bottom as the bailer is lowered, then prevents its release as the bailer is raised (see Fig. 11-15).

This device is particularly useful when samples must be recovered from depths greater than the range (or capability) of suction lift pumps, when volatile strip-

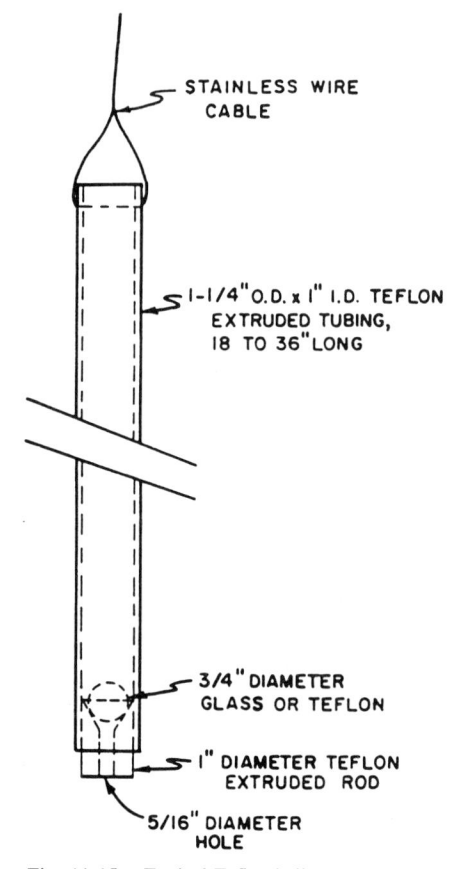

Fig. 11-15. Typical Teflon bailer (EPA, 1977).

ping is of concern, or when well casing diameters are too narrow to accept submersible pumps (Dunlap et al., 1977). It is the method of choice for the collection of samples which are susceptible to volatile component stripping or degradation, due to the aeration associated with most other recovery systems. Samples can be recovered with a minimum of aeration if care is taken to gradually lower the bailer until it contacts the water surface and is then allowed to sink as it fills. The primary disadvantages of bailers are their limited sample volume and inability to collect discrete samples from a depth below the water surface.

When collecting samples with a bailer, the following procedures should be used (Dunlap et al., 1977):

- Using clean, uncontaminated equipment, i.e., an electronic level indicator (avoid indicating paste), determine the water level in the well, then calculate the fluid volume in the casing.

- Purge well (as previously discussed).
- Attach bailer to cable or line for lowering.
- Lower bailer slowly until it contacts water surface.
- Allow bailer to sink and fill with a minimum of surface disturbance.
- Slowly raise bailer to surface. Do not allow bailer line to contact ground.
- Tip bailer to allow slow discharge from top to flow gently down the side of the sample bottle with minimum entry turbulence.
- Repeat the above steps as needed to acquire sufficient volume.
- Preserve the sample, if necessary, according to guidelines.
- Check that a Teflon® liner is present in cap if required. Secure the cap tightly.
- Label the sample bottle with an appropriate tag. Be sure to complete the tag with all necessary information. Record the information in the field logbook and complete all chain-of-custody documents.
- Place the properly labeled sample bottle in an appropriate carrying container maintained at 4°C through the sampling and transportation period.
- Rinse bailer with deionized water before reuse. In some cases, especially where trace analysis is desired, it may be prudent to use a separate bailer for each well or, if possible, to thoroughly decontaminate the bailer after each use according to specific laboratory instructions. After use, place in plastic bag for return to lab.

REFERENCES

Barfield, B., Warner, R., and Haan, C. 1981. *Applied Hydrology and Sedimentology for Disturbed Areas*. Stillwater, OK: Oklahoma Technical Press.

Bates, R., and Jackson, J., Editors. 1980. *Glossary of Geology*, 2nd Ed. Falls Church, VA: American Geological Institute.

Dunlap, W., McNabb, J., Scalf, M., and Crosby, R. 1977. *Sampling for Organic Chemicals and Microorganisms in the Subsurface*. U.S. Environmental Protection Agency 600/2-77-176. Washington, DC: U.S. Govt. Printing Office.

EPA. 1977. *Procedures Manual for Groundwater Monitoring at Solid Waste Disposal Facilities*. U.S. Environmental Protection Agency 530/SW-611. Washington, DC: U.S. Govt. Printing Office.

EPA. 1983. *Procedures Manual for Groundwater Monitoring at Solid Waste Disposal Facilities*. Office of Water and Waste Management, U.S. Environmental Protection Agency SW-611. Washington, DC: U.S. Govt. Printing Office.

Fetter, C. W., Jr. 1980. *Applied Hydrogeology*. Columbus, OH: Charles E. Merril Publishing Co.

Gillham, R., and Cherry, J. 1982. Contaminant Migration in Saturated Unconsolidated Geologic Deposits. In *Recent Trends in Hydrogeology*, pp. 31-62. Geological Society of America Special Paper 189, Narasimhan, T., Editor. Boulder, CO: Geological Society of America.

Houston, W., and Kasim, A., 1982. Physical Properties of Porous Geologic Materials. In *Recent Trends in Hydrogeology*, pp. 143-162. Geological Society of America Special Paper 189, Narasimhan, T., Editor. Boulder, CO: Geological Society of America.

Manning, J. 1988. Lecture to author, January 1988.

Viessman, W., Jr., Knapp, J., Lewis, G., and Harbaugh, T. 1977. *Introduction to Hydrology*, 2nd Ed. New York, NY: Harper and Row.

12
Emergency Procedures

INTRODUCTION

Emergency response procedures are a necessary part of a safe hazardous waste remediation project. These procedures should focus on any emergency occurring on the immediate site that is being cleaned up and in the communities surrounding the site. EPA governs emergencies affecting the community, while OSHA regulates emergencies at the remediation site.

After reviewing these regulatory standards, procedures for fire fighting, protection, and spill containment are discussed. Practical suggestions are made concerning these activities for best implementation. Finally, medical emergency response is discussed only briefly, because it is our belief that the American Red Cross should be utilized as a resource for this type of training, they have many reference books and manuals available at reasonable prices.

OBJECTIVES

- To become familiar with Federal regulations concerning required procedures and workers' responsibilities in the event of an emergency.
- To introduce the main types of emergencies, what hazards they pose, and what workers can do to combat them.
- To introduce first-aid actions for injuries which can occur at an incident.

LAWS

Community Emergencies

Superfund Amendments Reauthorization Act (SARA) Title III, Emergency Planning and Community Right to Know. Before the 1986 Superfund Amendments and Reauthorization Act (SARA), there were no mandatory requirements for local governments, industries, or workers regarding hazardous substance emergencies. The Environmental Protection Agency had a voluntary Chemical Emergency Preparedness Program (CEPP), but after Union Carbide's disasters in Bhopal, India and Institute, West Virginia, it became apparent that this program was not effectively addressing the problem.

In response, the SARA legislation established specific requirements for Federal, State, and local governments and industry to plan, notify, and report on

EMERGENCY PROCEDURES

hazardous and toxic chemicals storage, emissions, and spills. Fig. 12-1 describes the intended flow of assistance and communication in the emergency contingency plan, from the Federal through the State and local levels. Fig. 12-2 also describes, in flow chart fashion, what sections of Title III apply to whom, from the Federal level down to the facility which must comply. It also addresses what emergency plans, procedures, and training workers must receive in order to operate more safely. The following is an outline of the relevant sections of the law and their provisions.

SARA Title III Sections 301–303, Emergency Planning.

These sections are designed to develop government response and preparedness capabilities. In SARA, Local Emergency Planning Committees (LEPC) are designated and emergency response or contingency plans developed.

The plan must include:

- identification of facilities
- transportation routes of hazardous substances to and from facilities
- emergency response procedures
    i. on site
    ii. off site
- designation of coordinators
    i. at facility
    ii. in community to implement plan

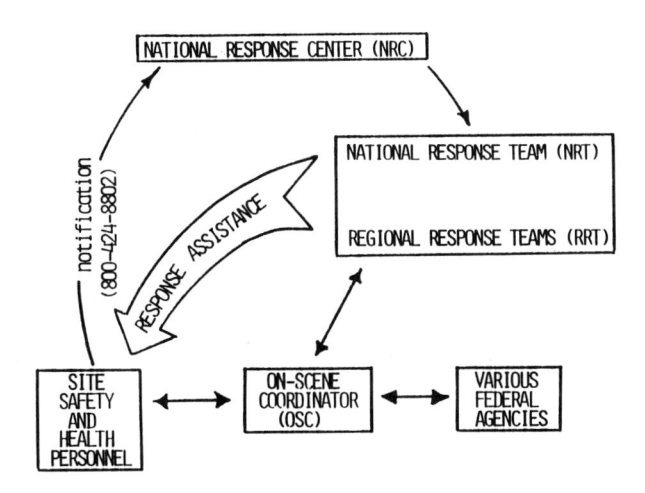

Fig. 12-1. The National Contingency Plan (NCP) response procedure.

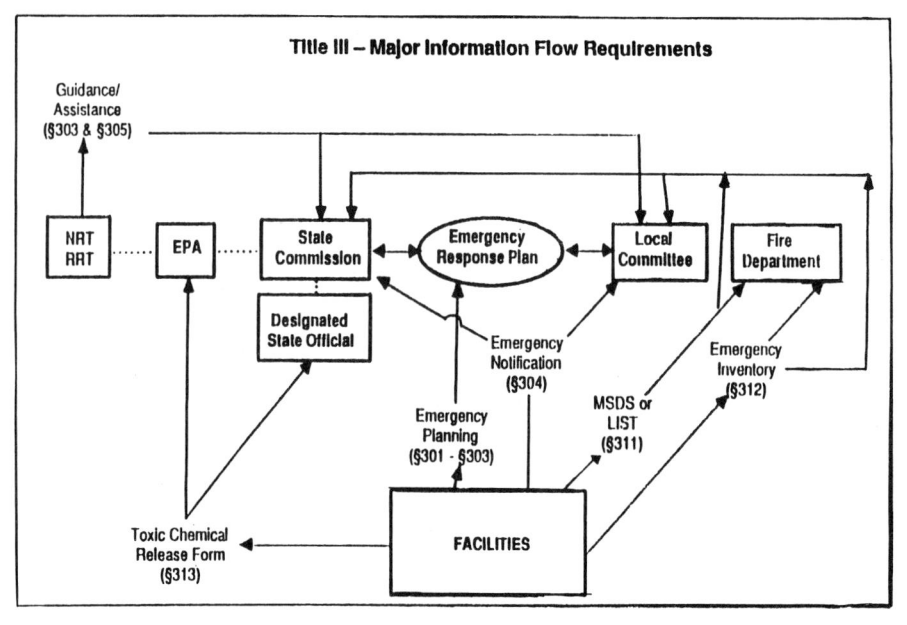

278 WORKER PROTECTION DURING HAZARDOUS WASTE REMEDIATION

Fig. 12-2. SARA Title III activities flow chart.

- emergency notification procedures
- methods of determining
    i. how release occurred
    ii. what area was affected
    iii. what population was affected
- who is responsible for facility or community owned emergency equipment
- evacuation plans
- information concerning response personnel training
- methods and schedules for exercising the plan

The NRT, or National Response Team, has published the *Hazardous Materials Emergency Planning Guide* to assist local planners in constructing their contingency plans. After their implementation these plans must be reviewed at least annually by the State Commission and local committee, with the assistance of the Regional Response Team. The focus of planning will include the type and amounts of substances found on the Extremely Hazardous Substances (EHS) list. If your site contains substances on this list in quantities exceeding the "threshold planning quantity" (TPQ) then you will need to know how to deal with these substances in the event of an emergency.

SARA Title III Section 304—Emergency Notification. Any release of an extremely hazardous substance exceeding its "reportable quantity" (RQ) must be reported to the Local Emergency Planning Commission (LEPC), the State Emergency Response Commission (SERC), and the National Response Center (NRC) immediately. This emergency notification must include:

- The chemical name.
- Whether or not it is extremely hazardous.
- How much was released.
- The time and duration of the release.
- What the chemical spilled into.
- Possible health risks.
- Precautions such as evacuation.
- Name and number of contact person.

A followup written notice must be made confirming actions taken during the emergency.

SARA Title III Section 305. This section establishes $5 million/year through 1990 for training grants so workers are better qualified to handle emergencies. The grants are administered through the Federal Emergency Management Agency (FEMA). The EPA is required to review emergency systems to determine their effectiveness.

SARA Title III Sections 311–312—Community Right-to-Know Reporting Requirements. Section 311 requires facilities to have Material Safety Data Sheets (MSDS) for chemicals it has, and to provide MSDS copies or a list of the chemicals to the Local Emergency Planning Commission (LEPC), the State Emergency Response Commission, and the local fire department. Section 312 requires submission of a chemical inventory form (known as Tier I or Tier II) to the above groups, describing the amounts of hazardous chemicals stored, used, and their locations. Upon request, further information must be submitted to a commission, department, or the public.

Section 313. Establishes an EPA inventory of toxic chemicals emissions along with a form (EPA Form R) for certain facilities (depending on the emission size and quantities of hazardous chemicals used). These emissions are reported so that government officials and the public are aware and may evaluate potential environmental damage that may result.

SARA Title III Section 322. Addresses trade secrets of chemical formulations which companies do not wish to divulge in the Material Safety Data Sheet (AEMA, 1987).

Figures 12-1 and 12-2 describe the intended sequence in which the law must be followed, and Table 12-1 the Extremely Hazardous Substances (EHS) list, with the RQs and TPQs of each substance.

Table 12-1. List of Extremely Hazardous Substances and Their Substances and Threshold Planning Quantities

CAS No.	Chemical name	Notes	Reportable quantity* (pounds)	Threshold planning quantity (pounds)
75-86-5	Acetone Cyanohydrin	e	10	1,000
1752-30-3	Acetone Thiosemicarbazide	e	1	1,000/10,000
107-02-8	Acrolein		1	500
79-06-1	Acrylamide	d, l	5,000	1,000/10,000
107-13-1	Acrylonitrile	d, l	100	10,000
814-68-6	Acrylyl Chloride	e, h	1	100
111-69-3	Adiponitrile	e, l	1	1,000
116-06-3	Aldicarb	c	1	100/10,000
309-00-2	Aldrin	d	1	500/10,000
107-18-6	Allyl Alcohol		100	1,000
107-11-9	Allylamine	e	1	500
20859-73-8	Aluminum Phosphide	b	100	500
54-62-6	Aminopterin	e	1	500/10,000
78-53-5	Amiton	e	1	500
3734-97-2	Amiton Oxalate	e	1	100/10,000
7664-41-7	Ammonia		100	500
16919-58-7	Ammonium Chloroplatinate	a, e	1	10,000
300-62-9	Amphetamine	e	1	1,000
62-53-3	Aniline	d, l	5,000	1,000
88-05-1	Aniline, 2,4,6-Trimethyl-	e	1	500
7783-70-2	Antimony Pentafluoride	e	1	500
1397-94-0	Antimycin A	c, e	1	1,000/10,000
86-88-4	ANTU	d	100	500/10,000
1303-28-2	Arsenic Pentoxide	d	5,000	100/10,000
1327-53-3	Arsenous Oxide	d, h	5,000	100/10,000
7784-34-1	Arsenous Trichloride	d	5,000	500
7784-42-1	Arsine	e	1	100
2642-71-9	Azinphos-Ethyl	e	1	100/10,000
86-50-0	Azinphos-Methyl		1	10/10,000
1405-87-4	Bacitracin	a, e, m	1	10,000
98-87-3	Benzal Chloride	d	5,000	500
98-16-8	Benzenamine, 3-(Trifluoromethyl)-	e	1	500
100-14-1	Benzene, 1-(Chloromethyl)-4-Nitro-	e	1	500/10,000
98-05-5	Benzenearsonic Acid	e	1	10/10,000
98-09-9	Benzenesulfonyl Chloride	a	100	10,000
3615-21-2	Benzimidazole, 4,5-Dichloro-2-(Trifluoromethyl)-	e, g	1	500/10,000
98-07-7	Benzotrichloride	d	1	100
100-44-7	Benzyl Chloride	d	100	500
140-29-4	Benzyl Cyanide	e, h	1	500
15271-41-7	Bicyclo[2.2.1]Heptane-2-Carbonitrile, 5-Chloro-6-(((Methylamino)Carbonyl)Oxy)Imino)-, (1s-(1-alpha, 2-beta, 4-alpha, 5-alpha, 6E))-	e	1	500/10,000

EMERGENCY PROCEDURES 281

CAS No.	Chemical	Notes	Col1	Col2
534-07-6	Bis(Chloromethyl) Ketone	e	1	10/10,000
4044-65-9	Bitoscanate	e	1	500/10,000
10294-34-5	Boron Trichloride	e	1	500
7637-07-2	Boron Trifluoride	e	1	500
353-42-4	Boron Trifluoride Compound With Methyl Ether (1:1)	e	1	100/10,000
28772-56-7	Bromadiolone	e	1	500
7726-95-6	Bromine	e	1	1,000
106-99-0	Butadiene	e, i	1	10,000
109-19-3	Butyl Isovalerate	a, e	1	10,000
111-34-2	Butyl Vinyl Ether	a, e	1	10,000
1306-19-0	Cadmium Oxide	c, e	1	100/10,000
2223-93-0	Cadmium Stearate	d	1,000	1,000/10,000
7778-44-1	Calcium Arsenate	d	1	500/10,000
8001-35-2	Camphechlor	d	1	100/10,000
56-25-7	Cantharidin	e	1	100/10,000
51-83-2	Carbachol Chloride	e	1	500/10,000
26419-73-8	Carbamic Acid, Methyl-, O-((2,4-Dimethyl-1,3-Dithiolan-2-yl)Methylene)Amino)-		10	10/10,000
1563-66-2	Carbofuran	i	100	10,000
75-15-0	Carbon Disulfide	e	1	500
786-19-6	Carbophenothion	a, e	1	10,000
2244-16-8	Carvone	a, e	1	1,000
57-74-9	Chlordane	e	1	500
470-90-6	Chlorfenvinfos	e	1	100
7782-50-5	Chlorine	e	10	500
24934-91-6	Chlormephos	e, h	1	100/10,000
999-81-5	Chlormequat Chloride	a	1,000	10,000
107-20-0	Chloroacetaldehyde	e	1	100/10,000
79-11-8	Chloroacetic Acid	e	1	500
107-07-3	Chloroethanol	e	1	1,000
627-11-2	Chloroethyl Chloroformate	d, i	5,000	10,000
67-66-3	Chloroform	d, i	1	100
542-88-1	Chloromethyl Ether	d, c, d	1	100
107-30-2	Chloromethyl Methyl Ether	c, e	1	100/10,000
3691-35-8	Chlorophacinone	e	1	500/10,000
1982-47-4	Chloroxuron	e	1	500
21923-23-9	Chlorthiophos	e, h	1	1/10,000
10025-73-7	Chromic Chloride	e	1	10,000
7440-48-4	Cobalt	a, e	1	100/10,000
62207-76-5	Cobalt, ((2,2'-(1,2-Ethanediylbis (Nitrilomethylidyne))Bis(6-Fluorophenolato))(2-)-N,N',O,O')-	e	1	100/10,000
10210-68-1	Cobalt Carbonyl	e, h	1	10/10,000

282 WORKER PROTECTION DURING HAZARDOUS WASTE REMEDIATION

Table 12-1. (Continued)

CAS No.	Chemical name	Notes	Reportable quantity* (pounds)	Threshold planning quantity (pounds)
64-86-8	Colchicine	e, h	1	10/10,000
117-52-2	Coumafuryl	a, h	1	10,000
56-72-4	Coumaphos		10	10/10,000
5836-29-3	Coumatetralyl		1	500/10,000
95-48-7	Cresol, o-	e, d	1,000	1,000/10,000
535-89-7	Crimidine	e, d	1	100/10,000
4170-30-3	Crotonaldehyde	e	1	100
123-73-9	Crotonaldehyde, (E)-		100	1,000
506-68-3	Cyanogen Bromide		100	1,000
506-78-5	Cyanogen Iodide		1,000	500/10,000
2636-26-2	Cyanophos		1	1,000/10,000
675-14-9	Cyanuric Fluoride		1	1,000
66-81-9	Cycloheximide	e	1	100
108-91-8	Cyclohexylamine	e	1	100/10,000
287-92-3	Cyclopentane	e, i	1	10,000
633-03-4	C. I. Basic Green 1	a, e	1	10,000
17702-41-9	Decaborane(14)	e	1	10,000
8065-48-3	Demeton	e	1	500/10,000
919-86-8	Demeton-S-Methyl	e	1	500
10311-84-9	Dialifor	e	1	100/10,000
19287-45-7	Diborane	a, E	100	100
84-74-2	Dibutyl Phthalate	a, e	10	10,000
8023-53-8	Dichlorobenzalkonium Chloride	d	1	10,000
111-44-4	Dichloroethyl Ether	e	1	10,000
149-74-6	Dichloromethylphenylsilane		1	1,000
62-73-7	Dichlorvos	e	10	1,000
141-66-2	Dicrotophos		1	100
1464-53-5	Diepoxybutane	e, d	1	500
814-49-3	Diethyl Chlorophosphate	e, h	1	500
1642-54-2	Diethylcarbamazine Citrate	e	1	100/10,000
93-05-0	Diethyl-p-Phenylenediamine	a, e	1	10,000
71-63-6	Digitoxin	c, e	1	100/10,000
2238-07-5	Diglycidyl Ether	e	1	1,000
20830-75-5	Digoxin	e, h	1	10/10,000
115-26-4	Dimefox	e	10	500
60-51-5	Dimethoate		1	500/10,000
2524-03-0	Dimethyl Phosphorochloridothioate		1	500
131-11-3	Dimethyl Phthalate	a, E	5,000	10,000
77-78-1	Dimethyl Sulfate	d	1	500
75-18-3	Dimethyl Sulfide	e	1	100
75-78-5	Dimethyldichlorosilane	e, h	1	500
57-14-7	Dimethylhydrazine	d	1	1,000
99-98-9	Dimethyl-p-Phenylenediamine	e	1	10/10,000
644-64-4	Dimetilan		1	500/10,000
534-52-1	Dinitrocresol		10	10/10,000
88-85-7	Dinoseb		1,000	100/10,000
1420-07-1	Dinoterb		1	500/10,000
117-84-0	Dioctyl Phthalate	a, E	5,000	10,000

EMERGENCY PROCEDURES 283

CAS	Chemical	Notes	col1	col2
78-34-2	Dioxathion	e	—	500
646-06-0	Dioxolane		—	10,000
82-66-6	Diphacinone	a, e	—	10/10,000
152-16-9	Diphosphoramide, Octamethyl-	e	100	100
298-04-4	Disulfoton		—	500
514-73-8	Dithiazanine Iodide		—	500/10,000
541-53-7	Dithiobiuret	e	100	100/10,000
316-42-7	Emetine, Dihydrochloride	e, h	—	1/10,000
115-29-7	Endosulfan		—	10/10,000
2778-04-3	Endothion	e	—	500/10,000
72-20-8	Endrin		—	500/10,000
106-89-8	Epichlorohydrin	d, l	1,000	1,000
2104-64-5	EPN	e	—	100/10,000
50-14-6	Ergocalciferol	c, e	—	1,000/10,000
379-79-3	Ergotamine Tartrate	e	—	500/10,000
1622-32-8	Ethanesulfonyl Chloride, 2-Chloro-	e	—	500
10140-87-1	Ethanol, 1,2-Dichloro-, Acetate	e	—	1,000
563-12-2	Ethion		—	1,000
13194-48-4	Ethoprophos	e	—	500
538-07-8	Ethylbis(2-Chloroethyl)Amine	e, h	—	500
371-62-0	Ethylene Fluorohydrin	c, e, h	10	10
75-21-8	Ethylene Oxide	d, l	—	1,000
107-15-3	Ethylenediamine		—	10,000
151-56-4	Ethyleneimine	d	5,000	500
2235-25-8	Ethylmercuric Phosphate	a, e	—	10,000
542-90-5	Ethylthiocyanate	e	—	10,000
22224-92-6	Fenamiphos	e	—	10/10,000
122-14-5	Fenitrothion	e	—	500
115-90-2	Fensulfothion	e	—	500
4301-50-2	Fluenetil	e, x	—	100/10,000
7782-41-4	Fluorine	j	10	500
640-19-7	Fluoroacetamide		100	100/10,000
144-49-0	Fluoroacetic Acid	c, e	—	10/10,000
359-06-8	Fluoroacetyl Chloride	e	—	10
51-21-8	Fluorouracil	e	—	500/10,000
944-22-9	Fonofos	d, l	—	500
50-00-0	Formaldehyde	e, h	1,000	500
107-16-4	Formaldehyde Cyanohydrin	e, h	—	1,000
23422-53-9	Formetanate Hydrochloride	e	—	500/10,000
2540-82-1	Formothion	e	—	100

284 WORKER PROTECTION DURING HAZARDOUS WASTE REMEDIATION

Table 12-1. (Continued)

CAS No.	Chemical name	Notes	Reportable quantity (pounds)	Threshold planning quantity (pounds)
17702-57-7	Formparanate	e	1	100/10,000
21548-32-3	Fosthietan	e	1	500
3878-19-1	Fuberidazole	e	1	100/10,000
110-00-9	Furan		100	500
13450-90-3	Gallium Trichloride	e	1	500/10,000
77-47-4	Hexachlorocyclopentadiene	d, h	1	100
1335-87-1	Hexachloronaphthalene	a, e	1	10,000
4835-11-4	Hexamethylenediamine, N,N′-Dibutyl-	e	1	500
302-01-2	Hydrazine	d	1	1,000
74-90-8	Hydrocyanic Acid		10	100
7647-01-0	Hydrogen Chloride (Gas Only)	e, l	1	500
7664-39-3	Hydrogen Fluoride		100	100
7722-84-1	Hydrogen Peroxide (Conc >52%)	e, l	1	1,000
7783-07-5	Hydrogen Selenide	e, l	1	10
7783-06-4	Hydrogen Sulfide	e	100	500
123-31-9	Hydroquinone		1	500/10,000
53-86-1	Indomethacin	a, e	1	10,000
10025-97-5	Indium Tetrachloride	a, e	1	10,000
13463-40-6	Iron, Pentacarbonyl	e	1	100
297-78-9	Isobenzan	e, h	1	100/10,000
78-82-0	Isobutyronitrile	e	1	1,000
10025-36-3	Isocyanic Acid, 3,4-Dichlorophenyl Ester	e	1	500/10,000
465-73-6	Isodrin		1	100/10,000
55-91-4	Isofluorphate	c	100	100
4098-71-9	Isophorone Diisocyanate	b, e	1	100
108-23-6	Isopropyl Chloroformate	e	1	1,000
625-55-8	Isopropyl Formate	e	1	500
119-38-0	Isopropylmethylpyrazolyl Dimethylcarbamate	e	1	500
78-97-7	Lactonitrile	e	1	1,000
21609-90-5	Leptophos	e	1	500/10,000
541-25-3	Lewisite	c, e, h	1	10
58-89-9	Lindane	d	1	1,000/10,000
7580-67-8	Lithium Hydride	b, e	1	100
109-77-3	Malononitrile	e	1	500/10,000
12108-13-3	Manganese, Tricarbonyl Methylcyclopentadienyl	e, h	1	100
51-75-2	Mechlorethamine	c, e	1	10
950-10-7	Mephosfolan	e	1	500
1600-27-7	Mercuric Acetate	e	1	500/10,000
7487-94-7	Mercuric Chloride	e	1	500/10,000
21908-53-2	Mercuric Oxide	e	1	500/10,000
108-67-8	Mesitylene		1	10,000
10476-95-6	Methacrolein Diacetate	a, e	1	1,000
760-93-0	Methacrylic Anhydride	e	1	500
126-98-7	Methacrylonitrile	e, h	1	500
920-46-7	Methacryloyl Chloride	e	1	100
30674-80-7	Methacryloyloxyethyl Isocyanate	e, h	1	100/10,000
10265-92-6	Methamidophos	e	1	100/10,000
558-25-8	Methanesulfonyl Fluoride	e	1	1,000

EMERGENCY PROCEDURES 285

CAS	Chemical	Notes	TPQ	RQ
950-37-8	Methidathion	e	—	500/10,000
2032-65-7	Methiocarb		10	500/10,000
16752-77-5	Methomyl		100	500/10,000
151-38-2	Methoxyethylmercuric Acetate	h	—	500/10,000
80-63-7	Methyl 2-Chloroacrylate	e	1	500
74-83-9	Methyl Bromide	e	1,000	1,000
79-22-1	Methyl Chloroformate	d, e	1,000	500
624-92-0	Methyl Disulfide	e	1	100
60-34-4	Methyl Hydrazine		10	100
624-83-9	Methyl Isocyanate	b, e	1	500
556-61-6	Methyl Isothiocyanate	e	—	500
74-93-1	Methyl Mercaptan	e	100	500
3735-23-7	Methyl Phenkapton	b, e	—	500
676-97-1	Methyl Phosphonic Dichloride	e	1	100
556-64-9	Methyl Thiocyanate	e	1	10,000
78-94-4	Methyl Vinyl Ketone	e, h	1	10
502-39-6	Methylmercuric Dicyanamide	e, e	—	500/10,000
75-79-6	Methyltrichlorosilane		—	500
1129-41-5	Metolcarb		—	100/10,000
7786-34-7	Mevinphos		10	500
315-18-4	Mexacarbate	d	1,000	500/10,000
50-07-7	Mitomycin C	e	1	500/10,000
6923-22-4	Monocrotophos	a, h	1	10/10,000
2763-96-4	Muscimol	e, d	1,000	10,000
505-60-2	Mustard Gas	a, d	1	500
7440-02-0	Nickel	c, e	—	10,000
13463-39-3	Nickel Carbonyl		1	1
54-11-5	Nicotine		100	100
65-30-5	Nicotine Sulfate	c, e	—	100/10,000
7697-37-2	Nitric Acid		1,000	1,000
10102-43-9	Nitric Oxide	e	10	100
98-95-3	Nitrobenzene		1,000	10,000
1122-60-7	Nitrocyclohexane	d, h	—	500
10102-44-0	Nitrogen Dioxide	e	10	100
62-75-9	Nitrosodimethylamine	a, e	1	1,000
991-42-4	Norbornide	e	—	100/10,000
0	Organorhodium Complex (PMN-82-147)	c, e	—	10/10,000
65-86-1	Orotic Acid		—	10,000
20816-12-0	Osmium Tetroxide		1,000	10,000
630-60-4	Ouabain		—	100/10,000

286 WORKER PROTECTION DURING HAZARDOUS WASTE REMEDIATION

Table 12-1. (Continued)

CAS No.	Chemical name	Notes	Reportable quantity* (pounds)	Threshold planning quantity (pounds)
23135-22-0	Oxamyl	e	1	100/10,000
78-71-7	Oxetane, 3,3-Bis(Chloromethyl)-	1		500
2497-07-6	Oxydisulfoton	e, h	e	500
100028-15-6	Ozone	e	1	100
1910-42-5	Paraquat	e	1	10/10,000
2074-50-2	Paraquat Methosulfate	e	1	10/10,000
56-38-2	Parathion	e	1	100
298-00-0	Parathion-Methyl	c, d	100	100/10,000
12002-03-8	Paris Green	c	100	500/10,000
19624-22-7	Pentaborane	d	1	500
76-01-7	Pentachloroethane	e	1	10,000
87-86-5	Pentachlorophenol	a, d	10	10,000
2570-26-5	Pentadecylamine	a, d	1	100/10,000
79-21-0	Peracetic Acid	e	1	500
594-42-3	Perchloromethylmercaptan	e	1	500
108-95-2	Phenol		100	500/10,000
97-18-7	Phenol, 2,2'-Thiobis(4,6-Dichloro-		1,000	100/10,000
4418-66-0	Phenol, 2,2'-Thiobis(4-Chloro-6-Methyl)-	e	1	100/10,000
64-00-6	Phenol, 3-(1-Methylethyl)-, Methylcarbamate	e	1	500/10,000
58-36-6	Phenoxarsine, 10,10'-Oxydi-	e	1	500/10,000
696-28-6	Phenyl Dichloroarsine	d, h	1	500
59-88-1	Phenylhydrazine Hydrochloride	e	1	1,000/10,000
62-38-4	Phenylmercury Acetate		100	500/10,000
2097-19-0	Phenylsilatrane	e, h	1	100/10,000
103-85-5	Phenylthiourea		108	100/10,000
298-02-2	Phorate		10	10
4104-14-7	Phosacetim	e	1	100/10,000
947-02-4	Phosfolan	e	1	100/10,000
75-44-5	Phosgene	—	10	10
732-11-6	Phosmet	e	1	10/10,000
13171-21-6	Phosphamidon	e	1	100
7803-51-2	Phosphine		100	500
2703-13-1	Phosphonothioic Acid, Methyl-, O-Ethyl O-(4-(Methylthio)Phenyl) Ester	e	1	500
50782-69-9	Phosphonothioic Acid, Methyl-, S-(2-(Bis(1-Methylethyl)Amino)Ethyl O-Ethyl Ester	e	1	100
2665-30-7	Phosphonothioic Acid, Methyl-, O-(4-Nitrophenyl) O-Phenyl Ester	e	1	500
3254-63-5	Phosphoric Acid, Dimethyl 4-(Methylthio) Phenyl Ester	e	1	500
2587-90-8	Phosphorothioic Acid, O,O-Dimethyl-S-(2-Methylthio) Ethyl Ester	c, e, g	1	500
7723-14-0	Phosphorus	b, h	1	100

EMERGENCY PROCEDURES 287

CAS Number	Chemical Name	Notes	TPQ1	TPQ2
10025-87-3	Phosphorus Oxychloride	d	1,000	500
10026-13-8	Phosphorus Pentachloride	b, e		500
1314-56-3	Phosphorus Pentoxide	b, e		10
7719-12-2	Phosphorus Trichloride		1,000	1,000
84-80-0	Phylloquinone			10,000
57-47-6	Physostigmine	a, e		100/10,000
57-64-7	Physostigmine, Salicylate (1:1)	e		100/10,000
124-87-8	Picrotoxin	e		500/10,000
110-89-4	Piperidine	e		1,000
5281-13-0	Piprotal	e		100/10,000
23505-41-1	Pirimifos-Ethyl			1,000
10025-65-7	Platinous Chloride	e		10,000
13454-96-1	Platinum Tetrachloride	a, e		10,000
10124-50-2	Potassium Arsenite	a, e	1,000	500/10,000
151-50-8	Potassium Cyanide	d	10	100
506-61-6	Potassium Silver Cyanide	b		500
2631-37-0	Promecarb	b, e		500/10,000
106-96-7	Propargyl Bromide	e, h		10
57-57-8	Propiolactone, Beta-	e		500
107-12-0	Propionitrile	e	10	500
542-76-7	Propionitrile, 3-Chloro-		1,000	1,000
70-69-9	Propiophenone, 4-Amino-	e, g		100/10,000
109-61-5	Propyl Chloroformate	e, e	1	500
1331-17-5	Propylene Glycol, Allyl Ether	e		10,000
75-56-9	Propylene Oxide	e, i		10,000
75-55-8	Propyleneimine	d		10,000
2275-18-5	Prothoate	e, e		100/10,000
95-63-8	Pseudocumene	e, c	5,000	10,000
129-00-0	Pyrene	e, u		1,000/10,000
140-76-1	Pyridine, 2-Methyl-5-Vinyl-	e		500
504-24-5	Pyridine, 4-Amino-	e, f	1,000	500/10,000
1124-33-0	Pyridine, 4-Nitro-, 1-Oxide	e		500/10,000
53558-25-1	Pyriminil	e, h		100/10,000
10049-07-7	Rhodium Trichloride	e, a, e		10,000
14167-18-1	Salcomine	e, e, h		500/10,000
107-44-8	Sarin		10	10
7783-00-8	Selenious Acid	e		1,000/10,000
7791-23-3	Selenium Oxychloride	e		500
563-41-7	Semicarbazide Hydrochloride	e		1,000/10,000
3037-72-7	Silane, (4-Aminobutyl)Diethoxymethyl-	a, e		1,000
128-56-3	Sodium Anthraquinone-1-Sulfonate	d		10,000
7631-89-2	Sodium Arsenate	b	1,000	1,000/10,000
7784-46-5	Sodium Arsenite	b	1,000	500/10,000
26628-22-8	Sodium Azide (Na(N3))	b	1,000	500
124-65-2	Sodium Cacodylate	e		100/10,000
143-33-9	Sodium Cyanide (Na(CN))	b	10	100
62-74-8	Sodium Fluoroacetate		10	10/10,000
131-52-2	Sodium Pentachlorophenate	e		100/10,000
13410-01-0	Sodium Selenate	e		100/10,000
10102-18-8	Sodium Selenite	e, h	100	100/10,000

288 WORKER PROTECTION DURING HAZARDOUS WASTE REMEDIATION

Table 12-1. (Continued)

CAS No.	Chemical name	Notes	Reportable quantity* (pounds)	Threshold planning quantity (pounds)
10102-20-2	Sodium Tellurite	e	1	500/10,000
900-95-8	Stannane, Acetoxytriphenyl-	e, g	1	500/10,000
57-24-9	Strychnine	c	10	100/10,000
60-41-3	Strychnine, Sulfate	e		100/10,000
3689-24-5	Sulfotep		100	500
3569-57-1	Sulfoxide, 3-Chloropropyl Octyl	e	1	500
7446-09-5	Sulfur Dioxide	e, l	1	500
7783-60-0	Sulfur Tetrafluoride	e	1	100
7446-11-9	Sulfur Trioxide	e	1	100
7664-93-9	Sulfuric Acid	b, e	1,000	1,000
77-81-6	Tabun	c, e, h	1	10
13494-80-9	Tellurium	e	1	500/10,000
7783-80-4	Tellurium Hexafluoride	e, k		100
107-49-3	TEPP		10	100
13071-79-9	Terbufos	e, h	1	100
78-00-2	Tetraethyllead	c, d	10	100
597-64-8	Tetraethyltin	c, e	1	100
75-74-1	Tetramethyllead	c, e, l		100
509-14-8	Tetranitromethane		10	500
1314-32-5	Thallic Oxide	a	100	10,000
10031-59-1	Thallium Sulfate	h	100	100/10,000
6533-73-9	Thallous Carbonate	c, h	100	100/10,000
7791-12-0	Thallous Chloride	c, h	100	100/10,000
2757-18-8	Thallous Malonate	c, e, h	1	100/10,000
7446-18-6	Thallous Sulfate		100	100/10,000
2231-57-4	Thiocarbazide	e		1,000/10,000
21564-17-0	Thiocyanic Acid, 2-(Benzothiazolylthio)Methyl Ester	a, e	1	10,000
39196-18-4	Thiofanox		100	100/10,000
640-15-3	Thiometon	a, e	1	10,000
297-97-2	Thionazin		100	500
108-98-5	Thiophenol		100	500
79-19-6	Thiosemicarbazide		100	100/10,000
5344-82-1	Thiourea, (2-Chlorophenyl)-		100	100/10,000
614-78-8	Thiourea, (2-Methylphenyl)-	e	1	500/10,000

EMERGENCY PROCEDURES

CAS	Chemical	Notes	RQ	TPQ
7550-45-0	Titanium Tetrachloride	e	1	100
584-84-9	Toluene 2,4-Diisocyanate	e	100	500
91-08-7	Toluene 2,6-Diisocyanate	e	100	100
110-57-6	Trans-1,4-Dichlorobutene	e	1	500
1031-47-6	Triamiphos	e	1	500/10,000
24017-47-8	Triazofos	e	1	500
76-02-8	Trichloroacetyl Chloride	e, h	1	500
115-21-9	Trichloroethylsilane	e, k, h	1	500
327-98-0	Trichloronate	e, h	1	500
98-13-5	Trichlorophenylsilane	e	1	500
52-68-6	Trichlorophon	a	100	10,000
1558-25-4	Trichloro(Chloromethyl)Silane	e	1	100
27137-85-5	Trichloro(Dichlorophenyl)Silane	e	1	500
998-30-1	Triethoxysilane	e	1	500
75-77-4	Trimethylchlorosilane	e	1	1,000
824-11-3	Trimethylolpropane Phosphite	e, h	1	100/10,000
1066-45-1	Trimethyltin Chloride	e	1	500/10,000
639-58-7	Triphenyltin Chloride	e	1	500/10,000
555-77-1	Tris(2-Chloroethyl)Amine	e, h	1	100
2001-95-8	Valinomycin	c, e	1	1,000/10,000
1314-62-1	Vanadium Pentoxide	d, l	1,000	100/10,000
108-05-4	Vinyl Acetate Monomer	a, e	5,000	1,000
3048-64-4	Vinylnorbornene		1	10,000
81-81-2	Warfarin		100	500/10,000
129-06-6	Warfarin Sodium	e, h	1	100/10,000
28347-13-9	Xylylene Dichloride	e	1	100/10,000
58270-08-9	Zinc, Dichloro(4,4-Dimethyl-5(((Methylamino) Carbonyl)Oxy)lmino)Pentanenitrile)-, (T-4)-	e	1	100/10,000
1314-84-7	Zinc Phosphide	b	100	500

*Only the statutory or final RQ is shown. For more information, see 40 CFR Table 302.4

Notes:
a This chemical does not meet acute toxicity criteria. Its TPQ is set at 10,000 pounds.
b This material is a reactive solid. The TPQ does not default to 10,000 pounds for non-powder, non-molten, non-solution form.
c The calculated TPQ changed after technical review as described in the technical support document.
d Indicates that the RQ is subject to change when the assessment of potential carcinogenicity and/or other toxicity is completed
e Statutory reportable quantity for purposes of notification under SARA sect 304(a)(2).
f The statutory 1 pound reportable quantity for methyl isocyanate may be adjusted in a future rulemaking action.
g New chemicals added that were not part of the original list of 402 substances.
h Revised TPQ based on new or re-evaluated toxicity data.
i TPQ is revised to its calculated value and does not change due to technical review as in proposed rule.
j The TPQ was revised after proposal due to calculation error.
k Chemicals on the original list that do not meet toxicity criteria but because of their high production volume and recognized toxicity are considered chemicals of concern ("Other Chemicals").
m Chemicals delisted by EPA in 1987.

Source: 40 CFR 355, 1987.

Site Emergencies

Emergency Response Requirements under the OSHA Standard 29 CFR 1910.120. The OSHA Standard, 29 CFR 1910.120, also addresses the issue of emergency response. An emergency response plan is a required part of the site-specific safety and health plan, and must be established prior to site cleanup, as opposed to the community focus under Title III law. Many similarities exist between the two. Some of the issues the Emergency Response Plan must address in writing are:

- Predetermined criteria for the recognition of an emergency situation.
- Emergency lines of authority.
- The security and control of contaminants at a site during an emergency.
- The site layout.

Specific procedures to be developed by personnel are also an integral requirement of the plan. These include:

- An emergency medical plan, which addresses injuries in the exclusion zone and decontamination procedures governing those with life threatening injuries (see the flow chart, Fig. 12-5, on medical emergency response operations and accident/injury events).
- Emergency equipment list and equipment stocked.
- Evacuation routes with an emergency alarm.
- A post-emergency critique of the actions taken by workers.
- The reporting of emergency incidents.
- The compatibility of the Emergency Response Plan with those of fire, police, nearby medical personnel, and the community.

The emergency response plan also requires specific training for workers who must combat site disasters. It must be adequate for an effective response and incorporate regular drills. Other specific issues addressed by the Emergency Response Plan can be found in Chapter 2.

UNDERSTANDING AND RESPONDING TO EMERGENCIES

Emergencies: Overview

As an introduction to emergencies, it is appropriate to look at a national emergency list, which (except for nuclear war) portrays (on average) what types of emergencies occur and the order of their severity. The ten greatest hazards are:

EMERGENCY PROCEDURES

- **Hazardous material highway incidents.**
- Winter storms.
- Floods.
- **Hazardous material railroad incidents.**
- Tornadoes.
- Power failures.
- **Stationary hazardous materials incidents (storage facilities).**
- Urban fires.
- Wild fires.
- Pipeline or pump station incidents.

Notice that three of the ten most severe nationwide hazards involve hazardous materials, and pipeline incidents closely resemble many hazardous materials spill incidents. Some statistics can further describe the nature and scope of hazardous materials emergencies:

- 50% occur on land.
- 25% in air or water.
- 25% other.
  (After Al Emergency Mgmt Agency 1987)

As of 1986, only one-third of all full-time emergency responders were knowledgeable and properly equipped for hazardous materials emergencies. The EPA considers that only 200 out of the nation's 1200 emergency response teams have "good" hazardous materials capability. Concerning transportation incidents:

- 750,000 people in the U.S. are evacuated from their homes each year due to releases of hazardous substances.
- 20,000 such incidents occur annually
  (from Ansul, Inc. 1987)

Concerning stationary facilities:

- 75% of incidents occur at fixed facilities.
- 25% are transportation related.

5000 injuries occurred at fixed facilities between 1980–1985.

Major fixed facility releases have involved PCBs, chlorine, and sulfuric acid, among others. Faced with such a grim record, emergency response personnel realize that better training and planning are essential to future improvement. Having been introduced to specific chemical hazards in previous chapters, we will now turn our attention to the two other greatest dangers in most emergencies: *fires* and *spills* (AEMA, 1987).

Fire Extinguishing, Suppressants and Protection

Ways to Fight Fire. There are two ways to combat a fire: isolation and suppression. When flammable materials are isolated by distance (buffer zones) or barriers (earthen berms) then time is no longer critical. The danger to fire fighters is reduced or eliminated if the fire can be allowed to burn itself out. Here is where planning can save lives. If a fire must be fought it must be suppressed by inhibiting the combustion reaction; that is, removing one of the components of the fire triangle (heat, air, fuel); see Chapter 4. Common sense, engineering judgment, and fast action come into play. Some questions must be answered:

- What is burning?
- How much is burning?
- Where is it and what is near it?
- What are the exposure hazards?
- What suppressants and resources are available?

Sometimes a fire should be left to burn until expert advice is obtained since most portable extinguishers, on average, only have a 30 second duration.

Basic Considerations. There are two essential considerations to be made before attempting to suppress a fire. One involves the physical/chemical properties of the burning substance. For example, ordinary foam would not be good on a fire involving a water-soluble liquid. The second essential consideration is the method of application of the extinguishing agent. For example, water streams directed against a burning liquid with a very high flash point could be deadly to a firefighter.

It is crucial to know the basic physical and chemical properties of burning substances before combating them. These include:

- Ignition temperature.
- Solubility and density.
- Reactivity—incompatibilities.
- Combustion products.
- Flashpoint (see Table 12-2).

These properties dictate the application method and are specified for each chemical listed under the fire fighting phases column of Table 12-2. For instance, if a spill of isopentyl alcohol burst into flames during a cleanup, Table

Table 12-2. Part 1. Fire Fighting Phases.

The numbers at the left of the paragraphs below correspond to the numbers in the right-hand column of the list of properties which follows.

1. Stop flow of gas. Use water to keep fire-exposed containers cool and to protect men effecting the shutoff. If a leak or spill has not ignited, use water spray to disperse the gas of vapor and to protect workers attempting to stop a leak.
2. Use dry chemical, "alcohol" foam, or carbon dioxide; water may be ineffective, but should be used to keep fire-exposed containers cool. If a leak or spill has not ignited, use water spray to disperse the vapors and to protect workers attempting to stop a leak. Water spray may be used to flush spill away from exposures and to dilute spills to nonflammable mixtures.
3. Use water spray, dry chemical, "alcohol" foam, or carbon dioxide. Use water to keep fire-exposed containers cool. If a leak or spill has not ignited, use water spray to disperse the vapors and to protect workers attempting to stop a leak. Water spray may be used to flush spills away from exposures and to dilute spills to nonflammable mixtures.
4. Use dry chemical, foam, or carbon dioxide. Water may be ineffective, but water should be used to keep fire-exposed containers cool. If a leak or spill has not ignited, use water spray to disperse the vapors and to protect workers attempting to stop a leak. Water spray may be used to flush spills away from exposures.
5. Use water spray, dry chemical, foam, or carbon dioxide. Use water to keep fire-exposed containers cool. If a leak or spill has not ignited, use water spray to disperse the vapors and to provide protection for workers attempting to stop a leak. Water spray may be used to flush spills away from exposures.
6. Use water spray, dry chemical, "alcohol" foam, or carbon dioxide. Water or foam may cause frothing. Use water to keep fire-exposed containers cool. Water spray may be used to flush spills away from exposures and to dilute spills to noncombustible mixtures.
7. Use water spray, dry chemical, foam, or carbon dioxide. Water or foam may cause frothing. Use water to keep fire-exposed containers cool. Water spray may be used to flush spills away from exposures.

Table 12-2. Part 2. Properties of Common Flammable Chemicals.

	Health	Flammability	Reactivity	Flash Pt. °F	Water Soluble	Fire Fighting Phases
Acetophenone	1	2	0	180 (oc)	No	5
Acetone	1	3	0	0	Yes	2
Adipic Acid	1	1	0	385	No	7
Asphalt (typical)	0	1	0	400 + (oc)	No	7
1-Bromopentane	1	3	0	90	No	4
Butane	1	4	0	Gas	No	1
1-Butane	1	3	0	84	Yes	2
2-Butanol	1	3	0	75	Yes	2
Butenes	1	4	0	Gas	No	1
Butyl Acetate	1	3	0	72	Slightly	2
Butyl Acetate (iso)	1	3	0	64	No	4
Butyl Benzoate (n)	1	1	0	225 (oc)	No	7
Chlorohexane	0	3	0	95	No	4
1-Chloropentane	1	3	0	55 (oc)	No	4
Cod Liver Oil	0	1	0	412	No	7
Corn Oil	0	1	0	490	No	7
Cottonseed Oil	0	1	0	486	No	7
Cyclohexane	1	3	0	minus 4	No	4
Cyclohexanone	1	2	0	111	Slightly	3
Cyclohexyl Alcohol	1	2	0	154	Slightly	3
Cyclopentane	1	3	0	less than 20	No	4
Cyclopropane	1	4	0	Gas	No	1
Decane (n)	0	2	0	115	No	5

Table 12-2. Part 2. (Continued)

	Health	Flam-mability	Reac-tivity	Flash Pt. °F	Water Soluble	Fire Fighting Phases
Decanol	0	2	0	180	No	5
Dibutyl Phthalate	0	1	0	315	No	7
Diesel Fuel Oil No. 1-D	0	2	0	100 min. or legal	No	5
Diesel Fuel Oil No. 2-D	0	2	0	125 min. or legal	No	5
Diesel Fuel Oil No. 4-D	0	2	0	130 min. or legal	No	5
Diethanolamine	1	1	0	305	Yes	6
Diethylene Glycol	1	1	0	225	Yes	6
Diethyl Phthalate (o)	0	1	0	325 (oc)	No	7
Dioctyl Phthalate	0	1	0	425 (oc)	No	7
Dipropylene Glycol	0	1	0	280	Yes	6
Ethyl Acetate	1	3	0	24	Slightly	2
Ethyl Alcohol	0	3	0	55	Yes	2
Ethyl Benzoate	1	1	0	greater than 204	No	7
2-Ethylbutanol	1	2	0	135 (oc)	No	5
Ethylbutyl Acetate	1	2	0	130 (oc)	No	5
Ethylbutyl Ketone	1	2	0	115 (oc)	No	5
Ethylene Glycol	1	1	0	232	Yes	6
Ethel Methyl Ketone	1	3	0	21	Yes	2
Fuel Oil No. 1 (Range Oil, Kerosine)	0	2	0	100 min. or legal	No	5
Fuel Oil No. 2	0	2	0	100 min. or legal	No	5
Fuel Oil No. 4	0	2	0	130 min. or legal	No	5
Fuel Oil No. 5	0	2	0	130 min. or legal	No	5
Fuel Oil No. 6	0	2	0	150 min. or legal	No	5
Gasoline	1	3	0	minus 45	No	4
Glycerine	1	1	0	320	Yes	6
Heptane (n)	1	3	0	25	No	4
Hexane (n)	1	3	0	minus 7	No	4
1-Hexanol	1	2	0	145	Slightly	3
Hydrogen	0	4	0	Gas	Slightly	1
Isopentyl Alcohol	1	2	0	109	Slightly	3
Isopropyl Alcohol	1	3	0	53	Yes	2
Lanolin	0	1	0	460	No	7
Lard Oil (commercial or animal)	0	1	0	395	No	7
Linseed Oil (boiled)	0	1	0	403	No	7
Lubricating Oil (mineral)	0	1	0	300–450	No	7
Methanol	1	3	0	52	Yes	2
Methyl Acetate	1	3	0	14	Yes	2
2-Methyl-2-Butanol	1	2	0	103	Slightly	3
2-Methyl-2-Propanol	1	3	0	52	Yes	2
Methyl Salicylate	1	1	0	214	No	7
Mineral Oil	0	1	0	380 (oc)	No	7
Octane	0	3	0	56	No	4
Octyl Alcohol (n)	1	2	0	178	No	5
Oleic Acid	0	1	0	372	No	7
Oleo Oil	0	1	0	450	No	7
Peanut Oil	0	1	0	540	No	7
Pentane	1	4	0	Less than minus 40	No	1
1-Pentanol(n)	1	3	0	91	Slightly	2
2-Pentanol	1	3	0	94	Slightly	2
Pentanol-3	1	2	0	105	Slightly	3
Pentyl Acetate(n)	1	3	0	77	Slightly	2
Pentyl Benzene	1	2	0	150 (oc)	No	5
Pentyl Laurate	0	1	0	300 (oc)	No	7
Pentylnaphthalene	0	1	0	255 (oc)	No	7
Petroleum (crude)	1	3	0	20–90	No	4
Propane	1	4	0	Gas	No	1
Propyl Acetate	1	3	0	58	Yes	2
Propylene Glycol	0	1	0	210	Yes	6
Quenching Oil	0	1	0	365	No	1
Soy Bean Oil	0	1	0	540	No	7
Tallow Oil	0	1	0	492	No	7
Tetrahydronaphthalene	1	2	0	160	No	5

Table 12-2. Part 2. (*Continued*)

	Hazard Identification Health	Flam- mability	Reac- tivity	Flash Pt. °F	Water Soluble	Fire Fighting Phases
Transformer Oil	0	1	0	295 (oc)	No	7
2, 4, 5-Trichlorophenol in solvent	1	1	0	Solvent Flash Point	No	5
2, 4, 6-Trichlorophenol in solvent	1	1	0	Solvent Flash Point	No	5
Triethylene Glycol	1	1	0	350	Yes	6
Turpentine	1	3	0	95	No	4
Vegetable Oil (hydrogenated)	0	1	0	610 (oc)	No	7

12-2 says that the fire fighting phase appropriate would be 3: "Use water spray, dry chemical, 'alcohol' foam or carbon dioxide. Use water to keep fire-exposed containers cool. If a leak or spill has not ignited, use water spray to dispense the vapors, and to protect the workers attempting to stop the leak. Water spray may be used to flush spills away from exposures and to dilute spills to non-flammable mixtures" (NFPA, 1986). Variations in application include sprays, streams, coordinated multiple streams, and whether the suppressant is applied over the top, into, or at the base of the flames (see Tables 12-3 and 12-4). With proper nozzles, even gasoline spill fires of some types have been extinguished by using coordinated hose lines of water to sweep the flames off the surface of the liquid (NFPA, 1986).

Water. Water is the most available and generally applicable agent for suppressing fires. Many hazardous substances, however, may react violently with water (see the list of water-reactive substances in Chapter 4, Table 4-2). It is generally recommended only for class A fires. When used in a stream water cools, but when used as a spray it also smothers. Several treatments are used to improve the effectiveness of water as an extinguisher (Hammer, 1985):

- *Thickening agents*, which produce a slurry, gel, or water foam (for control of viscosity), reduce runoff, and increase blanketing.
- *Salts* depress the freezing point of water in subfreezing applications and help inhibit combustion, but corrosive gas may be generated if salts are used against strong oxidizers.
- *Surfactants (detergents)* decrease the surface tension of water, thereby increasing the cooling ability.

Some specifics about water's properties and applications need mentioning. When dealing with burning liquids:

- The lower the flashpoint, the less effective water is; but spray applications will absorb heat and reduce fire damage.

Table 12-3. Types of Fire Extinguishers.

KIND OF FIRE	APPROVED TYPE OF EXTINGUISHER						
	Foam	Carbon Dioxide	Soda Acid	Pump Tank	Gas Cartridge	Multi Purpose Dry Chemical	Ordinary Dry Chemical
Class A Fires Ordinary Combustibles • Wood • Paper • Cloth, etc.	•		•	•	•	•	
Class B Fires Flammable Liquids, Grease • Gasoline • Paints • Oils, etc.	•	•				•	•
Class C Fires Electrical Equipment • Motors • Switches, etc.		•				•	•

HOW TO OPERATE

FOAM: Don't spray stream into burning liquid. Allow foam to fall lightly on fire.

CARBON DIOXIDE: Direct discharge as close to fire as possible. First at edge of flames and gradually forward and upward.

SODA-ACID, GAS CARTRIDGE: Direct stream at base of flame.

PUMP TANK: Place foot on footrest and direct stream at base of flames.

DRY CHEMICAL: Direct at the base of flames. In case of Class A fires, follow up by directing the dry chemical at remaining material that is burning.

(NIOSH, 1979)

- Water causes "frothing" (a dangerous reaction) when applied against liquids with flashpoints above 212°F (water's boiling point).
- Water will extinguish fires when the fuel is heavier than water and is not water soluble; the application (as a spray applied gently) is significant.
- Water will extinguish fires when the fuel is water-soluble by cooling and diluting, but must be applied at sufficient flooding rates.
- Water spray displaces air and reduces the fire and health risk associated with released substances which vaporize readily.

Note: Water may be ineffective in these applications unless used under favorable conditions by experienced, trained fire fighters (NFPA, 1986; Meyer, 1977).

Table 12-4. Suggested Methods of Extinguishing Chemical Fires.

Chemical	Extinguishing Methods
Cyanides	Water. Do not use CO_2 extinguishers. Avoid toxic fumes.
Chromic acid	Use water. Caution should be exercised against possibility of steam explosion.
Hydrofluoric acid	Use water. Neutralize with soda ash or lime. If water is ineffective, use "alcohol" foam.
Hydrochloric acid	Use water. Neutralize with soda ash or slaked lime.
Nitric acid	Use a water spray. Neutralize with soda ash or lime.
Sulfuric acid	Use large amounts of water. Reaction may occur. Neutralize with ash or lime; sand or gravel will also help.
Acetic acid	Use water spray, dry chemical, "alcohol" foam or carbon dioxide.
Ferric chloride	Use water.
Ammonium persulfate	Use water spray or water flooding. Avoid toxic fumes.
Caustics	Flood with water. Avoid spattering or splashing.
Ammonia	Stop flow of material. Use water to keep container cool. Avoid fumes.
Alkaline wastes	Use water. Neutralize with dilute acid (acetic) if necessary.
Mercury	Use water. Avoid mercury's toxic vapor.
Tetraethyl lead and lead oxide mixed	Fight fires from explosion-restraint location. Use water, dry chemical, foam, or carbon dioxide.
Lead compounds and oxides	Use flooding amounts of water.
Zinc compounds	Smother with suitable dry powder.
Sodium Compounds	Use water, dry powder. Neutralize with appropriate chemical if necessary.
Aluminum, phosphorus, and sulfur compounds	Do not use water. Smother with suitable dry powder.

Source: NFPA, 1986.

Other Suppressants. Many other types of fire suppressants are used. Carbon dioxide (CO_2), is widely used on class A, B, and C fires. CO_2 inhibits combustion by cooling and dissipating oxygen, and is used in portable and automatic extinguishing systems. One difficulty experienced with CO_2 extinguishers is that in dry air, the friction generated by the discharging extinguishing agent can create static electricity which can build up in the air and reignite a flammable substance. Plastic nozzles on portable CO_2 extinguishers reduce this problem. Carbon dioxide can also be a problem when high flooding rates could create an oxygen-deficient atmosphere for the fire fighter. This problem is overcome when CO_2 is used as a chemical foam.

Halons are another group of suppressants which are very popular. Halon is short for halogenated hydrocarbon, which is a compound of carbon in which one or more hydrogen atoms have been substituted with halogen atoms (chlorine, fluorine, bromine, iodine); see the Chemistry section of Chapter 4. Carbon tetrachloride was the first halon used, but it has been outlawed in U.S. Federal installations because under the wrong conditions its use generates toxic phosgene gas. Halons are named by a unique system of four or five digits, the position and quantity of each referring to the specific halogens employed in the molecular configuration of that halon (see Table 12-5). Although halons are expensive, and may generate toxic gas during the wrong applications, they have several distinct advantages. Being much heavier than air they blanket fires effectively even in very low concentrations. They are considered "clean" extinguishers because they leave little residue and do not damage electrical equipment or other valuables. They are also effective against class A, B, and C fires.

Other types of fire extinguishers include dry chemicals such as sodium chloride, sodium bicarbonate, and potassium bicarbonate. These substances and graphite-based dry chemical extinguishing materials have very high melting points which make them effective as blanketing agents and as conductors of heat. These properties are good against class D metal fires (Meyer, 1977). Pound for pound, dry chemicals are generally more effective suppressants than CO_2 for class A, B, and C fires.

Essential Prevention Measures

When dealing with fire at a hazardous waste facility or where hazardous substances are involved, awareness of the danger of combustion products is critical so that proper protective measures may be taken. Phosgene was mentioned as a combustion product of carbon tetrachloride; it is also released from other chlorinated hydrocarbons. Inhaling a 25 ppm concentration of phosgene can cause death in less than 60 minutes. Other extremely toxic combustion products include hydrogen sulfide gas, hydrogen cyanide gas, oxides of nitrogen, acrolein, metal fumes, and ammonia. Fire protection measures are, therefore, of paramount importance at hazardous waste sites. Some of the basics include:

Table 12-5. Halons: 3–5 Digit Nomenclature.

First Digit	Number of carbon atoms
Second Digit	Number of fluorine atoms
Third Digit	Number of chlorine atoms
Fourth Digit	Number of bromine atoms
Fifth Digit	Number of iodine atoms

- Extinguishers at clearly marked locations,
- Extinguishers always operable and accessible.
- Extinguishers inspected frequently.
- "Hot" work areas protected, particularly the Exclusion Zone and CRZ (see zoning section of Engineering Controls)
- Use of fire barriers.
- 50-foot buffer zones.
- Fire lanes.
- Electrical devices protected from flammables.
- Grounded and secured dispensing systems.
- Fire departments briefed per SARA Title III.

Spills and Spill Response

Spills of hazardous substances generally represent a level of severity just below that of fires. As with fires, time is a critical factor in dealing with this type of emergency; fast judgment and action are essential. The first questions are, what to do and how to do it?

Anatomy of a Spill Response In an effort to provide more practical than general information for use on an actual site, assumptions will be made to illustrate certain containment techniques and equipment. Fig. 12-3 illustrates the anatomy of a spill response. One scenario might be as follows:

- A worker finds a leaking drum in a staging area of approximately 100 drums and he is not sure if the compatibility tests have been done on the drums in this area.
- He notifies his supervisor, who in turn notifies the emergency response personnel via the command post. The material has not been tested. Because it's unknown, all precautions have to be taken with the spill. As mentioned in the Drum Handling section of Chapter 7, the material is hazardous until proven otherwise.
- In sizing up the situation, direct-read air monitoring equipment is used in determining the contents; see Chapter 8.
- After testing the spilled material is found to be a mixture of paint thinner, predominantly xylene. No additional or outside personnel need to be called in. This decision is made by the Emergency Response Coordinator, because all the necessary fire and containment equipment is on-hand in case the spill begins to burn or leak excessively.
- Next, the emergency response team designates personnel to retrieve fire extinguishing equipment, in case a fire occurs, while the released material is being contained.

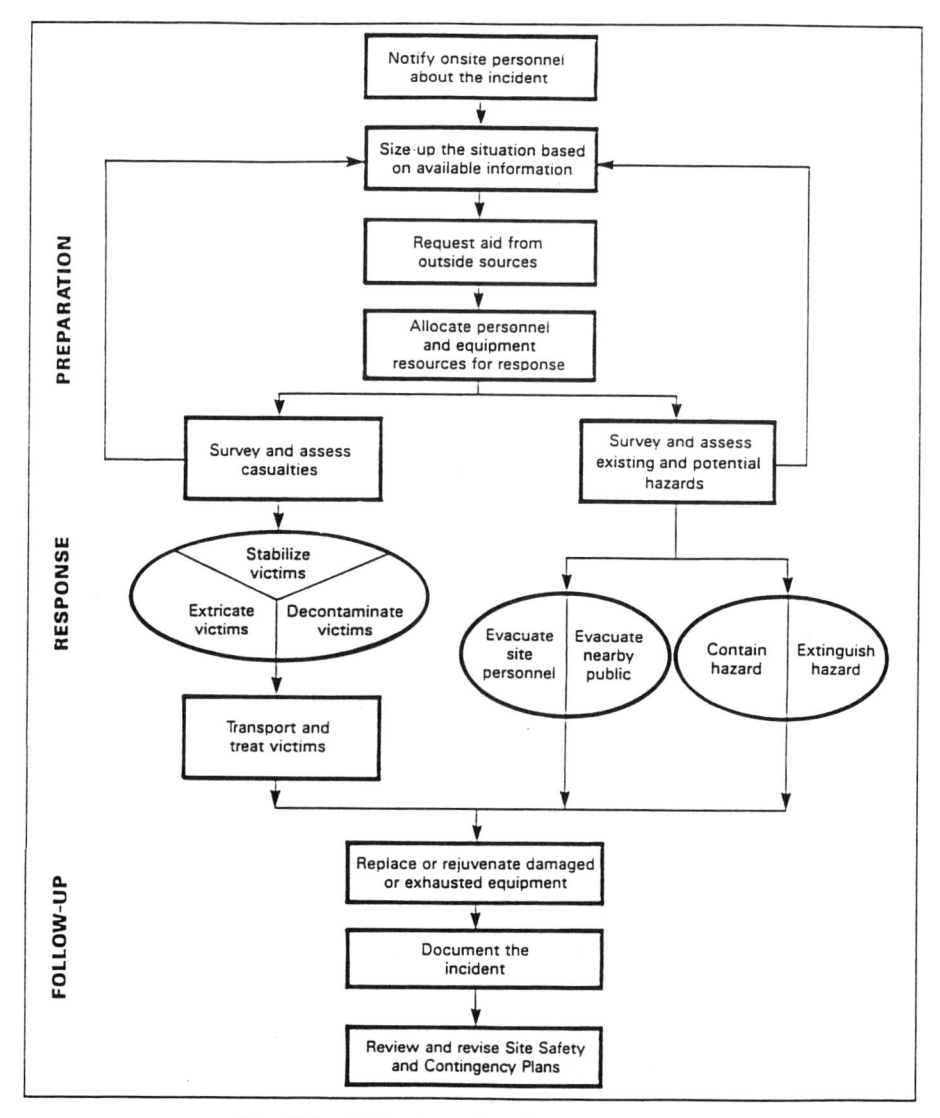

300 WORKER PROTECTION DURING HAZARDOUS WASTE REMEDIATION

Fig. 12-3. Spill response flow chart (NIOSH, 1985).

- No workers are injured, therefore this side of the response diagram (Fig. 12-3) is complete; see the section of this chapter on Medical Emergency Response for more details on procedures when injury occurs.
- The right side of the response diagram (Fig. 12-3) is to survey and assess existing and potential hazards. If additional information is found, backtracking to re-size the situation and/or requesting outside aid maybe in

order. For instance, while cleaning up the 55 gallon spill of xylene it is noticed that approximately 35 other drums are leaking and 10 are bulging.

These signs alert the emergency response coordinator to the fact that the local community could be in danger if an explosion occurred, and the hazard would be compounded by the presence of the spilled xylene and the 35 leaking drums. Therefore, the local fire department is brought in to fight the potential fire, but are kept at the command post until the pressurized drums are decompressed; see the Drum Handling section of Chapter 7 for procedures.
- While cleanup is beginning, evacuation of personnel is in order. In this example, only the people in the adjacent neighborhood and the convenience store personnel are evacuated until pressurized drums are decompressed.
- Once all members of the emergency response team are present and ready, the spill cleanup continues. The spilled material is diked using sorbent booms prior to any other cleanup activity. Then it is taken up (sorbed) using activated charcoal to minimized vapors and fire hazards. After that, the 35 other leaking drums are redrummed. The contaminated soil and all contaminated sorbent are removed and disposed of as hazardous waste.
- If any extinguishers or disposable equipment are used, they should be replenished or replaced immediately, and any damaged equipment should be repaired. Fortunately, only adsorbent booms and the charcoal absorbent would have to be replaced. No extinguishers are needed here. All other equipment goes through decontamination procedures and can be reused.
- Finally, the response is documented with sufficient detail so that the site safety plan can be modified to reflect changes in response activities that are more efficient.

As seen in the above analysis of Fig. 12-3, a spill response to even a straightforward incident can become quite involved. In summary, a threefold approach should be taken in response to spills:

- Assessment.
- Action.
- Followup.

Realize that all spill situations are different due to the ways it can occur, what the spilled substance is, where it goes, and how much there is.

Basic Containment Principles

There are some basic principles to follow in cleanup of a container spill no matter the volume. The first step is finding the source of the leak. Determine the volume of the leaked substance and identify the contaminants involved. Step two should be to stop the leak by repairing or replacing the container at fault, if possible. Then follows consolidation of the spilled material, packaging, labeling, and disposition. Finally, an assessment of the response and restoration of emergency equipment used will prepare the site for any future incidents. Here is a list of other critical guidelines to be aware of when responding to a spill emergency:

- Assume substances are hazardous-don't rush in
- approach from upwind if possible
- use references to familiarize yourself with the dangers of the substance
- remember that hazardous vapors and gases can travel
- let flammable gases burn until the source can be turned off
- work in pairs
- always have backup help

Specific Spill Containment Materials/Equipment. Knowledge of clean-up materials their abilities, availability, and costs are necessary for an expedient spill response.

Patch and plug products. Patch and plug products should be kept on hand for container spills, leaks, and ruptures. Once a spill is discovered and the source found, a patch or plug needs to be placed on the container, if applicable, to stop any additional release of chemicals to the environment.

The following patch and plug products have been used for container patching:

Material	*Application*
Clay	Small temporary patch
Teflon® sealant (example, "real tuff" brand)	Leaking threads, valves
Malleable sticks of adhesive materials (example: "Self Mold" brand)	For repairing cracks in metal vessels
Dry powder patch (example: "Plug n Dike" brand)	For patching cracks in metal or plastic vessels
Caulking compound (example: "DAP" brand)	
Tire repair patch	For repair of plastic vessels patches
Stainless steel putty (example: "Devcon" brand)	cracks in stainless steel vessels
Steel patch (with or without valve)	Bolt or epoxy to leak or rupture in tank

Sorbent materials. After the leak has been covered, plugged or patched, the spill must be contained, using sorbent materials and diking procedures. Sorbents can be used on spills of any size, but the sorbents have to be removed for disposal. This process can be quite cumbersome.

Sorbents come in several different configurations; as pillows, socks, rolls, sheets, and booms. Sorbents can consist of many different materials such as:

- Loose bentonite clay.
- Crushed corncobs.
- Sawdust.
- Feathers.
- Loose activated charcoal.
- Polypropylene shredded in pillows and socks or pressed into sheets and rolls.

Diking of spills in conjunction with the use of the sorbent materials makes for a more efficient response and cleanup. Diking can be a simple or a sophisticated process. Liquid polyfoam plastic (polyurethane), such as Mountains in Minutes® brand used for model train sets, is good for small dikes and is applied from a pressurized applicator. A backhoe can be mobilized and surrounding soil can be excavated. The soil can be placed in a dam configuration for quick response time. Straw bales and sandbags are also used to dike spills; they are economical and are typically available even in the most rural areas.

Sorbent materials can be placed in a dike configuration and this will provide a temporary containment until other means can be implemented. The above discussion applies to fairly small spills and should only be expected to be successful under calm or controlled conditions.

Contaminant Diversion and Containment in Waterways: Spills of contaminants into waterways pose a serious cleanup situation, as the *Exxon Valdez* disaster in Alaska, and the Ashland Chemical Corporation fuel oil spill into the Monongehela River will attest. Often in these large waterway spills a 50% collection is considered "good." Here are offered some tips in assessing a spill, and methods of impeding contaminants.

The assessment of a spill is critical in establishing the extent of the problem and thereby directing the appropriate resources to combat it. The two big questions at first are, how much and what is it? The physical and chemical properties of the contaminant will dictate everything from collection methods to evacuation routes. The amount of contaminant released will dictate the resources needed. Answers to these two questions will direct the immediate and ongoing assessment of human/environmental impact from the release. Further assessment must be made of the waterway's characteristics. Where are collection and agitation areas? Are there dams, ice formations, or other impediments? One waterway will affect the migration and dilution of a contaminant differently than others. After the information gathering, the mobilization of resources, and any emergency precautions are made, one general theme must be followed: An attack plan on a *timetable*.

Many techniques are applied to spills, attempting to consolidate, absorb, or diffuse them. No one method is usually completely effective, so a multiple tactics approach is desirable. Booms are commonly used to confine floating contaminants, their effectiveness depending on the water's flow rate. Siphon dams (see Fig. 12-4) also trap floating contaminants while allowing underlying water to continue down a channel. Filter fences are also used in channels. They involve stretching a fence across the flow and allowing sorbents (pressed against the fence) to collect the contaminants at whatever their level as they pass.

For heavier-than-water contaminants under-water dams, curtains, or dikes are employed, allowing for water overflow. If the contaminant's migration can be predicted and the element of time is not crucial, then downstream diversion trenches and capture pits can be constructed. When the contaminant arrives, the flow is drawn off until either the contaminant is removed from the waterway or the trench/pit is filled.

On occasion, spilled waterborne hydrocarbons have been ignited rather than

Fig. 12-4. Siphon construction for spill containment (EPA, 1987).

face other consequences from the release. Recent developments include the use of chemical agents which flocculate or disperse specific types of contaminants (such as oil). Internationally, the Soviet Union has huge suction apparatuses on their ships for ocean spills, capable of skimming 2 million gallons an hour. In spite of these advances, there is still significant room for improvement of present spill containment technologies in the future.

MEDICAL EMERGENCY RESPONSE AND FIRST AID

First aid is the immediate care given to a person who is injured or suddenly becomes ill. The principal aims are:

- To care for life-threatening conditions.
- To minimize further injury and complications.
- To minimize infection.
- To make the victim as comfortable as possible.
- To transport the victim to medical facilities, when necessary, in such a manner as not to complicate the injury or subject the victim to unnecessary discomfort.

There is a particular need for first aid training in hazardous waste assessments and remediations because medical treatment is often not immediately available at the site. Also, medical personnel are hesitant to treat "contaminated" workers. Therefore, an expedient plan for injured personnel decontamination (workers with severe injuries may only require gross contaminant removal) must be a part of first aid training. When a person is injured, someone must (1) take charge, (2) provide necessary decon, (3) administer first aid, and (4) arrange for medical assistance. It is recommended that all hazardous waste workers become certified in *American Red Cross CPR* and *First Aid*. Companies engaged in this work can arrange in-house training for their employees. The following response activities are summarized in Fig. 12-5.

Primary Survey

Several conditions are considered life-threatening, but three in particular require immediate action:

- Respiratory arrest.
- Heart failure.
- Severe bleeding.

Check for these three conditions while being careful not to move the victim any

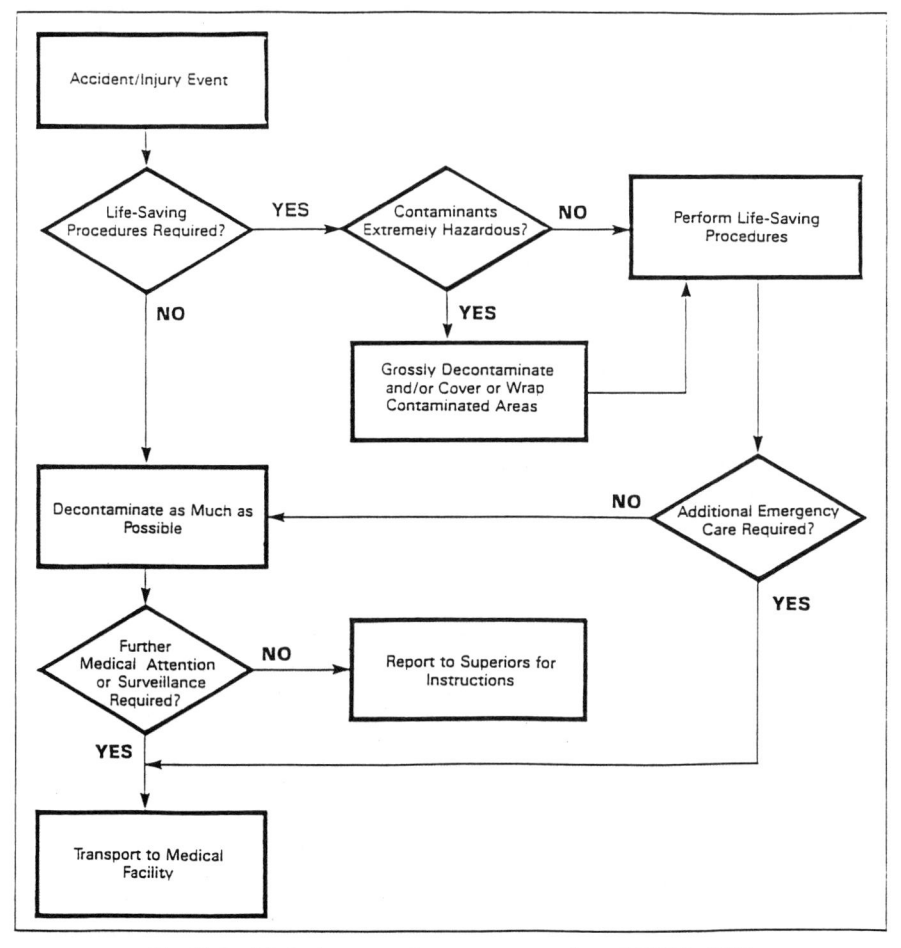

306 WORKER PROTECTION DURING HAZARDOUS WASTE REMEDIATION

Fig. 12-5. Medical emergencies response procedures (NIOSH, 1985).

more than is necessary to support life. Control these life-threatening conditions first.

Respiratory arrest requires immediate action. Victim recovery rate is 98% if artificial ventilation begins within one minute after breathing stops, but drops to 75% after two minutes and declines rapidly after three minutes.

Bleeding control is often very simple. Most external bleeding can be controlled by applying direct pressure to the open wound, permitting normal blood clotting to occur. Place gauze or the cleanest material available against the bleeding point and apply firm pressure with the hand until a bandage can be applied. The bandage knot should be tied over the wound and this, like all bandages, should be snug but not uncomfortably tight. If bleeding continues, apply more pressure. Elevate the bleeding part of the body above the level of the heart to slow the flow of blood and speed clotting.

In cases of severe bleeding where direct pressure is not controlling the bleeding, finger or hand pressure on an artery at a pressure point should be applied in addition to direct pressure. Direct pressure bandages should not be removed when bleeding stops, but pressure point contact should be stopped when bleeding stops. The pressure points most often used in first aid are inside the elbow and at the groin. A tourniquet, which may be very dangerous, should be used only when large arteries are cut or in cases of limb severance. Training in tourniquet application is complex, and is provided by the American Red Cross.

Loss of blood or other factors may lead to *shock*, a collapse or depression of the cardiovascular system due to accident or sudden illness. Circulation is reduced and organs begin to die. Symptoms are:

- Face: pale, dazed look.
- Eyes: dull; pupils dilated (larger).
- Breathing: shallow, irregular.
- Pulse: rapid, weak.
- Skin: cold, clammy.

First aid for shock includes:

- Open airway; start artificial ventilation or CPR if necessary.
- Ensure fresh air.
- Loosen clothing.
- Keep victim warm and dry.
- Give nothing to eat or drink.
- Calm and reassure the victim; keep onlookers away.

Anaphylactic shock, a sensitivity reaction to ingested, inhaled, or injected substances (including insect venom) is a true emergency and requires medication.

Secondary Survey

When the life-threatening conditions have been controlled, the secondary survey should begin. The secondary survey is a head-to-toe examination to check carefully for any additional injuries that can cause serious complications.

- Scalp cuts and skull injury. Without moving the head, check for blood in the hair, and gently feel for bone fragments or depressions.
- Spine fractures, especially in the neck area. Gently feel; if a spinal injury is suspected, stabilize the head immediately: Paralysis of arms or legs indicates spinal injury.

- Chest fractures and penetrating chest wounds. Observe chest movement; look for sides not rising together or one side not moving.
- Fractures. Check for grating, discoloration, swelling, tenderness, bony protrusions, lumps.
- Abdominal spasms and tenderness. Gently feel the abdomen.
- Burns and wounds. Visually examine the victim.

First aid treatment of injuries identified during the secondary survey involves application of a wide range of information and can be referenced through the American Red Cross Library. The goals of first aid are stated at the beginning of this chapter, and can be summarized as follows: prevention of further harm (both injury and infection specifically from moving the worker or in his decontamination) and to stabilize the victim for transportation to a medical facility.

Basic guidelines of treatment for secondary injuries are as follows:

Fractures. Do not move the victim until trained help arrives with proper stabilizing equipment.

Wounds. Cover cuts and scrapes which are not bleeding severely with a sterile bandage, after gently cleaning them if they are small. If a foreign object is in the wound, remove it only if it is small and near the surface. Puncture wounds present the added danger of tetanus infection (lockjaw) and a tetanus shot may be advised by the medical personnel. When a person is impaled on a fixed object, if possible cut or remove the object from its attachment and leave it in the victim as he is transported for treatment.

Burns. Determine the severity of a burn by depth (called *degree*), size, and location. Third degree burns are the deepest. Critical locations are hands, feet, face, and genitals. Cool thin, but not deep, burns in water; small deep burns may be cooled with cold packs. Cover a burn with a sterile dressing, covering charred clothing which is stuck to the burn. Wash all chemical burns with water immediately, and remove chemical-saturated clothing even if it is touching a burn. For chemical burns in the eye, hold the eye open and wash it under running water for 15 minutes.

REFERENCES

AEMA. 1987. "Memo EPA Title III Fact Sheet." Alabama Emergency Management Agency, The State of Alabama.

American Red Cross. 1979. *Standard First Aid and Personal Safety*, 2nd Edition.

Ansul, Inc. 1987. *The Facts about Spill Control* Marinette, WI

EPA. 1987. Title 40 CPR 355. List of Extremely Hazardous Substances. U.S. Environmental Protection Agency. Washington, DC: Office of the Federal Register.

Hammer, W. 1985. Chapter 20 in *Occupational Safety Management and Engineering* 3rd Edition. Englewood Cliffs, NJ: Prentice Hall, Inc.

Meyer, E. 1977. *Chemistry of Hazardous Materials.* Englewood Cliffs, NJ: Prentice Hall, Inc.

NFPA. 1986. *Fire Protection Guide on Hazardous Materials*, 9th Edition. Quincy, MA: National Fire Protection Association.

NIOSH/OSHA/CG/EPA. 1985. Chapter 12, *Occupational Safety and Health Guidance Manual for Hazardous Waste Site Activities.* NIOSH Publication No. 85-115. Washington, DC: U.S. Govt. Printing Office.

OSHA. 1989. Title 29 CFR 1910.120. Hazardous Waste Operations and Emergency Response; Final Rule. U.S. Occupational Safety and Health Administration. Washington, DC: Office of the Federal Register.

13

Transportation of Hazardous Wastes

INTRODUCTION

Public pressure has led to increasingly strict governmental regulations on the transportation of hazardous materials. The hazardous waste stored at an abandoned or uncontrolled site is subject to these regulations when it is shipped to a treatment, storage, and disposal facility.

The cleanup contractor, as the generator of waste for shipment, and the transporter who carries it are subject to shipping regulations set by several federal agencies. Workers who are engaged in labeling drums and other shipping containers or loading and decontaminating trucks need specialized training in these tasks.

OBJECTIVES

- The reader will become familiar with applicable government regulations
- The reader will be able to identify labels and placards, and understand their meaning.
- The reader will learn to read Hazardous Material Table 172.101.
- The reader will become knowledgeable about manifests.
- The reader will be familiar with proper waste loading procedures and decontamination of vehicles.

DIFFERENTIATION OF TERMS

Three government agencies define and classify hazardous chemicals and other substances, each from a different point of view. The terminology is important in a discussion of workplace, environmental, and transportation regulations, since each term connotes subjectivity to a different set of regulations.

Hazardous chemicals, as defined by OSHA, are chemicals which are hazardous to people in the workplace or to the community if released. A Material Safety Data Sheet for each hazardous chemical used must be prepared by a responsible person in a company, and kept on file in a location accessible to workers and members of the community.

Hazardous materials are those materials which may present a danger during shipment by truck, rail, air, or water as determined by the Secretary of Transportation. If a hazardous material listed in the DOT Hazardous Materials Table is transported in one package in a quantity equal to or greater than the reportable quantity for that material, it is subject to DOT regulation.

Hazardous substances are substances determined by the Environmental Protection Agency to present a danger to the environment. If a hazardous substance is spilled or otherwise released into air, ground, or water in excess of EPA's listed reportable quantity, the release must be reported to EPA.

Hazardous wastes are defined by EPA as hazardous substances which have no commercial value. Unable to sell or recycle these substances, a contractor consigns hazardous wastes for transport to an appropriate treatment, storage, and disposal (TSD) facility. Although hazardous waste accounts for only 0.2% of all hazardous material transported annually in the United States, most substances removed from hazardous waste cleanup sites are classified as hazardous wastes. Any carrier engaged in the United States transportation of any hazardous waste must obtain an EPA Transporter Identification Number.

GOVERNMENT REGULATIONS

The Department of Transportation has been the primary regulatory agency for hazardous materials interstate transport since the enactment of the Transportation Safety Act in 1974. The Hazardous Materials Transportation Act in 1975 gave DOT the authority to impose stiff financial penalties for violations. In 49 CFR (Code of Federal Regulations), DOT regulations are spelled out which establish criteria for packaging, labeling, placarding, shipping papers, and training and responsibilities of personnel involved in the shipment of hazardous materials.

The Resource Conservation and Recovery Act (RCRA) of 1976 authorized the Environmental Protection Agency to institute a program of hazardous waste management initiating "cradle-to-grave" tracking of hazardous wastes through a prescribed manifest system; worker training and management of environmental releases were added. EPA regulations are set forth in 40 CFR; however, since Congress instructed EPA to coordinate closely with DOT, the agency relied heavily on the department's regulations already in effect and 49 CFR was expanded to include EPA's list of hazardous substances.

The Comprehensive Environmental Response, Compensation and Liability Act of 1980 (CERCLA), commonly referred to as the Superfund Act, increased penalties for unreported release incidents including spills during transportation.

EPA categorizes hazardous wastes in one way; DOT groups hazardous material in another and requires labeling by DOT hazard class. United Nations hazard class numbers must appear on international shipments, and are frequently seen in the lower corner of DOT labels and placards.

Hazardous waste shippers must determine the DOT classification of hazardous wastes; the relationship is clear from a comparison of Tables 13-1 and 13-2.

Table 13-1. EPA Hazard Categories.

Ignitable	Burnable solids, liquids, gases
Corrosive	Having pH of 2 or below or equal to or above 12.5
Reactive	Explosive, or violently reacting with air or water
Toxic	Poisonous to living organisms

Table 13-2. DOT Hazard Classes.

Blasting agent	Designated for blasting; little probability of initiating an explosion
Class A explosives	Function by detonation
Class B explosives	Function by rapid combustion
Class C explosives	Manufactured articles containing restricted quantities of A or B
Combustible liquid	Flashpoint 100–200°F
Compressed gas	Absolute pressure exceeding 40 psi at 70°F, 104 psi at 130°F, or flammable and 40 psi at 100°F
(a) Flammable gas	Flammable or explosive
(b) Non-flammable gas	Non-flammable or not explosive, hazardous by virtue of being under pressure
Corrosive material	Visible destruction or irreversible damage to skin, severe corrosion rate on steel
Cryogenic liquid	Gas converted to liquid at extremely cold temperature (below −130°F)
Etiologic agents	May cause human disease
Flammable liquid	Flashpoint of less than 100°F
Flammable solid	Likely to cause fire through absorption of water or spontaneous chemical changes
Irritating materials	Give off dangerous or intensely irritating fumes when exposed to air or fire
Organic peroxide	Derivative of hydrogen peroxide in which part of the hydrogen has been replaced by an organic material
ORM-A, B, C, D, E	Other regulated materials that do not meet the definition of a hazardous material listed in the Hazardous Materials Table
Oxidizer	Yields oxygen readily to stimulate the combustion of certain other substances
Poison A	Specifically enumerated gases or liquids of such nature that a small quantity of gas or the vapor of liquid is dangerous to life
Poison B	So toxic to man as to create a hazard during transportation
Radioactive	Spontaneously emits radiation capable of penetrating and damaging living tissue
Toxic by inhalation	Gives off toxic vapors at normal temperature (68°F)

Table 13-3 is included for your information, but is not required for shipments inside the United States.

The *Hazardous Materials Table*, found in *Section 172.101 of 49 CFR*, is the source of information to be used in determining proper shipping name, packaging, labeling, and hazard class of materials for shipment. The table is also printed in books published by the American Trucking Association in which the table has been adapted to highway shipment by replacement of air and water regulation columns with a truck placarding column (Currie, 1987). Table 13-4 shows a section of the Hazardous Materials Table. Table 13-5 explains the contents of the Hazardous Materials Table and gives a key to the symbols used therein.

Special Situations

Situations may exist where mixtures of wastes in containers, or combinations of containers in trucks, complicate the determination of accurate hazard class for labeling and placarding. The following guidelines may be used in these situations.

Mixtures in a Container. If mixed flammable liquids form a compound not listed in the Hazardous Materials Table, the proper shipping name is Flammable Liquid, n.o.s. (not otherwise specified). The container is labeled Flammable Liquid, and the truck placarded similarly. A mixture or solution of hazardous and nonhazardous materials is labeled by the hazard class of the listed materials.

A Truck with a Mixed Load of Containers. If two or more hazard classes are being transported on a truck the "Dangerous" placard plus the placard de-

Table 13-3. UN Hazard Classes.

Class 1	Explosives
Class 2	Gases
Class 3	Flammable liquids
Class 4	Flammable solids
	Spontaneously combustible
	Dangerous when wet
Class 5	Oxidizers
	Organic peroxides
Class 6	Poisonous
	Etiological Agents
Class 7	Radioactives
Class 8	Corrosives
Class 9	Miscellaneous

314 WORKER PROTECTION DURING HAZARDOUS WASTE REMEDIATION

Table 13-4. Hazardous Materials Table.

(1) +/A/W	(2) Hazardous materials descriptions and proper shipping names	(3) Hazard class	(3A) Identification number	(4) Label(s) required (if not excepted)	(5) Packaging (a) Exceptions	(5) Packaging (b) Specific requirements	(6) Maximum net quantity in one package (a) Passenger carrying aircraft or railcar	(6) (b) Cargo only aircraft	(7) Water shipments (a) Cargo vessel	(7) (b) Passenger vessel	(7) (c) Other requirements
A	Sodium dichromate	ORM-A	NA1479	None	173.505	173.510	No limit	No limit	1,2	1,2	
A	Sodium fluoride, solid	ORM-B	UN1690	None	173.505	173.510	No limit	No limit	1,2	1,2	
	Sodium fluoride solution	Corrosive material	UN1690	Corrosive	173.244	173.245	1 quart	5 gallons	1,2	1,2	Stow away from acids
	Sodium hydrate. See Sodium hydroxide										
	Sodium hydride	Flammable solid	UN1427	Flammable solid and Dangerous when wet	None	173.198	Forbidden	25 pounds	1,2	5	Segregation same as for flammable solids labeled Dangerous When Wet
A	Sodium hydrogen sulfate, solid	ORM-B	UN1821	None	173.505	173.800	25 pounds	100 pounds	1,2	1,2	
	Sodium hydrogen sulfate solution	Corrosive material	UN2837	Corrosive	173.244	173.245	1 quart	1 gallon			
A	Sodium hydrogen sulfite, solid	ORM-B	NA2693	None	173.505	173.800	25 pounds	100 pounds	1,2	1,2	
	Sodium hydrogen sulfite, solution	Corrosive material	NA2693	Corrosive	173.244	173.245	1 quart	5 gallons	1,2	1,2	
	Sodium hydrosulfide, solid (with less than 25% water of crystallization)	Flammable solid	UN2318	Flammable solid	173.153	173.154	25 pounds	100 pounds	1,2	1,2	
	Sodium hydrosulfide, solid (with not less than 25% water of crystallization)	Corrosive material	NA2923	Corrosive	173.244	173.245b	25 pounds	100 pounds	1,2	1,2	
	Sodium hydrosulfide, solution	Corrosive material	NA2922	Corrosive	173.244	173.245	1 quart	5 gallons	1,2	1,2	
	Sodium hydrosulfite (*sodium dithionite*)	Flammable solid	UN1384	Flammable solid	173.153	173.204	25 pounds	100 pounds	1,2	1,2	Keep dry. Below deck stowage in metal drums only. Separate from flammable gases, liquids, oxidizing materials, or organic peroxides
	Sodium hydroxide, dry, solid, flake, bead, or granular	Corrosive material	UN1823	Corrosive	173.244	173.245b	25 pounds	200 pounds	1,2	1,2	Keep dry
	Sodium hydroxide, liquid *or* solution	Corrosive material	UN1824	Corrosive	173.244	173.249	1 quart	5 gallons	1,2	1,2	
	Sodium hypochlorite. See Hypochlorite solution or Hypochlorite solution containing not more than 7% available chlorine										
A	Sodium metabisulfite	ORM-B	NA2693	None	173.505	173.510	No limit	No limit	1,2	5	Segregation same as for flammable solids labeled Dangerous When Wet
	Sodium, metal *or* metallic	Flammable solid	UN1428	Flammable solid and Dangerous when wet	None	173.206	Forbidden	25 pounds			
	Sodium, metal dispersion in organic solvent	Flammable solid	UN1429	Flammable solid and Dangerous when wet	None	173.230	Forbidden	10 pounds	1,2	5	Segregation same as for flammable solids labeled Dangerous When Wet

Table 13-5. Hazardous Material Table, Key to Symbols.

Column 1	Notes the applicability of the regulations in special situations. • The plus (+) fixes the proper shipping name and hazard class for that entry without regard to whether the material meets the definition of that class. • A letter "A," "W" signifies the commodity subject to the regulations only when transported by air (A) and/or water (W). Exception: Hazardous wastes are subject to the regulations when transported by any means. • Letter "E" identifies materials classified as hazardous substances, regardless of mode of transportation.
Column 2	Lists the hazardous materials description and proper shipping name. It is the responsibility of the shipper to provide this information. *Note:* The numbers in italics (RQ 1000/454) following a shipping name of a material identified by the letter "E" in column 1 means that the reportable quantity is 1000 pounds or 454 kilograms.
Column 3	Shows the hazard class for each commodity listed. "Forbidden" means material prohibited from being offered or accepted for transportation, but does not apply if materials are diluted, stabilized, or incorporated in devices.
Column 3A	Lists the identification numbers assigned to hazardous materials: "UN"—international and domestic shipments "NA"—not recognized for international shipments, except to and from Canada
Column 4	Shows the hazardous materials label(s) that must be affixed to each package. For some, the table specifies that more than one label be used. (Arsine requires both Poison Gas and Flammable Gas labels, but is assigned the single hazard class of Poison A).
Column 5	If a rule number is shown, the rule sets forth the circumstances under which there is a "Limited Quantity" exception. If "none" appears, no "Limited Quantity" exception is provided for the material.
Column 5B	Lists the number of the rule which sets forth the packaging specifications for each hazardous material. *Note:* Information here is primarily of concern to the shipper, but it may contain information on packaging specification numbers which can be useful to the motor carrier in determining that a hazardous material is in the proper package or container.
Columns 6A, 6B	Information applicable to air transportation.
Columns 7A, 7B, 7C	Information applicable to water transportation.

scribing the highest hazard, may be used. (A list of hazard priorities is found under Shipment of Samples.)

Hazardous Wastes Which Are Also Hazardous Materials. These are properly named with the word "Waste" preceding the shipping name on the shipping paper and package markings, unless the word "waste" is included in the proper shipping name in the table (as "Hazardous Waste, liquid or solid, n.o.s.")—for example, "Waste Acetone."

LABELING CONTAINERS

Hazard class labels must be affixed to all containers with capacity of 110 gallons or less. Labels are strictly regulated regarding color, print type and size, and placement on containers. In addition to the proper label, the following markings should be stenciled or otherwise affixed to each container.

- Proper shipping name.
- UN or NA identification number.
- Gross weight if over 110 pounds.
- "Inhalation Hazard" if the package contains one (except in one liter or less in a lab pack).
- "Dangerous When Wet" if the material is water reactive.
- Name and address of consignee if it will be transferred during shipment.
- "This End Up" for liquids (not gas cylinders).
- Bung label on metal drums containing certain flammable liquids.
- "Hazardous Waste" label.
- Reportable quantity if it is an EPA Hazardous Substance.

Labels are not required on ORM class materials. ORM-E wastes must be listed on the manifest, and labeled if the package contains any other hazardous materials requiring a label.

Each package containing a single material meeting the definition of more than one hazard class must be labeled for each class. Table 13-6 shows hazard class combinations requiring multiple labels for a single material.

Multiple labels are required for multiple-class materials; however, only one

Table 13-6. Multiple Labels.

Hazard Class		Additional Class
Explosive	+	any other
Radioactive	+	any other
Poison A	+	any other
Poison B	+	flammable liquid
		flammable solid
		oxidizer
		corrosive
Corrosive	+	flammable liquid
		flammable solid

placard is used on the vehicle. Hydrogen sulfide, labeled Poison and Flammable Gas, is placarded Flammable Gas. The easiest way to determine appropriate placarding is to use the American Trucking Associations adaptation of the Hazardous Materials Table (49 CFR Section 172.101) from the transportation standard.

PLACARDING VEHICLES

All vehicles transporting hazardous materials must bear a placard on *each end* and *each side*, away from other writing and *clearly visible*. The size, color, and printing of placards is regulated in the same manner as has been described regarding labels. Carriers transporting radioactive materials, combustible liquids in containers less than 110 gallons (see the note below), etiologic agents, ORM class materials, and small hazardous loads which total less than 1,000 pounds are not required to placard these materials, except when the small load contains explosive A or B, poison A, water reactive flammable solids, or radioactive III materials.

Note: EPA and DOT are not in agreement on criteria for combustibles. DOT classifies materials with flashpoints of 0–100°F as flammable, and 100–200°F as combustible. EPA categorizes as ignitables materials with 0–140°F flashpoint. Hazardous wastes with flashpoints between 100°F and 140°F are ignitable by EPA standards, but are combustible when shipped and fall under the DOT labels requirement and placard exception in containers with capacities less than 110 gallons. A liquid with a flashpoint of above 140°F is not of concern on a site, where EPA hazard classes are the important classes, but when it leaves the site on a truck it is subject to DOT regulations and is placarded as combustible.

The placard imparts information in several ways simultaneously, as can be seen in Fig. 13-1. Placards on tank trucks must also include the UN identification number of the material, as shown below, and tankers with several compartments must display a placard for each.

It is important that placards be accurate. A shipper or carrier who overplacards, or uses placards indicating a hazard more serious than is actually present, is subject to large fines, as is one who underplacards his vehicles.

MANIFEST SYSTEM

Several types of shipping papers normally accompany transported hazardous materials; however, when hazardous waste is transported the only legally authorized document, which must accompany the waste at all times, is the original generator-prepared hazardous waste manifest.

The manifest must contain the proper shipping name from the Hazardous Materials Table, the hazard class, the identification number, and the total quan-

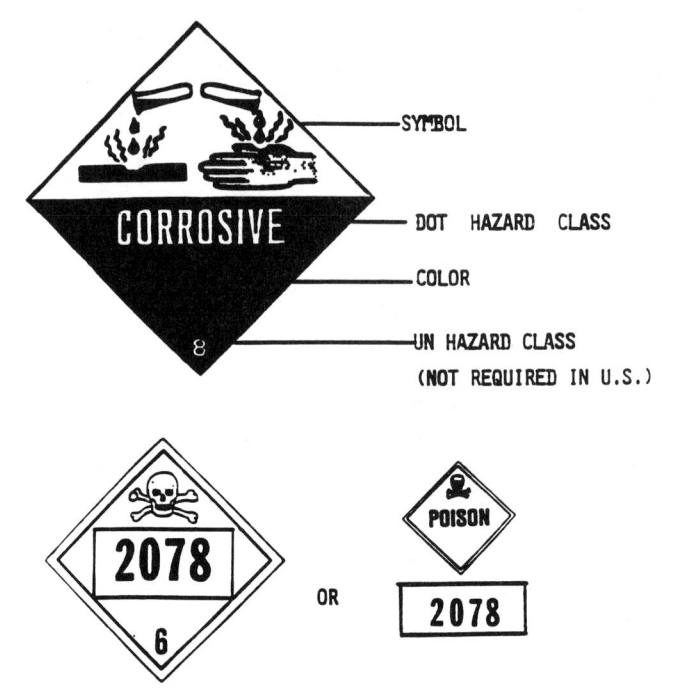

Fig. 13-1. DOT placard examples (DOT, 1976).

tity, typed or printed in English. These are not estimates: there must be no discrepancies! The transporter must verify the manifest listing upon receipt of the materials, as he too can be prosecuted for manifest errors or for unknowingly transporting hazardous wastes. The manifest is a multilayered form, with copies for the generator, transporter, and disposer (who keeps one, sends one back to the generator, and sends one to the state regulatory agency). During transportation, the manifest must be readily accessible in the cab of the truck, separate from other shipping papers or conspicuous by being tabbed or placed first, and must be within reach of a seatbelted driver and visible to a person entering the cab. A manifest is shown in Fig. 13-2.

TRUCKS ON SITE

One person should be responsible for routing transport trucks on a hazardous waste site in such a way as to provide the least hazard to human and vehicular traffic and to facilitate decontamination. Transporters should be furnished appropriate protective clothing and, of course, properly completed and signed manifest forms.

UNIFORM HAZARDOUS WASTE MANIFEST

1. Generator's US EPA ID No. | Manifest Document No. | 2. Page 1 of | Information in the shaded areas is not required by Federal law.

3. Generator's Name and Mailing Address

A. State Manifest Document Number

B. State Generator's ID

4. Generator's Phone ( )

5. Transporter 1 Company Name | 6. US EPA ID Number | C. State Transporter's ID | D. Transporter's Phone

7. Transporter 2 Company Name | 8. US EPA ID Number | E. State Transporter's ID | F. Transporter's Phone

9. Designated Facility Name and Site Address | 10. US EPA ID Number | G. State Facility's ID | H. Facility's Phone

11. US DOT Description *(Including Proper Shipping Name, Hazard Class, and ID Number)* HM	12. Containers No.	Type	13. Total Quantity	14. Unit Wt/Vol	I. Waste No.
a.					
b.					
c.					
d.					

J. Additional Descriptions for Materials Listed Above

K. Handling Codes for Wastes Listed Above

15. Special Handling Instructions and Additional Information

16. **GENERATOR'S CERTIFICATION:** I hereby declare that the contents of this consignment are fully and accurately described above by proper shipping name and are classified, packed, marked, and labeled, and are in all respects in proper condition for transport by highway according to applicable international and national governmental regulations.

If I am a large quantity generator, I certify that I have a program in place to reduce the volume and toxicity of waste generated to the degree I have determined to be economically practicable and that I have selected the practicable method of treatment, storage, or disposal currently available to me which minimizes the present and future threat to human health and the environment; OR, if I am a small quantity generator, I have made a good faith effort to minimize my waste generation and select the best waste management method that is available to me and that I can afford.

Printed/Typed Name | Signature | Month Day Year

GENERATOR

17. Transporter 1 Acknowledgement of Receipt of Materials

Printed/Typed Name | Signature | Month Day Year

18. Transporter 2 Acknowledgement of Receipt of Materials

Printed/Typed Name | Signature | Month Day Year

TRANSPORTER

19. Discrepancy Indication Space

20. Facility Owner or Operator: Certification of receipt of hazardous materials covered by this manifest except as noted in Item 19.

Printed/Typed Name | Signature | Month Day Year

FACILITY

ORIGINAL - RETURN TO GENERATOR

Fig. 13-2. The uniform manifest form used for hazardous waste shipment (EPA, 1982).

Truck loading of bulk waste or containers should be accomplished according to a plan which minimizes contamination of the truck itself and the surrounding ground area. Plastic ground covers and truck bed liners are used for this purpose, and covers should be added after loading to prevent loss of contaminated materials during transport.

Any part of a truck which touches contaminated materials or contamination on the ground must be decontaminated as the truck goes through the contamination reduction corridor. As with personnel and tool decontamination, all wash and rinse solutions must be contained and disposed of as hazardous waste. A list of recommended equipment for vehicle decontamination is provided in Chapter 14.

TRANSPORTATION ACCIDENTS

Transportation accidents are regulated by 40 CFR and 49 CFR. In the event of a discharge of hazardous waste during transportation, the transporter must take appropriate and immediate action to protect human health and the environment. It is the transporter's responsibility to clean up the discharge. A call to the National Response Center (NRC) must be made as promptly as possible if the accident results in a death or injury requiring hospitalization, estimated damage exceeding $50,000, fire or spill of radioactive materials or etiologic agents, or any other situation in which the driver feels a report is justified. A written report must follow within 15 days. Releases of reportable quantities of hazardous substances must be reported to the U.S. Coast Guard National Response Center and to the local emergency planning committee.

Emergency hotlines should be contacted for immediate advice:

CHEMTREC: 1-800-424-9300   NRC: 1-202-366-4488

Public relations following transportation accidents which result in a release of a hazardous substance or waste, or even the possibility of such a release as perceived by the community, can be handled well if prior planning is done and a properly prepared spokesperson is at the site quickly.

COOPERATION WITH THE TSD FACILITY

On arrival at the treatment, storage, and disposal facility, the load will be checked for the following:

- Batch count (number of containers) exact?
- Bulk load accurate within 10%?
- Physical condition of containers acceptable?
- Inventory of contents and Generator's Waste Material Profile Sheet on file,

having been submitted for approval in advance? At most facilities, delivery is limited to certain hours and days, and must be scheduled with the dispatcher in advance.

REFERENCES

Currie, J. V. 1987. *Driver's Guide to Hazardous Materials*. Alexandria, VA: American Trucking Association.
U.S. Department of Transportation, Title 49 CFR. Washington, DC: U.S. Govt. Printing Office.
U.S. Environmental Protection Agency. Title 40 CFR 261. Washington, DC: U.S. Govt. Printing Office.

14

Decontamination

INTRODUCTION

Decontamination is the process of removing or neutralizing contaminants that have accumulated on personnel and equipment and is critical to health and safety at hazardous waste sites. Decontamination protects workers from hazardous substances that may contaminate and eventually permeate the protective clothing, respiratory equipment, tools, vehicles, and other equipment used on site; it protects all site personnel by minimizing the transfer of harmful materials into clean areas; it helps prevent mixing of incompatible chemicals; and it protects the community by preventing uncontrolled transportation of contaminants from the site.

OBJECTIVES

On completion of this chapter, reader will:

- Know the goals and components of a good decontamination plan.
- Differentiate between physical and chemical decontamination methods and appropriate use of each.
- Be able to describe procedures for decontaminating personnel, clothing, tools, and vehicles.
- Be aware of the effectiveness of evaluation techniques.

PROGRAM DESIGN

A decontamination plan must be developed as part of the site safety plan and put into effect before any personnel or equipment may enter areas where the potential for exposure to hazardous substances exists. Factors which must be considered in planning include, but are certainly not limited to, those listed below. When variables in the factors change as the investigation or cleanup progresses, the plan must be changed to accommodate new conditions.

What Types of Contaminants Are Suspected to Be On Site?

Different methods of removal are used for different substances: chemical poisons versus infectious wastes; dusts or other particles versus liquids; acids versus bases; water soluble chemicals versus chemicals insoluble in water.

What Will Be the Level of Activity and the Potential for Worker Contamination?

Decontamination for investigations involving limited contact with contaminants during sampling may be less elaborate than decontamination of workers who handle wastes during the clean-up phase.

How Contaminated Is the Site?

Liquid waste contained in intact tanks or drums may be handled with less worker contamination than spilled liquid or leaking containers. Solid wastes may be limited to containers or spread throughout the site.

How Long Will Work Continue on the Site?

A long-term cleanup operation makes permanent, well equipped decontamination stations feasible. Brief sampling or cleanup activities, which may move from spot to spot, require portable equipment: a "decon kit" would be appropriate or, for companies with many workers, a mobile self-contained facility such as a trailer can be used. Emergency response to spills or accidents calls for immediate decontamination arrangements.

How Much Space is Available at the Appropriate Location on the Site?

The arrangement of decontamination stations will depend on the size, shape, and topography of the space available adjacent to the exclusion zone or hot zone.

What Utilities Will Be Available on the Site?

If clean water is not available it will have to be brought in. Some means of producing hot water or steam may be desired.

How Can Safe and Efficient Arrangements Be Made for Disposal of Contaminated Clothing, Equipment, and Solutions Which Cannot Be Cleaned?

All these materials must be containerized and properly disposed of. In some cases decontamination solutions may be disposed of at wastewater treatment plants. If not, it may have to be disposed of as a hazardous waste, as will the other decontamination materials.

Will Other Functional Units Be Incorporated into the Decontamination Corridor(s)?

Rest room facilities, an air tank change area, worker cooling stations, and an emergency medical station can safely be included at appropriate locations in the personnel decontamination corridor. A truck decontamination plan should include provisions for decontaminating drivers and helpers, and accommodations for waiting drivers.

How Much Decontamination Will Be Done on an Ill or Injured Worker?

This must be planned and practiced in advance of emergency so quick, efficient action may be taken. If it is evident that lifesaving procedures are not immediately required, decontaminate the worker as much as possible before he receives medical attention. If it is known or suspected that the victim requires immediate lifesaving measures, and he is contaminated with hazardous substances, transfer of contaminants can be reduced by quickly rinsing or wiping off gross contamination, or covering or wrapping contaminated areas of his clothing. Of course it is important to warn emergency and medical personnel of the nature of the contamination.

METHODS

Removal of contaminants takes place in two stages, with different methods used in each stage. Physical removal of bulk contaminants, such as spilled liquid, mud, or oily accumulation, precedes chemical removal of contaminants which tightly adhere to or permeate materials.

Physical Decontamination

Physical processes such as wiping, scraping, scrubbing, blowing, rinsing with flowing or pressurized water, or volatilizing with steam jets are used to remove heavy external contamination. Removal and disposal of inexpensive outer garments, use of disposable samplers and other tools, and disposal of outer tool protectors or truck bed liners are simple physical removal methods. Some adhesives and muds are easier to remove after being allowed to dry, and some adhesive wastes release their hold after being frozen or melted.

Following physical removal of bulk contaminants, chemical solutions may be used to neutralize, solidify, dissolve, or thermally degrade surface or permeated contaminants.

Many solutions which might react with a contaminant in such a way as to aid its removal are hazardous themselves, and therefore are not appropriate for use on or by humans. Some organic solvents selected for equipment and vehicle decontamination may not be suitable for protective clothing decontamination due to their potential toxicity or flammability. Also, some organic solvents which will not harm tools or vehicles may damage protective clothing made of polymer organics.

Another important consideration is whether the decontamination solution is compatible with the waste on which it will be used, and other solutions with which it may mix during decontamination.

A chemical solution chosen for use in decontamination must be readily available on the site. For this reason, water and detergents are commonly used. Easy storage and handling are additional factors in their favor. Waste solutions may need to be disposed of as hazardous waste.

Cleaning solutions selected for use will be effective if they cause one or more of the following results:

- Dissolve the contaminant so that it remains in the solvent, not on the equipment, and is rinsed away with it. A chemist or industrial hygienist should be consulted for determination of the effectiveness and risks of solvents.
- Rinse progressively with clean solution (preferably in a continuous shower) to maintain a concentration gradient at the contaminated surface. The gradient encourages migration of the contaminant into the solution.
- Reduce adhesion between the contaminant and the surface. Household and other detergents have this action.

DECONTAMINATION PROCEDURES

Lines of stations set up for decontamination of personnel and portable tools are usually combined; vehicles and heavy equipment should be decontaminated at a separate, physically removed location.

Decontamination of Protective Clothing

A series of stations is set up in an orderly fashion (a straight line is preferred, as it prevents misdirection of personnel and possible crosscontamination) such that workers leaving the exclusion zone move into the prevailing wind from greater to lesser contamination. Decontamination and removal of protective garments and other equipment takes place beginning with the most contaminated (outer gloves and boot covers), followed by less contaminated (chemical protective suits), and ending with least contaminated (inner gloves, air tanks). Figs. 14-1 through 14-4 show possible arrangements of stations in the decontamination line under different levels of protection.

Workers move from there to a body wash and redress area at the exit from the contamination reduction corridor into the clean support zone.

Each procedure is performed at a separate station, separated physically and arranged in an order of decreasing contamination. Setup at each station includes supplies appropriate to the station's procedure, and decontamination personnel to help the worker at stations where help is needed.

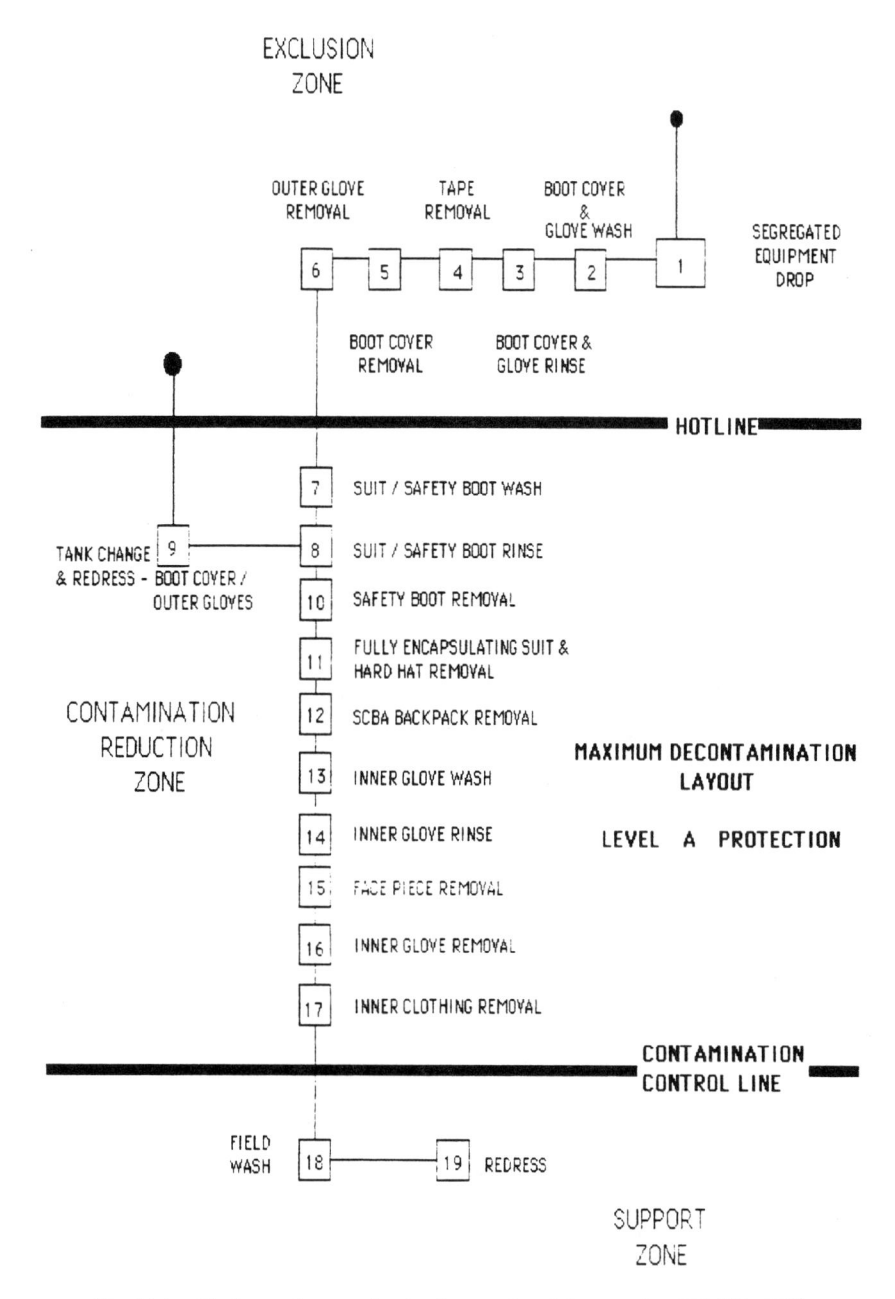

Fig. 14-1. Maximum decontamination layout—Level A protection (NIOSH, 1985).

DECONTAMINATION 327

EXCLUSION ZONE

| 6 | 5 | 4 | 3 | 2 | 1 |

- 6: OUTER GLOVE REMOVAL
- 5: TAPE REMOVAL
- 4: BOOT COVER REMOVAL
- 3: BOOT COVER & GLOVE RINSE
- 2: BOOT COVER & GLOVE WASH
- 1: SEGREGATED EQUIPMENT DROP

════ HOTLINE ════

| 7 | SUIT/SAFETY BOOT WASH |
| 8 | SUIT/SAFETY BOOT RINSE |

9: CANISTER OR MASK CHANGE & REDRESS - BOOT COVER/ OUTER GLOVES

10	SAFETY BOOT REMOVAL
11	SPLASH SUIT REMOVAL
12	INNER GLOVE WASH
13	INNER GLOVE RINSE
14	FACE PIECE REMOVAL
15	INNER GLOVE REMOVAL
16	INNER CLOTHING REMOVAL

CONTAMINATION REDUCTION ZONE

MAXIMUM DECONTAMINATION LAYOUT

LEVEL C PROTECTION

════ CONTAMINATION CONTROL LINE ════

| 17 | FIELD WASH |
| 18 | REDRESS |

SUPPORT ZONE

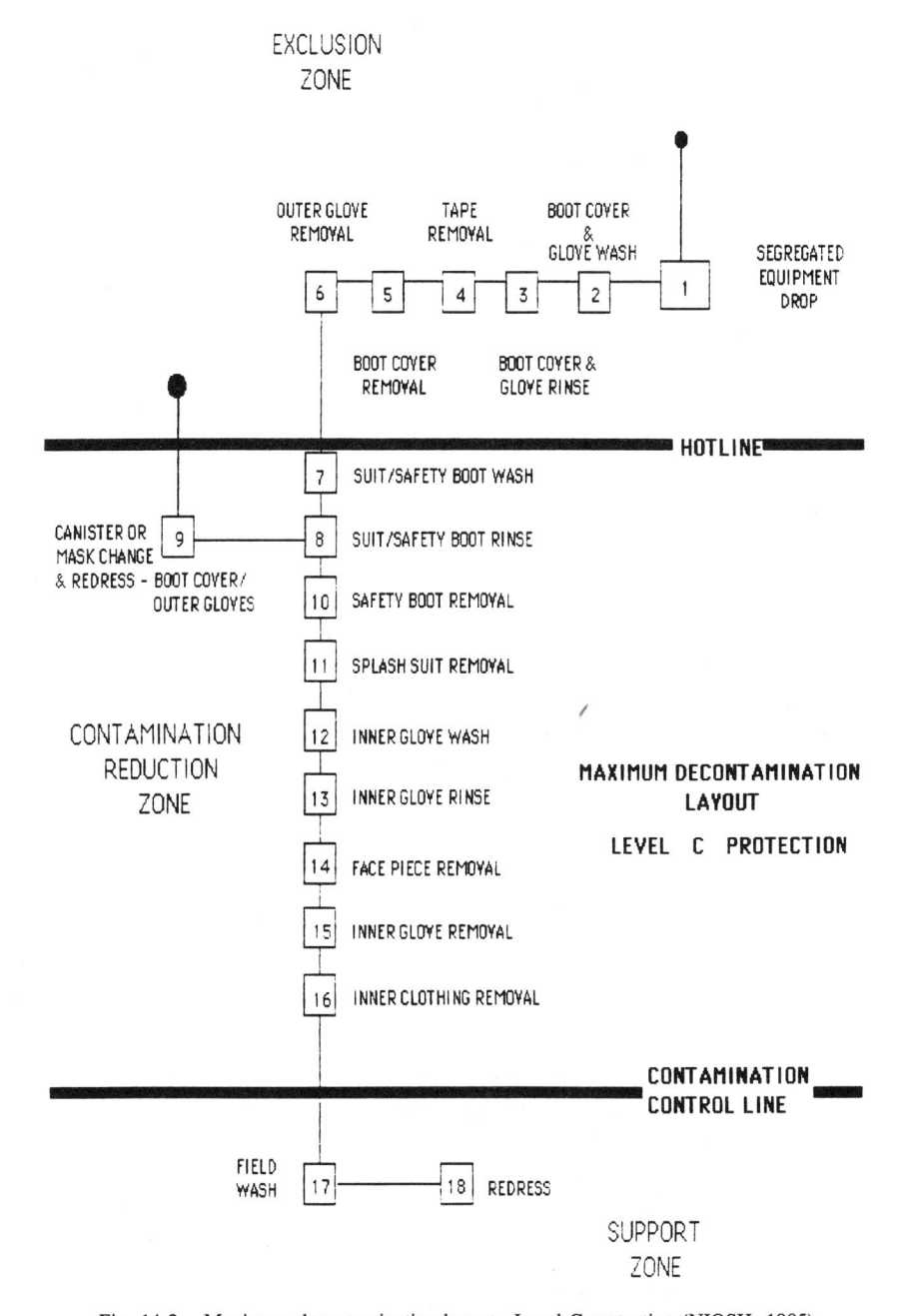

Fig. 14-2. Maximum decontamination layout—Level C protection (NIOSH, 1985).

MINIMUM DECONTAMINATION LAYOUT

LEVELS A & B PROTECTION

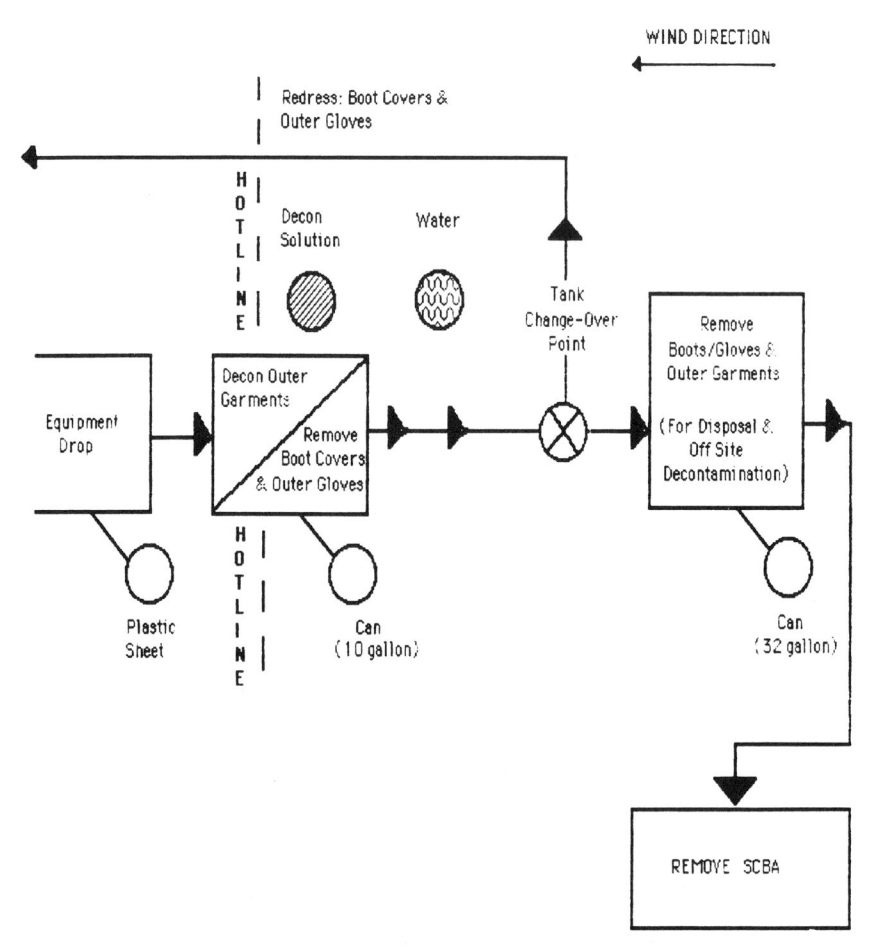

Fig. 14-3. Minimum decontamination layout—Levels A and B protection (NIOSH, 1985).

Some Recommended Equipment for Decontamination of Personnel and Personal Protective Clothing and Equipment.

- Drop cloths of plastic or other suitable materials on which heavily contaminated equipment and outer protective clothing may be deposited.
- Collection containers, such as drums or suitably lined trash cans, for storing disposable clothing and heavily contaminated personal protective clothing or equipment that must be discarded.

MINIMUM DECONTAMINATION LAYOUT

LEVEL C PROTECTION

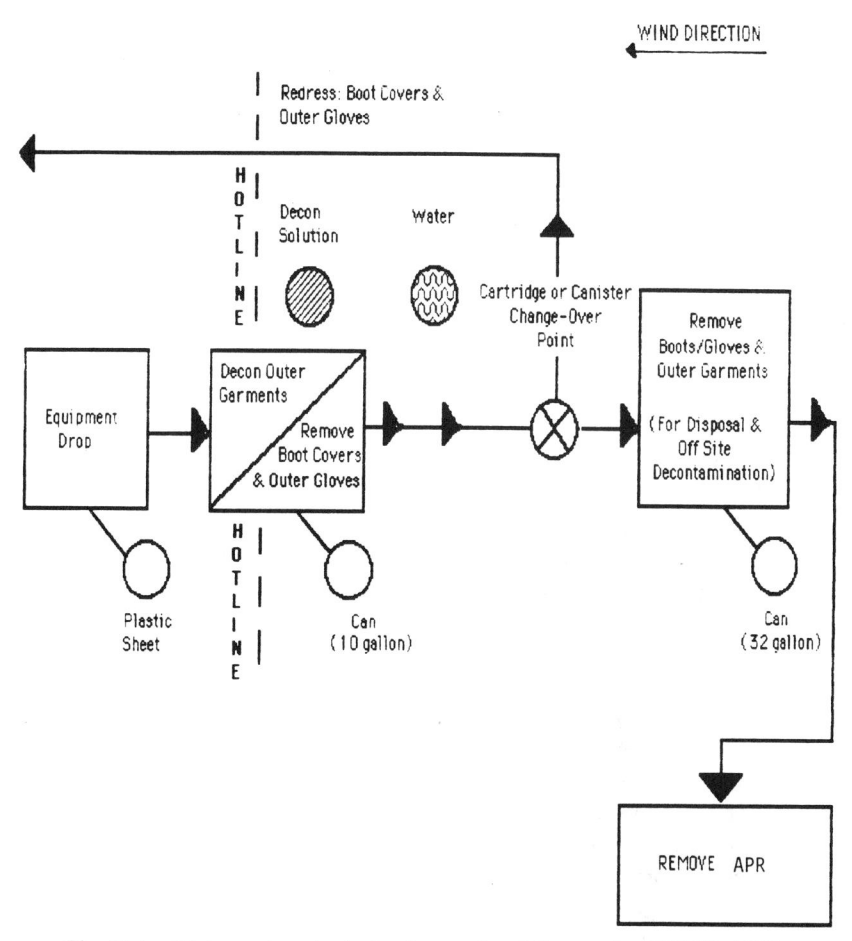

Fig. 14-4. Minimum decontamination layout—Level C protection (NIOSH, 1985).

- Lined box with absorbents for wiping or rinsing off gross contaminants and liquid contaminants.
- Large galvanized tubs, stock tanks, or children's wading pools to hold wash and rinse solutions. These should be at least large enough for a worker to place a booted foot in, and should have either no drain or a drain connected to a collection tank or appropriate treatment system.
- Wash solutions selected to wash off and reduce the hazards associated with the contaminants.

330 WORKER PROTECTION DURING HAZARDOUS WASTE REMEDIATION

- Rinse solutions selected to remove contaminants and contaminated wash solutions.
- Long-handled, soft-bristled brushes to help wash and rinse off contaminants.
- Paper or cloth towels for drying protective clothing and equipment.
- Lockers and cabinets for storage of decontaminated clothing and equipment.
- Metal or plastic cans or drums for storage of contaminated wash and rinse solutions.
- Plastic sheeting, sealed pads with drains, or other appropriate methods for containing and collecting contaminated wash and rinse solutions spilled during decontamination.
- Shower facilities for full body wash or, at a minimum, personal wash sinks (with drains connected to a collection tank or appropriate treatment system).
- Soap or wash solution, wash cloths, and towels for personnel.
- Lockers or closets for clean clothing and personal item storage.

The Decontamination Line. Entry and exit points to the decontamination line should be clearly marked. Workers entering the exclusion zone must pass through the contamination reduction corridor by a route which does not permit them to be contaminated by workers in the decon line. If workers from different parts of the hot zone require different decon solutions or procedures, more than one line of stations is arranged and clearly marked.

Special areas of consideration for inclusion in a decontamination line, if appropriate, are:

- Tank change station, arranged at a point in the line where removal by a helper of part of the worker's suit does not permit contamination of the worker's unprotected body, or of the helper at the tank change station.
- Cooling station in a shady area during hot weather, provided with cooling devices such as water hose or fans and plenty of potable water.
- Rest room beyond the point in the line where minimum decontamination deemed necessary has taken place.
- Emergency medical station, together with an emergency decontamination plan.

Protective clothing should be stored in such a way that air can circulate freely around the outside and inside of the garment to aid in evaporation of remaining solutions and contaminants. Permeated chemicals are extremely difficult to remove from fabrics, and garments found to be permeated should be disposed of if no method can be found which forces the contaminant to reverse its migration into the fabric.

Decontamination of Tools

Hand tools and some portable sampling equipment may be partially protected from gross contamination by being encased in materials such as plastics; this will reduce decontamination. Disposable tools, especially samplers such as the drum thief, are cost effective and enable workers to avoid the problem of decontamination.

In general, decontamination of samplers and other tools is difficult and likely to be less than completely effective. As with clothing, physical removal of liquids, residues, mud, and other thick materials is the important first step. Most tools can take stronger methods, such as chipping or steam jetting, than can be used on personnel in protective clothing. Chemical solutions which are stronger may also be used, as long as personnel using them are protected. The metals and gaskets of tools and equipment can be damaged by strong acids or caustic basic compounds.

All hand tools should be cleaned after each use and before storage.

Decontamination of Vehicles and Heavy Equipment

Normal site operation will probably make daily cleaning of heavy equipment impractical; since most of this equipment is metal and permeation is unlikely to occur, partial daily decontamination is adequate. Gross contamination should be removed daily to prevent buildup. Permeable surfaces such as seats, tires, and the inside of the operator cabin should be completely decontaminated every day. Air vents and ducts should be flushed. Heavy equipment whose exterior is grossly contaminated should remain parked in the exclusion zone when not in use.

Vehicles hauling hazardous waste away from the contaminated areas must be thoroughly decontaminated before leaving the contamination reduction zone.

Some Recommended Equipment for Heavy Equipment and Vehicle Decontamination.

- Storage tanks or appropriate treatment systems for temporary storage and/or treatment of contaminated wash and rinse solutions.
- Drains or pumps for collection of contaminated wash and rinse solutions.
- Long handled brushes for general exterior cleaning.
- Wash solutions selected to remove and reduce the hazards associated with the contamination.
- Rinse solutions selected to remove contaminants and contaminated wash solutions.
- Pressurized sprayers for washing and rinsing, particularly hard-to-reach areas.

- Curtains, enclosures, or spray booths to contain splashes from pressurized sprays.
- Long handled brushes, rods, and shovels for dislodging contaminants and contaminated soil caught in tires and the undersides of vehicles and equipment.
- Containers to hold contaminants and contaminated soil removed from tires and the undersides of vehicles and equipment.
- Wash and rinse buckets for use in the decontamination of operator areas inside vehicles and equipment.
- Brooms and brushes for cleaning operator areas inside vehicles and equipment.
- Containers for storage and disposal of contaminated wash and rinse solution, damaged or heavily contaminated parts, and equipment to be discarded.

PROTECTING DECONTAMINATION PERSONNEL

Decontamination workers may be exposed to hazardous substances from two sources:

- Workers, clothing, and equipment leaving the exclusion zone.
- Decontamination solutions before, during, and after their use.

Protective clothing and respirators which are appropriate to levels of specific contaminants with which they will come in contact must be selected for decontamination personnel. Consideration should be given to splashing and vaporizing of contaminants. Workers at the first several stations entered by contaminated personnel may need the most protection.

These workers must be decontaminated before leaving the contamination reduction corridor, and their clothing and solutions cleaned or disposed of properly.

MEASURING EFFECTIVENESS

It is difficult in many cases to be sure when decontamination procedures have been effective and clothing and equipment can be described with assurance as being clean. Site-specific methods for evaluating the effectiveness of decontamination may be created by qualified persons on the spot, and appropriate known methods included in the evaluation techniques.

A visual inspection in natural or artificial light will show some, but by no means all, staining contaminants. Stains, dirt, or changes in a fabric are detectable; nonstaining or permeating contaminants are not. Inspection by ultraviolet light aids in detection of certain fluorescent contaminants, such as some aromatic hydrocarbons. (Ultraviolet light itself is dangerous, and the user must not look directly at the source.)

Laboratory instruments are available which detect and analyze some surface contaminants, but they are not portable for field use and usually require destruction of the surface. Swabbing the surface with an appropriate solvent and having the swab sample analyzed is useful. The solvent and analytical method should be selected by a chemist. There may be a time lag before results are known. The cleaning solutions themselves may be analyzed after use; again, an inconvenient time lag may occur.

Due to the difficulty of certain determination of the effectiveness of decontamination, procedures should be repeated after obvious contaminants have been removed. This is especially important when the contaminant is an extremely hazardous one.

REFERENCES

NIOSH/OSHA/USCG/EPA. 1985. *Occupational Safety and Health Guidance Manual for Hazardous Waste Site Activities*. Washington, DC: U.S. Govt. Printing Office.

15

Site Safety Plan

INTRODUCTION

In this final chapter we will address an important document, the site safety plan. It is developed to define crucial actions required during any phase of a cleanup. It is used as a working plan and should be modified as changes are foreseen in site activities.

The first part of this chapter will deal with OSHA and EPA requirements with respect to the site safety plan, then a basic "generic" plan is illustrated, with a practical description of each of the major headings required by OSHA, 29 CFR 1910.120. A review of the benefits and advantages to implementing a good site safety plan sets the stage for the final comments concerning hazard prevention through use of the site safety plan and a checklist for all workers to use prior to site entry.

THE ROLE OF OSHA AND EPA

The specifications for a site safety plan required by SARA through OSHA are discussed in Chapter 2 of this book. In summary, OSHA requires that a site safety plan be developed for each site to be investigated, sampled, or remediated and that each and every worker understand its contents.

The EPA gets involved in the site safety plan activities by requiring that all potential contractors submit a skeletal site safety plan with their bid package for each site investigation and upon award complete the details. Selection of the construction phase contractor is made by the U.S. Army Corps of Engineers. They also require a proposed safety plan with each site's bid application. Both the EPA and the Corps review the bid packages on the basis of technical competence and on the financial proposal. Therefore, a low bid contractor may only be awarded the project if the technical proposal is sound and a thorough proposed site safety plan is included.

ANATOMY OF A BASIC PLAN

Plan Description

OSHA Standard 29 CFR 1910.120 contains minimum specifications which all site safety plans must meet. Under this standard, the plan must be a site-specific, written document which comprises a separate chapter of the employer's safety and health program (see Chapter 2). OSHA requires that *all* employees abide by the provisions of the plan. All site safety plans must include *at a minimum*, the following *specific provisions*:

Site *safety and health personnel* must be designated.

- The safety and health supervisor must be named.
- All other key safety and health personnel must be named.

A complete *hazard assessment*, based on a thorough site characterization, must be included for each task and phase of operation involved.

A written *sampling procedure* must be included and followed during all sampling operations.

Employee *training assignments* must be specified. These assignments must be sufficient to ensure that all employees are adequately trained to perform any task which their job assignment may require.

Personal protective equipment to be used by employees must be specified for each task or operation conducted onsite. A PPE program which addresses the following topics must be included:

- Selection.
- Use and limitations.
- Work mission duration.
- Maintenance.
- Storage.
- Decontamination.
- Training and proper fitting.
- Donning and doffing.
- Inspection.
- Limitations during temperature extremes.
- Program evaluation

Medical surveillance requirements for on-site work must be specified.

Site control measures to be used must be included. These measures must be sufficient to prevent the contamination of employees or the migration of contaminants into uncontaminated areas. Site control topics to be covered include;

- Site map.
- Site work zones.
- Standard operational safety procedures.
- Use of the buddy system.
- Communication systems (including alarm systems) to be used on site.
- Identification of the nearest medical assistance.

Decontamination procedures must be specified. These procedures must provide for the decontamination of *all* personnel and equipment exiting the exclusion zone.

Monitoring requirements, as needed to insure the adequate protection of on-site personnel, must be stated. The site monitoring program must specify:

- The frequency and types of air monitoring to be used.
- The frequency and types of personnel monitoring to be used.
- Environmental sampling techniques.
- Instrumentation to be used.
- Methods of instrument calibration and maintenance.

An *emergency response plan* sufficient to handle any anticipated site emergencies is required.

Confined space entry procedures must be specified for any areas of confined space hazard on the site.

A *spill containment program* must be included if major spills are possible.

Trenching and excavation procedures sufficient to comply with applicable OSHA regulations must be specified.

A new technologies program must be included, incorporating procedures for the implementation and evaluation of new technologies, equipment, and control measures designed to enhance employee protection.

Note: Many specific requirements listed in 29 CFR 1910.120 pertain to the provisions listed above. These requirements should be incorporated into the site safety plan in order to comply with the intent of the OSHA standard (see Chapter 2).

Putting The Site Safety Plan Into Action

The site safety plan should be implemented *before* on-site operations begin. An initial plan should be developed prior to the initial site entry and modified or refined, based on the information gathered.

The plan should be modified whenever;

- Site conditions change.
- Different phases of the operation begin.
- Additional information about the site becomes available.

The site safety and health supervisor is responsible for:

- Developing and implementing the plan.
- Verifying compliance with the plan.

- Conducting inspections to evaluate the effectiveness of the plan.
- Modifying the plan as needed.

Requirements for Informing Employees of Provisions of the Plan

OSHA requires that all employees and other involved parties (e.g., employee representatives, contractors, subcontractors, and OSHA officials) be kept informed of the provisions of the site safety plan.

Employee briefings are to be held:

- Prior to initial site entry.
- At other times as often as required to ensure compliance with the plan.

The plan must be posted on site so as to be available to all affected personnel.

Note: A generic site safety plan follows in Fig. 15-1. Please note that this plan is highly generalized and would require extensive modification to produce an acceptable site-specific safety and health plan. The generic plan predates 29 CFR 1910.120 and is out of compliance with several key provisions of the standard. Modification sufficient to eliminate these deficiencies would produce a much longer document.

Generic Site Safety Plan

```
A.  SITE DESCRIPTION
    Date_____    Location_____
    Hazards_____
    Area affected_____

    Surrounding population_____
    Topography_____
    Weather conditions_____

    Additional information_____
    _____
    _____
    _____
```

Fig. 15-1. Generic site safety plan created by U.S. Coast Guard. (NIOSH, 1985).

B. ENTRY OBJECTIVES - The objective of the initial entry to the contaminated area is to __(describes actions, tasks to be accomplished; i.e., identify contaminated soil; monitor conditions, etc.)__

_____
_____
_____

C. ONSITE ORGANIZATION AND COORDINATION - The following personnel are designated to carry out the stated job functions on site. (Note: One person may carry out more than one job function.)

PROJECT TEAM LEADER_____
SCIENTIFIC ADVISOR_____
SITE SAFETY OFFICER_____
PUBLIC INFORMATION OFFICER_____
SECURITY OFFICER_____
RECORDKEEPER_____
FINANCIAL OFFICER_____
FIELD TEAM LEADER_____
FIELD TEAM MEMBERS_____
_____
_____
_____
_____
_____

FEDERAL AGENCY REPS __(i.e., EPA, NIOSH)_____
_____
_____
_____

STATE AGENCY REPS _____
_____
_____

LOCAL AGENCY REPS _____
_____
_____

CONTRACTOR(S) _____
_____
_____

All personnel arriving or departing the site should log in and out with the Recordkeeper. All activities on site must be cleared through the Project Team Leader.

D. ONSITE CONTROL

__(Name of individual or agency)__ has been designated to coordinate access control and security on site. A safe perimeter has been established at __(distance or description of controlled area)__
_____

No unauthorized person should be within this area.

The onsite Command Post and staging area have been established at _____
_____

Fig. 15-1. (*Continued*)

SITE SAFETY PLAN 339

The prevailing wind conditions are _____. This location is upwind from the Exclusion Zone.

Control boundaries have been established, and the Exclusion Zone (the contaminated area), hotline, Contamination Reduction Zone, and Support Zone (clean area) have been identified and designated as follows: __(describe boundaries and/or attach map of controlled area)__
_____
_____
_____

These boundaries are identified by: __(marking of zones, i.e., red boundary tape - hotline; traffic cones - Support Zone; etc.)__
_____
_____

E. HAZARD EVALUATION

The following substance(s) are known or suspected to be on site. The primary hazards of each are identified.

Substances Involved	Concentrations (If Known)	Primary Hazards
(chemical name)		(e.g., toxic on inhalation)

The following additional hazards are expected on site: __(i.e., slippery ground, uneven terrain, etc.)__

Hazardous substance information form(s) for the involved substance(s) have been completed and are attached.

F. PERSONAL PROTECTIVE EQUIPMENT

Based on evaluation of potential hazards, the following levels of personal protection have been designated for the applicable work areas or tasks:

Location	Job Function	Level of Protection
Exclusion Zone	_____	A B C D Other
	_____	A B C D Other
	_____	A B C D Other
	_____	A B C D Other
Contamination Reduction Zone	_____	A B C D Other
	_____	A B C D Other
	_____	A B C D Other
	_____	A B C D Other

Specific protective equipment for each level of protection is as follows:

Level A	Fully-encapsulating suit	Level C	Splash gear (type)
	SCBA		Full-face canister resp.
	(disposable coveralls)		
	_____		_____
	_____		_____

Fig. 15-1. (*Continued*)

Level B   Splash gear (type) _____   Level D   _____
          SCBA _____   _____
          _____   _____
          _____   _____
          _____   _____

Other _____
       _____
       _____

The following protective clothing materials are required for the involved substances:

Substance	Material
(chemical name)	(material name, e.g., Viton)
_____	_____
_____	_____
_____	_____

If air-purifying respirators are authorized, __(filtering medium)__ is the appropriate canister for use with the involved substances and concentrations. A competent individual has determined that all criteria for using this type of respiratory protection have been met.

NO CHANGES TO THE SPECIFIED LEVELS OF PROTECTION SHALL BE MADE WITHOUT THE APPROVAL OF THE SITE SAFETY OFFICER AND THE PROJECT TEAM LEADER.

G. ONSITE WORK PLANS

Work party(s) consisting of ____ persons will perform the following tasks:

Project Team Leader ___(name)___   _____(function)_____

Work Party #1   _____   _____
                                         _____
                                         _____
                                         _____

Work Party #2   _____   _____
                                         _____
                                         _____
                                         _____

Rescue Team   _____   _____
(required for   _____
entries to IDLH   _____
environments)   _____

Decontamination
Team   _____   _____
                                         _____
                                         _____
                                         _____

The work party(s) were briefed on the contents of this plan at _____

Fig. 15-1. (*Continued*)

SITE SAFETY PLAN   341

H. COMMUNICATION PROCEDURES

Channel _____ has been designated as the radio frequency for personnel in the Exclusion Zone. All other onsite communications will use channel _____.

Personnel in the Exclusion Zone should remain in constant radio communication or within sight of the Project Team Leader. Any failure of radio communication requires an evaluation of whether personnel should leave the Exclusion Zone.

__(Horn blast, siren, etc.)__ is the emergency signal to indicate that all personnel should leave the Exclusion Zone. In addition, a loud hailer is available if required.

The following standard hand signals will be used in case of failure of radio communications:

```
Hand gripping throat  --------------- Out of air, can't breathe
Grip partner's wrist or ------------ Leave area immediately
  both hands around waist
Hands on top of head  --------------- Need assistance
Thumbs up  ------------------------- OK, I am all right, I understand
Thumbs down ------------------------ No, negative
```

Telephone communication to the Command Post should be established as soon as practicable. The phone number is _____.

I. DECONTAMINATION PROCEDURES

Personnel and equipment leaving the Exclusion Zone shall be thoroughly decontaminated. The standard level _____ decontamination protocol shall be used with the following decontamination stations: (1) _____
(2) _____ (3) _____ (4) _____ (5) _____
(6) _____ (7) _____ (8) _____ (9) _____
(10) _____ Other _____

Emergency decontamination will include the following stations: _____
_____
_____

The following decontamination equipment is required: _____
_____
_____

__(Normally detergent and water)__ will be used as the decontamination solution.

J. SITE SAFETY AND HEALTH PLAN

1. ____(name)____ is the designated Site Safety Officer and is directly responsible to the Project Team Leader for safety recommendations on site.

2. Emergency Medical Care

__(names of qualified personnel)__ are the qualified EMTs on site.
__(medical facility names)__, at __(address)__,
phone _____ is located _____ minutes from this location.
__(name of person)__ was contacted at __(time)__ and briefed on the situation, the potential hazards, and the substances involved. A map

Fig. 15-1. (*Continued*)

of alternative routes to this facility is available at ___(normally Command Post)___.

Local ambulance service is available from _____ at phone _____. Their response time is _____ minutes. Whenever possible, arrangements should be made for onsite standby.

First-aid equipment is available on site at the following locations:

    First-aid kit   _____
    Emergency eye wash   _____
    Emergency shower   _____
    (other)_____   _____

Emergency medical information for substances present:

Substance	Exposure Symptoms	First-Aid Instructions

List of emergency phone numbers:

Agency/Facility	Phone #	Contact
Police		
Fire		
Hospital		
Airport		
Public Health Advisor		

3. Environmental Monitoring

   The following environmental monitoring instruments shall be used on site (cross out if not applicable) at the specified intervals.

   Combustible Gas Indicator - continuous/hourly/daily/other _____
   O$_2$ Monitor   - continuous/hourly/daily/other _____
   Colorimetric Tubes   - continuous/hourly/daily/other _____
   ____(type)_____   _____
   _____   _____

   HNU/OVA   - continuous/hourly/daily/other _____
   Other _____   - continuous/hourly/daily/other _____
   _____   - continuous/hourly/daily/other _____

4. Emergency Procedures (should be modified as required for incident)

   The following standard emergency procedures will be used by onsite personnel. The Site Safety Officer shall be notified of any onsite emergencies and be responsible for ensuring that the appropriate procedures are followed.

   <u>Personnel Injury in the Exclusion Zone</u>: Upon notification of an injury in the Exclusion Zone, the designated emergency signal _____ shall be sounded. All site personnel shall assemble at the decontamination line. The rescue team will enter the Exclusion Zone (if required) to remove the injured person to the hotline. The Site Safety

Fig. 15-1. (*Continued*)

Officer and Project Team Leader should evaluate the nature of the injury, and the affected person should be decontaminated to the extent possible prior to movement to the Support Zone. The onsite EMT shall initiate the appropriate first aid, and contact should be made for an ambulance and with the designated medical facility (if required). No persons shall reenter the Exclusion Zone until the cause of the injury or symptoms is determined.

Personnel Injury in the Support Zone: Upon notification of an injury in the Support Zone, the Project Team Leader and Site Safety Officer will assess the nature of the injury. If the cause of the injury or loss of the injured person does not affect the performance of site personnel, operations may continue, with the onsite EMT initiating the appropriate first aid and necessary follow-up as stated above. If the injury increases the risk to others, the designated emergency signal _____ shall be sounded and all site personnel shall move to the decontamination line for further instructions. Activities on site will stop until the added risk is removed or minimized.

Fire/Explosion: Upon notification of a fire or explosion on site, the designated emergency signal _____ shall be sounded and all site personnel assembled at the decontamination line. The fire department shall be alerted and all personnel moved to a safe distance from the involved area.

Personal Protective Equipment Failure: If any site worker experiences a failure or alteration of protective equipment that affects the protection factor, that person and his/her buddy shall immediately leave the Exclusion Zone. Reentry shall not be permitted until the equipment has been repaired or replaced.

Other Equipment Failure: If any other equipment on site fails to operate properly, the Project Team Leader and Site Safety Officer shall be notified and then determine the effect of this failure on continuing operations on site. If the failure affects the safety of personnel or prevents completion of the Work Plan tasks, all personnel shall leave the Exclusion Zone until the situation is evaluated and appropriate actions taken.

The following emergency escape routes are designated for use in those situations where egress from the Exclusion Zone cannot occur through the decontamination line: _(describe alternate routes to leave area in emergencies)_ _____
_____
_____

In all situations, when an onsite emergency results in evacuation of the Exclusion Zone, personnel shall not reenter until:

1. The conditions resulting in the emergency have been corrected.
2. The hazards have been reassessed.
3. The Site Safety Plan has been reviewed.
4. Site personnel have been briefed on any changes in the Site Safety Plan.

5. Personal Monitoring

The following personal monitoring will be in effect on site:

Personal exposure sampling: _(describe any personal sampling programs_

Fig. 15-1. (*Continued*)

```
being carried out on site personnel. This would include use of sampling
pumps, air monitors, etc.)
Medical monitoring: The expected air temperature will be    (  °F)  . If
it is determined that heat stress monitoring is required (mandatory if
over 70°F) the following procedures shall be followed:
   (describe procedures in effect, i.e., monitoring body temperature, body
weight, pulse rate)
_____
_____
_____
_____

_____

All site personnel have read the above plan and are familiar with its
provisions.

Site Safety Oficer  _____(name)_____    _____(signature)_____
Project Team Leader_____    _____
Other Site Personnel_____    _____
                    _____    _____
                    _____    _____
                    _____    _____
                    _____    _____
                    _____    _____
```

Fig. 15-1. (*Continued*)

Methods of Interpretation

Taking each major topic required in the site safety plan and emphasizing its importance will provide an understanding for each section of the plan. By understanding the plan, implementation will be facilitated.

Safety and health personnel are listed in the site safety plan so that all personnel on that site will known from whom to request information and to whom to report emergencies. Workers will comply with directives concerning safety given by the individuals the workers know to be responsible and in charge. Also, the designated workers in charge of safety and health will know their responsibilities because they are detailed in the plan.

The hazard assessment for the specific site reveals the potential exposure to the worker. This topic will likely change several times throughout the duration of the project as the hazards change.

Sampling procedures and site monitoring are separate and very important factors to the outcome of a site cleanup. Monitoring (direct-read) and analysis of results from site sampling determine what personal protective equipment and cleanup methods are to be used. Following and understanding the specific details of monitoring (including calibration) and sampling are obviously crucial elements to the plan.

Knowledge of the specifics of the site operation and safety measures are *required training* for all site workers. The training is documented in the plan as a check and balance for the site compliance. If a topic listed under training in the plan has not been addressed in actual worker training, then the Safety and Health Manager should be contacted to remedy this situation.

Knowledge of certain topics provide obvious benefits. They are:

- PPE. This section will provide information about where, what and how to wear protective equipment to limit worker exposures.
- Decontamination procedures. Solutions/procedures vary in effectiveness with different contaminants; therefore, correct decon leads to containment of contaminants. These procedures and solutions should be reviewed prior to any decontamination activity.
- Medical surveillance. This part outlines the medical data to be obtained for any worker exposed to contaminants and the procedures to keep track of the individual's health.
- Site control.
- Trenching and excavation.
- Confined space entry.

The above three topics all provide the worker with a description of day-to-day tasks for safe operation on a site.

- Emergency response.
- Spill containment.

These last two procedures should be *second nature* to all workers designated to respond to any emergency, by defining the personnel and the procedures, the first step in implementation is done.

- New technologies. This section alerts the worker that new and possibly "untried" methodologies are being employed on site and to use due caution.

The site safety plan should not necessarily be limited to the OSHA requirements since sites have unique hazards; they should also be addressed. These topics for the site safety plan might include such procedures for working safely around lagoons, ponds, and impoundments, explosion or biohazard waste handling procedures.

BENEFITS ANALYSIS

There are obviously many benefits to incorporating a good site safety plan into any phase of activity at a Superfund site. A review of the obvious and the subtle benefits is important since the implications of poor safety can disrupt not only site workers but an entire community. The well known explosion in Elizabeth,

NJ is a case in point. After the worst materials, nerve-gas type pesticides, radioactive materials, known and suspected carcinogens, explosives, and compressed gases had been removed, the site exploded in April 1980 (see Fig. 15-2). As a result of extensive preparation based on the site safety plan by all workers (firemen and police included) and the members of the adjoining communities, no-one was killed. All the extremely poisonous materials had been removed and this reduced the level of danger dramatically. The contractor had a *good* site safety plan and because of this, the surrounding cities, were spared from the exposure to extremely toxic substances.

This site emergency was handled very well because all personnel involved were informed and well trained in the specific response techniques required for this site. There are still more advantages to having a sound, practical, "tried and true" site safety plan:

- Since the site safety plan must be implemented before operations begin, operation methods can be reviewed and flaws corrected.
- As a written document, the plan sets a standard for all workers to follow with no exceptions. Therefore one worker will not endanger another at the expense of his ignorance.

Fig. 15-2. Explosion at a well-known Superfund site (EPA, 1980).

If the site safety plan is followed and understood by all workers, it has been proven to:

- Reduce worker stress.
- Reduce time away from work.
- Increase job efficiency.
- Increase worker health significantly.
- Reduce job costs dramatically.
- Responsibilities for each job description with respect to safety are outlined in the plan, and therefore, will aid in preventing any procedures left unexecuted because erroneous assumptions were made.

Hazard Awareness Review

The potential hazards on a hazardous waste remediation project, protective measures and safe work methods will be summarized as an introduction and motivation for use of a checklist for workers to use for safe site entry. The amount of risk to the worker varies with the type and degree of exposure to hazards at a site. Exposure can result in any of the following:

- Asphyxiation.
- Poisoning.
- Cancer.
- Infertility.
- Damage to liver, kidneys, nerve cells, etc.
- Harmful effects to the unborn child.
- Loss of limbs.
- Skin diseases.
- Loss of hearing.
- Eye injuries.

The specific hazards that may be encountered by any hazardous waste site worker are:

- Toxic substances.
- Flammable materials.
- Explosive materials.
- Excessive noise.
- Corrosive materials.
- Biologically active materials such as bacteria and viruses.
- Heat or cold stress.
- Air in the work area, such as in a tank or ditch, that is deficient in oxygen.

- Accidents resulting in physical harm.
- Radioactive materials.
- Cancer-causing agents.

Failure to comply with safe work practices such as the situations below cause unnecessary and extremely dangerous exposures:

- Lack of qualified personnel and/or the proper equipment to evaulate hazards and define the levels of protection needed. Improper selection of or insufficient training in the maintenance and use of personal protective equipment (such as respirators, special clothing, or safety glasses) before entering a work site.
- Failure to follow instructions or wear prescribed protective equipment.
- Failure or lack of engineering controls such as shields or drum handling equipment.
- Unexpected hazards at the work site.
- Insufficient time to put on protective equipment in an emergency.
- Inadequate emergency procedures and/or protective equipment.
- Walking unnecessarily through puddles or into vapor mists, etc.
- Failure to decontaminate immediately after splashes or spills occur.

In determining or verifying the actual protective measures that you need to know prior to site entry, reference the sites safety plan and standard operating procedures. Examples, scenarios, or questions that alert a worker to consider protection measures are as follows:

- What protective clothing and equipment are required for the hazardous substances you may encounter at the work site?
- What potential explosive and/or flammable conditions may be present?
- What, if any confined spaces will you have to enter? (The air in these spaces MUST be checked for unsafe concentrations of airborne contaminants and for sufficient levels of oxygen.)
- What emergency equipment is available, where is it located, and how does one use it? (This equipment must be checked often and kept in good working condition.)
- What is the availability of standby personnel?
- What are the standard operating procedures (SOPs) for evacuation and rescue in the event of an emergency?
- If conditions or situations are likely to change during the work period, how will you be notified?
- What is the work/rest cycle for each task?

- What are the prescribed decontamination procedures?
- Will the buddy system be used and who will be your buddy?

Stay out of contaminated areas if you are not properly trained, equipped, or authorized to enter.

Do not take chances with life-threatening materials or situations.

Finally, in review, here is a list of things to be continually aware of at the site:

- Any weather changes. For example, when it gets hot or the air is calm, chemical concentrations in the air can increase. This may require additional protection.
- Wind direction. For example, avoid dust and vapors by working upwind if possible.
- Odors that may indicate the presence of chemicals.
- The location of someone who can help if an emergency arises.
- Your employers SOPs, and the specific site safety plan which you should follow for any necessary decontamination procedures, including cleaning and storing or disposing of contaminated equipment and clothing. You can expose your family or friends to toxic substances by carrying contaminants on your clothing, shoes, tools, etc.
- Washing your hands before eating, drinking, smoking, or using the restroom.
- Showering and changing into clean clothes and depositing your work clothes in the proper area before leaving the work site.
- Keeping food, drinks, smoking materials, and personal care items in clean areas only.
- Heavy equipment operating near you.
- The proper handling of drums and other equipment so as to prevent personal injury.
- The need for proper personal protective equipment and its limitations.
- Emergency procedures and the evacuation signal.
- Where and how to exit from every area.

Checklist for Safe Site Entry

By using the requirements for training under the OSHA standard 29 CFR 1910.120 the following checklist has been prepared to assist you, the worker, in evaluating and achieving a safe entry and maintaining a safe work environment at a hazardous waste site.

To use the following checklist, place a check in the blank next to the statement if the information activity/equipment is readied for worker site entry. This list is not totally inclusive; therefore, it should be used as a guideline in preparing your specific on-site checklist.

HAZARDOUS WASTE WORKER SITE ENTRY CHECKLIST

____ 1. Worker knows the names of the *on-site* site safety and health personnel.
____ 2. Worker knows the site hazards. (Reviewed reference materials)
____ 3. Personal protective equipment (PPE) selected is appropriate for specific job task.
    ____ User is familiar with equipment and has successfully completed training.
    ____ User can recognize symptoms of heat strain related to work in PPE and knows preventive measures to avoid heat injury.
____ 4. Chemical protective clothing selected is appropriate for hazards present and specific job task of user.
____ 5. Personal protective clothing has been checked for contamination, signs of chemical degradation, tears, pinholes, or other defects, and replaced if faulty or cleaned if not decontaminated.
____ 6. Respirators inspected for use (see chapter 9).
    ____ Facepiece inspected for damage and to check fit.
    ____ Respirator decontaminated and disinfected since previous use
    ____ Fresh cartridges or canister installed for APRs.
    ____ Air tanks full, and all system components checked for proper function for SCBA.
    ____ Airlines and escape air bottles inspected for SARs.
    ____ User has been successfully fit tested with the appropriate respirator facepiece.
____ 7. Worker knows safe work practices procedures for this project.
    ____ Confined space entry.
    ____ Trenching and excavation.
    ____ Drilling activities.
    ____ Use of heavy equipment.
    ____ Bulking of drummed wastes.
    ____ Handling of containers.
____ 8. Worker is familiar with all communication systems used on site.
____ 9. Worker is familiar with use of the buddy system on site.
____ 10. Worker is familiar with site layout, site zoning system, zone boundaries, and the zone barrier or boundary marking system use.
____ 11. Worker knows what additional engineering controls are being used and why.
    ____ Dikes.

___ Berms (earthern walls to segregate incompatible materials).
___ Ditches and excavations.
___ 12. Medical examinations have been conducted in compliance with medical surveillance requirements (29 CFR 1910.120).
___ 13. Bodily symptoms which will alert worker to overexposure of chemicals, oxygen-deficiency, and other site hazards are known.
___ 14. Familiar with decontamination procedures.
___ Decon station locations for equipment and personnel are known.
___ Contaminated equipment disposal locations are known.
___ 15. The latest revision of the site emergency response plan has been reviewed during on-site training.
___ Site Emergency Response personnel, and notification procedures are known.
___ Worker is familiar with their specific role in a response.
___ Worker is aware of potential emergencies.
___ Worker can recognize a developing emergency (i.e., bulging drums, bubbling liquids, or heat generation) and knows appropriate preventive measures.
___ Emergency exit locations known.
___ Evacuation signals, and emergency alert signals are known.
___ Emergency decon procedures, if different from normal procedures, are known.
___ Site-specific procedures for responding in the event of injury to a worker, including decon and first aid, are known.
___ 16. Spill containment procedures are known.
___ Know what equipment is available on site.
___ Know location the large quantities of materials on site, and variety of containers.
___ 17. Worker is familiar with safe trenching and excavation procedures, if applicable, on site.
___ 18. Worker is familiar with hazard monitoring procedures (including calibration and maintenance procedures for field equipment) which they are required to use on site.
___ 19. Worker knows location of command post, and is familiar with the site safety plan.

REFERENCE

US/NIOSH/OSHA/CG/EPA. 1985. Appendix B. *Occupational Safety and Health Guidance Manual for Hazardous Waste Site Activities*. NIOSH Publication No. 85-115. Washington, DC: U.S. Govt. Printing Office.

GLOSSARY

AA *Atomic absorption* spectrophotometry. Refers to the analytical method or apparatus used for metals analysis.

AAPCO *Association of American Pesticide Control Officials, Inc.* This association consists of officials charged by law with active execution of the laws regulating sale of economic poisons, and of deputies designated by these officials employed by State, Territorial, Dominion, or Federal agencies. The group objective is to promote uniform and effective legislation, definitions, rulings, and enforcement of laws relating to control of sale and distribution of economic poisons.

AAR *Association of American Railroads.*

Absorption (a) Penetration of a substance into the body of another.
(b) Transformation into other forms suffered by radiant energy passing through a material substance.

Absorbed dose The energy imparted to matter by ionizing radiation per unit mass of irradiated material at the place of interest. A special unit for this quantity is the rad.

ACGIH *American Conference of Governmental Industrial Hygienists.* An organization of professional personnel in governmental agencies or educational institutions engaged in occupational safety and health programs ACGIH develops and publishes recommended occupational exposure limits (see TLV) for hundreds of chemical substances and physical agents.

Acid A hydrogen-containing compound that reacts with water to produce hydrogen. Acids chemicals are corrosive. (See also pH.)

Activity (A) The number of disintegrations of a quantity of radionuclide per unit time. Care must be exercised in a measurement of activity to insure that the time is short enough not to be influenced by the radioactive decay and that the number of disintegrations is large enough to be statistically significant.

Acute effect An adverse effect on a human or animal, generally after a single significant exposure, with severe symptoms developing rapidly and coming quickly to a crisis. (See also Chronic effect.)

Acute toxicity The adverse (acute) effects resulting from a single dose of, or exposure to, a substance.

Aerosols Liquid droplets or solid particles dispersed in air, that are of fine enough particle size (0.01–100 microns) to remain so dispersed for a period of time.

AIHA American Industrial Hygiene Association.

Aliphatic hydrocarbon Major group of organic compounds with a straight or branched molecular chain structure of carbon atoms. There are three (3) groups: alkanes, alkenes, alkynes.

Alkali Any substance that in water solution is bitter, more or less irritating, or caustic to the skin. Strong alkalies in solution are corrosive to the skin and mucous membranes. (See also pH.)

Alkanes Saturated hydrocarbons—those containing only single bonds, are relatively unreactive; paraffins.

Alkenes Unsaturated hydrocarbons—those containing one or more double bonds; olefins.

Alkynes Unsaturated hydrocarbons—those containing one or more triple bonds; acetylenes.

Alpha particles or alpha rays Particulate ionizing radiation consisting of helium nuclei traveling at high speed.

Anhydrous Free from water.

ANSI American National Standards Institute.

Anorexia Lack or loss of the appetite for food.

Aromatic hydrocarbons A major group of unsaturated cyclic hydrocarbons typified by benzene which has a 6 carbon ring with 3 resonating double bonds.

Asbestos Any material containing more than 1% asbestos in any form.

Asbestosis A disease of the lungs caused by the inhalation of fine airborne fibers of asbestos.

Asphyxiant A vapor or gas which can cause unconsciousness or death by suffocation (lack of oxygen). Most simple asphyxiants are harmful to the body only when they become so concentrated that they reduce oxygen in the air (normally about 21%) to dangerous levels (18% or lower). Asphyxiation is one of the principal potential hazards of working in confined spaces.

ASTM American Society for Testing and Materials.

Atrophy Arrested development or wasting away of cells and tissue.

Attenuation Decrease in exposure rate of radiation caused by passage through material.

Auto-ignition temperature The minimum temperature at which a material will ignite without a spark or flame being present. Along with the flashpoint, auto-ignition temperature gives an indication of relative flammability.

Background radiation Radiation arising from sources other than the one directly under consideration. Background radiation due to cosmic rays and natural radioactivity is always present. There may also be additional background radiation due to the presence of sources of radiation in other parts of the building and/or area.

BAT *Best available technology* or most stringent type of control for existing discharges and applies to toxic pollutants as well as conventional and some nonconventional pollutants.

BADCT *Best available demonstrated control technology.* Applies only to new industrial sources of pollution. Pollution control is built into the entire facility.

BCT *Best conventional technology* for discharges of conventional pollutants; more stringent than BPT.

BEJ *Best engineering judgment* or type of control for pollution sources for which EPA has not issued regulations.

Beta particles or beta rays Particulate ionizing radiation consisting of electrons or positrons traveling at high speed.

Bioassay A term used to describe the technique by which a toxic agent, such as an insecticide, is detected and measured for potency. The technique involves testing of the toxicant at different dosage levels for ability to cause a physiological response (often death) in a test organism (e.g., insect, rat). In bioassay, chemicals are not identified individually. Bioassay may be used to determine the rate of loss after application of an insecticide to crop or soil, as confirmation of chemical assays of residues, for detection of insecticides as a cause of honeybee losses, etc.

Biocide A substance that, when absorbed by eating, drinking, or breathing, or otherwise consumed in relatively small quantities, causes illness or death, or even retardation of growth or shortening of life.

Biohazard A combination of the words "biological hazard;" infectious agents presenting a risk or potential risk to the well being of man or other animals, either directly through infection or indirectly through disruption of the environment.

Biohazard area Any area (a complete operating complex, a single facility, a room within a facility, etc.) in which work has been, or is being performed with biohazardous agents or materials.

Biological half-life The time required for a given species, organ, or tissue to eliminate half of a substance which it takes in.

Biological magnification The concentration of certain substances up a food chain. A very important mechanism in concentrating pesticides and heavy metals in organisms such as fish.

Biological treatment A process by which hazardous waste is rendered nonhazardous or is reduced in volume by relying on the action of microorganisms to degrade through organic waste.

Biological hazardous wastes (infectious) Any substance of human or animal origin, other than food wastes, which are to be disposed of and could harbor or transmit pathogenic organisms including, but not limited to, pathological specimens such as tissues, blood elements, excreta, secretions, and related substances. This category includes wastes from health care facilities and laboratories, and biological and chemical warfare agents. Wastes from hospitals would include malignant or benign tissues taken during autopsies, biopsies, or surgery; hypodermic needles; and bandaging materials. Although the production of biological warfare agents has been restricted and production of chemical agents discontinued, some quantities still remain and must be disposed of. See Title 9 CFR Part 102 (licensed veterinary biological products), Title 21 CFR Part 601 (licensing), or Title 42 CFR Part 72.

Biological wastewater treatment A type of wastewater treatment in which bacterial or biochemical action is intensified to stabilize, oxidize, and nitrify the unstable organic matter present. Intermittent sand filters, contact beds, trickling filters, and activated sludge tanks are examples of the equipment used.

Blasting agent A material designed for blasting that has been evaluated according to one of the tests described in Title 49 CFR 173.114a of the Department of Transportation and found to be so insensitive that there is very little probability of accidental initiation of explosion or of transition from deflagration to detonation.

BLEVE *Boiling liquid expanding vapor explosion.* In addition to its technical meaning, this acronym has acquired a common usage definition that has come to stand for virtually any rupture of a tank of liquid or liquefied compressed gas and has been expanded to include all vapor explosions. The technicle definition of BLEVE presents the hypothesis that rapid depressurization of a hot, saturated liquid may result in an explosion. The temperature of the hot liquid must be above the superheat limit temperature at 1 atmosphere, and the drop in tank pressure must be very rapid. This requires instantaneous homogeneous nucleation of the hot liquid. This phenomenon has *not* been observed as the cause of failure of a transportation container.

BMP *Best management practices.*

BOE *Bureau of Explosives*, Association of American Railroads.

Boiling point (B. P.) The temperature at which a liquid changes to a vapor state at a given pressure, usually expressed in degrees Fahrenheit at sea level. Flammable materials with low boiling points generally present special fire hazards.

BPT *Best practicable technology.* Minimum acceptable level of treatment for existing plants.

Breathing zone sample An air sample collected in the breathing area (around the nose) of a worker to assess his exposure to airborne contaminants.

°C or C *Degrees Centigrade (Celsius).*

C or Ceiling The maximum allowable human exposure limit for an airborne substance, not to be exceeded even momentarily. (See also PEL and TLV.)

CAA *Clean Air Act.*

Canister (air purifying) A container filled with sorbents and catalysts that remove gases and vapors from air drawn through the unit. The canister may also contain an aerosol (particulate) filter to remove solid or liquid particles.

Capacitor A device for accumulating and holding a charge of electricity and consisting of conducting surfaces separated by a dielectric.

Carcinogen A substance capable of causing cancer.

cc *Cubic centimeter.* A volume measurement in the metric system equal in capacity to one milliliter (ml). One quart is about 946 cubic centimeters.

CDC *Center for Disease Control.*

Centigrade (Celsius) The internationally used scale for measuring temperature, in which 100 is the boiling point of water at sea level (1 atmosphere), and 0 is its freezing point.

CEQ Council on Environmental Quality.

CERCLA *Comprehensive Environmental Response, Compensation and Liability Act* (1980). Superfund.

CFC *Chlorofluorocarbons.* A class of halon chemical compounds containing both chlorine and fluorine, used as refrigerants or cleaning solvents and commonly referred to as Freons.

CFR Code of Federal Regulations.

CGA *Compressed Gas Association.*

CGNRC *Coast Guard National Response Center.*

Chemical-resistant materials Materials that inhibit or protect against penetration of certain chemicals.

CHEMTREC *Chemical Transportation Emergency Center*, operated by the Chemical Manufacturers Association (CMA).

CHRIS *Chemical Hazards Response Information System* published by the United States Coast Guard.

Chronic effect Adverse effects resulting from repeated doses of, or exposures to, a substance over a relatively prolonged period of time.

CMA *Chemical Manufacturers Association.*

Concentration The relative amount of a substance when combined or mixed with other substances. Examples: 2 ppm hydrogen sulfide in air or a 50% caustic solution.

Combustible liquid class IIIA and IIIB (OSHA usage) Class II liquids include those with flashpoints at or above 100°F (37.8°C), and below 140°F (60°C) except any mixture having components with flashpoints of 200°F (93.3°C) or higher, the volume of which make up 99% or more of the total volume of the mixture (Title 29 CFR 1910.106).

Combustible liquid (DOT usage) Flashpoint 100°F to 200°F.

Compressed gas Material packaged in a cylinder, tank, or aerosol under pressure exceeding 40 psi at 70°F or other pressure parameters identified by DOT.

Consignee The addressee to whom the item is shipped.

Container Any portable device in which a material is stored, transported, disposed of, or otherwise handled. (See Title 40 CFR 260.10 (a) (9).)

Container, intermodal ISO An article of transport equipment that meets the standards of the International Organization for Standardization (ISO) designed to facilitate and optimize the carriage of goods by one or more modes of transportation without intermediate handling of the contents and equipped with features permitting ready handling and transfer from one mode to another. Containers may be fully enclosed with one or more doors, open top, tank, refrigerated, open rack, gondola, flatrack, and other designs. Included in this definition are modules or arrays that can be coupled to form an integral unit regardless of intention to move singly or in multiplex configuration.

Containerization The use of transport containers—container express (CONEX), military-owned demountable containers (MILVAN), commercially or Government-owned (or leased) shipping containers (SEAVAN), and roll on/roll off (RORO) trailers to unitize cargo for transportation, supply, and storage. Containerization aids carriage of goods by one or more modes of transportation without the need for intermediate handling of the contents, and incorporates supply, security, packaging, storage, and transportation into the distribution system from source to user.

Contamination (radioactive) Deposition or presence of radioactive material in any place where it is not desired, and particularly in any place where its presence can be harmful. The harm may be in vitiating the validity of an experiment or a procedure or in being a source of danger to persons.

Corrosive acid A liquid or solid, excluding poisons, that causes visible destruction or irreversible alterations in human skin tissue at the site of contact, or has a severe corrosion rate on steel. Liquids show a pH of 6.0 or less. See Title 49 CFR 173.240.

Corrosive alkaline A liquid or solid, excluding poisons, that causes visible destruction or irreversible alteration in human skin tissue at the site of contact, or has a severe corrosion rate on steel. Liquids show a pH of 8.0 or above. See Title CFR 173.240.

CPR *Cardiopulmonary resuscitation.*

CPSA *Consumer Product Safety Act*, Title 16 CFR 1500 series.

Curie (Ci) The special unit of activity; $1 \text{ Ci} = 3.7 \times 10^{10}$ disintegrations/second.

CPSC *Consumer Products Safety Commission.*

CWA *Clean Water Act*, Title 40 CFR.

Cyanosis Blue appearance of the skin, especially on the face and extremities, indicating a lack of sufficient oxygen in the arterial blood.

Dangerous when wet A label required for certain materials being shipped under US DOT, ICAO, and IMO regulations. Any of this labeled material that is in contact with water or moisture may produce flammable gases. In some cases, these gases are liable to spontaneous combustion.

Daughter An isotope formed by the decay of a given radioactive isotope. The daughter may be either radioactive or stable.

DCM Dangerous cargo manifest. (See Title 49 CFR 176.30.)

Dermal toxicity Adverse effects resulting from skin exposure to a substance.

Dermatitis Inflammation of the skin from any cause. There are two types of skin reaction: primary irritation dermatitis and sensitization dermatitis. (See Irritant and Sensitizer.)

Desiccant A substance such as silica gel that removes moisture (water vapor) from the air and is used to maintain a dry atmosphere in containers of food or chemical packagings.

Disposal drum A nonprofessional reference to a drum used to overpack damaged or leaking containers of hazardous materials for shipment; the proper shipping name is Salvage Drum as cited in Title 49 CFR 173.3

Distribution system (supply) A complex of facilities, equipment, methods, patterns and procedures designed to receive, store, maintain, distribute, and control the flow of items from one point to another.

DOC *Department of Commerce.*

DOD *Department of Defense.*

DOE *Department of Energy.*

DOJ *Department of Justice.*

DOL *Department of Labor.*

DOS *Department of State*

Dose The amount of energy or substance absorbed in a unit volume or an organ or individual. Dose rate is the dose delivered per unit of time. (See also Roentgen, RAD, REM.)

Dose equivalent (DE) A concept used in radiation protection work to permit the summation of doses from radiations having varying linear energy transfers, distributions of dose, etc. It is equal numerically to the product of absorbed dose in rads and arbitrarily defined quality factors, dose distribution factors and other necessary modifying factors. In the case of mixed radiations, the dose equivalent is assumed to be equal to the sum of the products of the absorbed dose of each radiation and its factors.

DOT *Department of Transportation.*

dps *Disintegrations per second.* A unit of measure relating to the breakdown of a radioactive material.

Dust Solid particles generated by handling, crushing, grinding, rapid impact, detonation, and decrepitation of organic or inorganic materials, such as rock, ore, metal, coal, wood, and grain. Dusts do not tend to flocculate except under electrostatic forces; they do not diffuse in air but settle under the influence of gravity.

Dyspnea Shortness of breath, difficult or labored breathing.

Ecology A branch of science concerned with interrelationship or organisms and their environments; the totality or pattern of relations between organisms and their environment.

Economic poison As defined in the Federal Insecticide, Fungicide, and Rodenticide Act (FIFRA), an economic poison is "any substance or mixture of substances intended for preventing, destroying, repelling, or mitigating any insects, rodents, nematodes, fungi, or weeds or any other forms of life declared to be pests ... any substance intended for use a plant regulator, defoliant, or desiccant." As so defined, economic poisons are known generally as *pesticides.*

Edema A swelling of body tissues as a result of fluid retention.

Effluent guidelines (CWA) Minimum, technology-based levels of pollution reduction that point sources must attain.

Effluent limitations (CWA) Specific control requirements directed at a specific discharge site.

Empty packagings As related to Title 49 CFR:

1. The description on the shipping paper for a package containing the residue of a hazardous may include the words *"RESIDUE: Last Contained Material"* in association with the basic description of the hazardous material last contained in the packaging.
2. For a tank car containing the residue (as defined in Title 49 CFR 171.8) a hazardous material, the requirements of Title 49 CFR 172.203 (e) and 174.25 (c) apply.
3. If a packaging, including a tank car, contains a residue that is a hazardous substance, the description on the shipping appears must be with the phrases *"RESIDUE: Last Contained"* and the letters "RQ" must be entered on the shipping paper either before or after the description.

EPA United States Environmental Protection Agency.

Epidemiology The science that deals with the study of disease in a general population. Determination of the incidence (rate of occurrence) and distribution of a particular disease (as by age, sex, or occupation) may provide information about the cause of the disease.

Etiological agent A viable microorganism or its toxin, which causes or may cause human disease; limited to the agents identified in Title 42 CFR part 72.

Etiology The study of the causes of disease.

Evaporation rate The rate at which a particular material will vaporize (evaporate) when compared with the rate of vaporization of a known material. The evaporation rate can be useful in evaluating the health and fire hazards of a material. The known material is usually normal butyl acetate (NBUAC or n-BuAc), with a vaporization rate designated as 1.0. Vaporization rates of other solvents or materials have three classifications:

1. *Fast* evaporating if greater than 3.0. *Examples*: methyl ethyl ketone (MEK) = 3.8; acetone = 5.6; hexane = 8.3
2. *Medium* evaporating if 0.8–3.0. *Examples*: 190 proof (95%) ethyl alcohol = 1.4; VM&P naphtha = 1.4; MIBK = 1.6.
3. *Slow* evaporating if less than 0.8. *Examples*: xylene = 0.6; isobutyl alcohol = 0.6; normal butyl alcohol = 0.4; water = 0.3; mineral spirits = 0.1.

Exotoxin A toxin produced and delivered by a microorganism into the surrounding medium.

Explosionproof equipment Apparatus enclosed in a case capable of withstanding an explosion of a specified gas or vapor that may occur and of preventing the ignition of a specified gas or vapor surrounding the enclosure by sparks, flashes, or explosion of the gas or vapor within; and that operates at an external temperature such that a surrounding flammable atmosphere will not be ignited.

Explosive class A Any of nine types of explosives as defined in Title 49 CFR 173.53, and listed in Title 49 CFR 172.101. Any chemical compound, mixture, or device having the primary or common purpose to function by detonation (i.e, with substantial instantaneous release of gas and heat, unless such compound, mixture, or device is otherwise classified for storage or transportation).

Explosive, class B Explosives that, in general, function by rapid combustion rather than detonation and include some explosive devices such as special fireworks, flash powders, some pyrotechnic signaling devices, and solid or liquid propellant exlosives including some smokeless powders. These explosives are defined in Title 49 CFR 172.101 and Title 49 CFR 173.88 of the Department of Transportation, respectively.

Explosive, class C Certain types of manufactured articles that contain Class A or Class B explosives, or both, as components but in restricted quantities; and certain types of fireworks. These explosives are defined in Title 49 CFR 172.101 and Title 49 CFR 173.100 of the Department of Transportation, respectively.

Explosive limits Some items have a minimum and maximum concentration in air which can be detonated by spark, shock, fire, etc. The lowest concentration is known as the lower explosive limit (LEL). The highest concentration is known as the upper explosive limit (UEL).

Exposure Subjection of a person to a toxic substance or harmful physical agent in the course of employment through any route of entry (e.g., inhalation, ingestion, skin contact, or absorption); includes past exposure and potential (e.g., accidental or possible) exposure, but does not include situations where the employer can demonstrate that the toxic substance or harmful physical agent is not used, handled, stored, generated, or present in the workplace in any manner different from typical non-occupational situations. An exposure to a substance or agent may or may not be an actual health hazard to the worker. An industrial hygienist evaluates exposures and determines if permissible exposure levels are exceeded.

Exposure rate absorbed dose rate The time rate at which an exposure or absorbed dose occurs; that is, exposure or absorbed dose per unit time. It implies a uniform or short-term average rate, unless expressly qualified (e.g., peak dose rate). In protection work it is usually expressed in mR/hr, mrads/hr.

°F or F *Degrees Fahrenheit.*

Fahrenheit The scale of temperature in which 212 is boiling water at 760 mm Hg and 32 is the freezing point.

FFDCA *Federal Food, Drug, and Cosmetic Act.* (See Title 21 USC 301–392.)

FHSLA (CPSC usage) *Federal Hazardous Substances Labeling Act.* (See Title 15 USC 1261–1275.)

FIFRA *Federal Insecticide, Fungicide, and Rodenticide Act* (See Title 40 CFR.)

Film badge A pack of appropriate photographic film and filters used to determine radiation exposure.

Fibrosis A condition marked by increase of interstitial fibrous tissue.

Flammable (DOT usage) Flashpoint < 100 F.

Flammable aerosol An aerosol which is required to be labeled "Flammable" under the United States Federal Hazardous Substances Labeling Act. For storage purposes, flammable aerosols are treated as Class IA liquids (NFPA 30, Flammable and Combustible Liquids Code).

Flammable gas Any compressed or liquified gas, except an aerosol, is flammable if either a mixture of 13% of less (by volume) with air forms a flammable mixture or the flammable range with air is wider than 12% regardless of the lower limit (at normal temperature and pressure). (ICAO Technical Instructions.)

Flammable limits Flammable liquids produce (by evaporation) a minimum and maximum concentration of flammable gases in air that will support combustion. The lowest concentration is known as the lower flammable limit (LFL). The highest concentration is known as the upper flammable limit (UFL).

Flammable liquid class IA (OSHA usage) Any liquid having a flashpoint below 73°F (22.8°C) and having a boiling point below 100°F (37.8°C) except any mixture having components with flashpoints of 100°F (37.8°C) or higher, the total of which comprise 99% or more or the total volume of the mixture (Title 29 CFR 1910.106).

Flammable liquid class IB (OSHA usage) Any liquid having a flashpoint below 73°F (22.8°C) and having a boiling point at or above 100°F (37.8°C), except at or above 100°F (37.8°C) or higher, the total of which make up 99% or more of the total volume of the mixture (Title 29 CFR 1910.106).

Flammable liquid class IC (OSHA usage) Any liquid having a flashpoint below at or above 73°F (22.8°C) and below 100°F (37.8°C), except any mixture having components with flashpoints of 100°F (37.8°C), or higher, the total of which make up 99% or more of the total volume of the mixture (Title 29 CFR 1910.106).

Flammable solid (DOT usage) Any solid material, other than one classed as an explosive, that under conditions normally incident to storage is liable to cause fire through friction or retained heat from manufacturing or processing; or that can be ignited readily, and when ignited burns so vigorously and persistently as to create serious storage hazard. Flammable solids, excluding Dangerous When Wet, are further defined in Title 49 Cfr 173.150.

Flashpoint The lowest temperature at which a liquid gives off enough vapor to form an ignitable mixture with air and produce a flame when a source of ignition is present. Two tests are used—*open cup* and *closed cup*.

FP or fl. pt. *Flashpoint.*

Friable Capable of being pulverized with hand pressure as relates to asbestos (Title 29 CFR 1910).

ft^3 *Cubic feet.* Calculated by multiplying length by width by depth of an item or space.

Full protective clothing Such units are typically recommended where high chemical gas, vapor, or fume concentrations in air may have a corrosive effect on exposed skin, and/or where the chemical in air may be readily absorbed through the skin to produce toxic effects. These suits are impervious to chemicals, offer full body protection, and include self-contained breathing apparatus (SCBA).

Fully encapsulating suits Full chemical protective suits that are impervious to chemicals, offer full body protection from chemicals and their vapors/fumes, and are to be used with self-contained breathing apparatus (SCBA).

Fume Gaslike emanation containing minute solid particles arising from the heating of a solid body such as lead. This physical change is often accompanied by a chemical reaction, such as oxidation. Fumes flocculate and sometimes coalesce. Odorous gases and vapors should not be called fumes.

FWPCA *Federal Water Pollution Control Act* (1972).

Gamma radiation, gamma rays Electromagnetic radiation of short wavelength and correspondingly high frequency, emitted by nuclei in the course of radioactive decay.

Gas A state of matter in which the material has very low density and viscosity; can expand and contract greatly in response to changes in temperature and pressure; easily diffuses into other gases; readily and uniformly distributes itself throughout any container. A gas can be changed to the liquid or solid state by the combined effect of increased pressure and/or decreased temperature.

Gastr-; gastro- (Prefix) Pertaining to the stomach.

GC/MS *Gas chromatography/mass spectrometry*. Refers to both analytical method and apparatus used for organics analysis.

Genetic effects Mutations or other changes which are produced by irradiation of the germ plasm.

g/kg *Grams per kilogram*, an expression of dose used in oral and dermal toxicity testing to indicate the grams of substance dosed per kilogram of animal body weight. (See also kg.)

GSA *General Services Administration*.

HAP *Hierarchical analytical protocol*. A procedure identified by the EPA to demonstrate the presence or absence of RCRA (Title 40 CFR) classes or Appendix VIII compounds in groundwater.

Hazardous air pollutant A pollutant to which no ambient air quality standard is applicable and that may cause or contribute to an increase in mortality or serious illness. For example, asbestos, beryllium, and mercury have been declared hazardous air pollutants.

Hazard assessment risk analysis A process used to qualitatively or quantitatively assess risk factors to determine mitigating actions.

Hazardous chemicals Chemicals or materials used in the workplace that are regulated under the OSHA Hazard Communication Standard or the "right-to-know" regulations in Title 29 CFR 1910.1200.

Hazard class A category of hazard associated with an HM/HW that has been determined capable of posing an unreasonable risk to health, safety, and property when transported (see Title 49 CFR 171.8). The hazard class used by the United States DOT and published in Title 49 CFR 172.101. The hazard classes used in the United States include: Explosive (class A, B, or C); flammable liquid; flammable solid; corrosive material; oxidizer; poison A; poison B; radioactive material; nonflammable gas; ORM-A, -B, -C, -D, and -E; etiologic agent; irritating material; organic peroxide; combustible liquid; flammable gas; and blasting agent.

Hazardous material In a broad sense, a hazardous material is any substance (HM) or mixture of substances having properties capable of producing adverse effects on the health and safety or the environment of a human being. Legal definitions are found in individual regulations.

Hazardous waste manifest, uniform (EPA usage) The shipping document, originated and signed by the waste generator or his authorized representative, that contains the information required by Title 40 CFR 262, Subpart B.

Hazardous substances Chemicals, mixtures of chemicals, or materials subject to the regulations contained in Title 40 CFR. For transportation purposes, means a material, and its mixtures or solution, identified by the letter E in column 2 of the Hazardous

Materials Table included in Title 49 CFR 172.101 when offered for transportation in one package, or in one transport vehicle if not packaged, and when the quantity of the material therein equals or exceeds the reportable quantity (RQ). For details, refer to Title 49 CFR 171.8 and Title 49 CFR 172.101.

Hazardous waste (HW) Any material listed as such in Title 40 CFR 261, Subpart D, that possesses any of the hazard characteristics of corrosivity, ignitability, reactivity, or toxicity as defined in Title 40 CFR 261, Subpart C, or that is contaminated by or mixed with any of the previously mentioned materials. (See Title 40 CFR 261.3.)

Hazardous waste generation The act or process of producing hazardous waste.

Hazardous waste landfill An excavated or engineered area on which hazardous waste is deposited and covered; proper protection of the environment from the materials to be deposited in such a landfill requires careful site selection, good design, proper operation, leachate collection and treatment, and thorough final closure.

Hazardous waste leachate The liquid that has percolated through or drained from hazardous waste emplaced in or on the ground.

Hazardous waste management Systematic control of the collection, source separation, storage, transportation, processing, treatment, recovery, and disposal of hazardous wastes.

Hazardous waste number The number assigned to each hazardous waste listed by EPA and to each hazardous waste characteristic.

Hazardous waste site A location where hazardous wastes are stored, treated, incinerated, or otherwise disposed of.

Hematology Study of the blood and the blood-forming organs.

Hepatitis Inflammation of the liver.

Herbicide A chemical intended for killing plants or interrupting their normal growth. A weed, grass, or brush killer. (See also Pesticides.)

HMTA *Hazardous Materials Transportation Act* (1975).

HPLC *High performance liquid chromatography*, used in organics analysis. Also called LC (liquid chromatography).

HSWA *Hazardous and Solid Waste Amendments* of 1984 (RCRA Jr.).

Hygroscopic Descriptive of a substance that has the property of absorbing moisture from the air, such as silica gel, calcium chloride, or zinc chloride.

Hypothermia Condition of reduced body temperature.

IATA *International Air Transport Association.*

IC Ion chromatography.

ICAO *International Civil Aviation Organization.*

ICP *Inductively coupled (argon) plasma.* Used with reference to both the analytical method and the apparatus.

Identification code for EPA The individual number assigned to each generator, transporter, and treatment, storage, or disposal facility by State or Federal regulatory agencies.

IDLH *Immediately dangerous to life and health.* An environmental condition which would immediately place a worker in jeopardy. Usually used to describe a condition existing where self-contained breathing apparatus must be used.

ID number Four-digit number by UN or NA, assigned to hazardous materials and dangerous goods. (See column 3a of the Hazardous Materials Table included in Title

49 CFR 172.102. Note also the cross-reference list for the number-to-name that follows the Hazardous Materials Table 102 as Appendix A.)

Ignitible (EPA usage) A liquid with a flashpoint less than 140%.

IMDG *International Maritime Dangerous Goods.*

IMDGC *International Maritime Dangerous Goods Codes.*

IMDG designation A hazardous material identifier published by the International Maritime Organization in their Dangerous Goods Code.

IMO *International Maritime Organization* (formerly IMCO).

Impermeability As applied to soil or subsoil, the degree to which fluids, particularly water, cannot penetrate in measurable quantities.

Impoundment See Surface impoundment.

Inactive facility The EPA designation for a treatment, storage, or disposal facility that has not accepted hazardous waste since November 19, 1980.

Inactive portion A portion of a hazardous waste management facility that has not operated since November 19, 1980, but is not yet a closed portion (no longer accepts wastes to that area).

Incineration An engineered process using controlled flame combustion to thermally degrade waste materials. Devices normally used for incineration include rotary kilns, fluidized beds, and liquid injectors. Incineration is used particularly for the destruction of organic wastes with a high BTU value. The wastes are detoxified by oxidations, and if the heat produced is high enough, they can sustain their own combustion and will not require additional fuel. EPA's draft regulations specify a recommended temperature of 1000° Celsius, with a residence time (the time the gases should stay in the combustion chamber) of 2 seconds.

Incompatible waste Waste unsuitable for commingling with another waste or materials, where the commingling might result in the following:

1. Extreme heat or pressure generation.
2. Fire.
3. Explosion or violent reaction.
4. Formation of substances that are shock sensitive, friction sensitive, or otherwise have the potential to react violently.
5. Formation of toxic dusts, mists, fumes, gases, or other chemicals.
6. Volatization of ignitable or toxic chemicals due to heat generation, in such a manner that the likelihood of contamination of groundwater or escape of the substances into the environment is increased.

Industrial wastes Unwanted materials produced in or eliminated from an industrial operation. They may be categorized under a variety of headings, such as liquid wastes, sludge wastes, and solid wastes. Hazardous wastes contain substances that, in low concentrations, are dangerous to life (especially human) for reasons of toxicity, corrosiveness, mutagenicity, and flammability.

Infectious waste Waste that contains pathogens or consists of tissues, organs, body parts, blood, and body fluids that are removed during surgery or other procedures. See Title 42 CFR Part 72. (See also Biologically hazardous waste).

Infiltration The flow of fluid into a substance through pores or small openings. The word is commonly used to denote the flow of water into soil material.

Ingestion The process of taking substances into the body, as in food, drink, medicine, etc.
Inhalation the breathing in of a substance in the form of gas, vapor, fume, mist, or dust.
Inhibitor A chemical added to another substance to prevent an unwanted occurrence of chemical change.
Injection The subsurface emplacement of a fluid or waste.
Injection well A well into which fluids are injected.
Inner liner A continuous layer or lining of material placed inside a tank or other container that protects the construction materials of the tank or container from the contents.
Inorganic compounds Chemical compounds that do not contain the element carbon.
Inorganic matter Chemical substances of mineral origin, not containing carbon to carbon bonding. Generally structured through ionic bonding.
Insecticide A chemical product used to kill and control nuisance insect species. (See also Pesticide.)
Institutional waste All solid waste emanating from institutions such as, but not limited to, hospitals, nursing homes, orphanages, schools, and universities.
Interim authorization The conditional permission from EPA that enables a State to operate its own hazardous waste management program.
Interim status A period of time, which began November 19, 1980, when hazardous waste storage and treatment facilities and hazardous waste transporters could continue to operate under a special set of regulations until the appropriate permit or license application is or was approved by EPA.
Intermunicipal agency An agency established by two or more municipalities with with responsibility for planning or administration of solid waste.
IPY *Inches per year* (as corrosion rate reference in Title 49 CFR 173.240 (a) (2) and 173.500 (b) (2) (i)).
Irritant Any material, liquid or solid substance, that upon contact with fire or when exposed to air gives off dangerous or intensely irritating fumes, such as tear gas, but not including poison class A or B material. (Materials named as irritants are presented in Title 49 CFR, 173.38).
ISO *International Organization for Standardization*
kg *Kilogram.* A metric unit of weight, about 2.2 United States pounds.
LC$_{50}$ *Lethal concentration$_{50}$*, the concentration of a material which on the basis of laboratory tests is expected to kill 50% of a group of test animals when administered as a single exposure (usually 1 or 4 hours). Also, other LC values can be expressed (e.g., LC$_{10}$ and LC$_{20}$).
LC$_{Lo}$ *Lethal concentration low.* The lowest concentration of a substance in air, other than LC$_{50}$, which has been reported to have caused death in humans or animals. The reported concentrations may be entered for periods of exposure that are less than 24 hours (acute) or greater than 24 hours (subacute and chronic).
LD$_{50}$ *Median lethal dose.* The dose which is required to produce death in 50% of the exposed species. Death is usually reckoned as occurring within the first 30 days.
LD$_{Lo}$ *Lethal dose low.* The lowest dose of a substance introduced by any route, other than inhalation, over any given period of time in one or more divided portions and reported to have caused death in humans or animals.

Label (DOT) Diamond-shaped, square-shaped, or rectangular-shaped attachment to a package that identifies the hazardous nature of a material. (See Title 49 CFR Part 172, Subpart E.)

Land treatment facility A facility or part of a facility where hazardous waste is applied or incorporated into the soil surface; such facilities are *disposal facilities* if the waste will remain after closure.

Latent period The time which elapses between exposure and the first manifestation of damage.

Leak or Leaking Any instance in which a article, container, or equipment has any hazardous material (e.g., PCB) on any part of its external surface or has released this substance to the surrounding environment.

LEL *Lower explosive limit.* The lowest concentration of the material in air that can be detonated by spark, shock, fire, etc.

LFL *Lower flammable limit.* The lowest concentration of the material in air that will support combustion from a spark or flame.

LUST *Leaking underground storage tanks.* (Now being called UST.)

m^3 *Cubic Meter or Stere.* A metric measure or volume, about 35.3 cubic feet or 1.3 cubic yards.

Macroencapsulation The isolation of a waste by embedding it in, or surrounding it with, a material that acts as a barrier to water or air (e.g., clay and plastic liners).

Magnetized material Any material which, when packed for air transport, has magnetic field strength of 0.159 A/M or more at a distance 2.1 m from any point on the surface of the assembled package. (See ICAAO Technical Instructions.)

Malaise Vague feeling of bodily discomfort.

Manifest, uniform hazardous waste When properly prepared and distributed, provides a tracking system that consists of forms originating with the generator or shipper and following from the generator to disposal in a permitted TSDF.

Manometer An instrument for measuring pressure that usually consists of a U-shaped tube containing a liquid, the surface of which in one end of the tube moves proportionally with pressure changes on the liquid in the other end. Also, a tube type of differential pressure gauge.

Marking Applying the required descriptive name, instructions, cautions, weight, or specifications or combination thereof on containers of HM/HW. (See Title 49 CFR 171.8.)

Material safety data sheet (OSHA usage) See MSDS.

Maximum permissible dose (MPD) The maximum RBE dose that the body of a person or specific parts thereof shall be permitted to receive in a stated period of time.

Melting point The temperature at which a material changes from a solid to a liquid.

mg *Milligram.* A metric unit of weight. There are 1,000 milligrams in one gram (g) of a substance.

mg/m^3 *Milligrams per cubic meter.* A unit for measuring concentrations of dusts, gases or mists in air.

MHE *Material handling equipment.*

Microorganism A living organism not discretely visible to the unaided eye. These organisms obtain nutrients from the discharge waste products (largely CO_2 or O_2) into the fluid in which they exist, thus serving to lower the nutrient level. Bacteria, fungi.

ml *Milliliter*. A metric unit of capacity, equal in volume to one cubic centimeter (cc), or about $1/16$ of a cubic inch. There are 1,000 milliliters in one liter (l).

mm *Millimeter*. A metric unit of length, equal to $1/1000$ of a meter, or about $1/25$ of an inch.

Monolithic Describing a structure that is without cracks or seams, self-supporting, and essentially homogeneous.

MSDS *Material safety data sheet*. An MSDS must be in English and include information regarding the specific identity of hazardous chemicals. Also includes information on health effects, first aid, chemical and physical properties, and emergency phone numbers.

MSHA *Mine Safety and Health Administration* of the United States Department of Interior.

MTB *Materials Transportation Bureau* (formerly of DOT); now the Research and Special Programs Administration (RSPA) of DOT.

Mutagen A substance capable of causing genetic damage.

NA number *North American* identification number. When NA precedes a four-digit number, it indicates that this identification number is used in the United States and Canada to identify a hazardous material (HM) or a group of HMs in transportation.

NAAQS *National Ambient Air Quality Standards*. CAA Section 109.

Narcosis Destruction of body tissue.

NEPA *National Environmental Policy Act* (1969).

NESHAPs *National Emission Standards for Hazardous Air Pollutants*. CAA Section 112 also refers to chemicals regulated under this program.

Neutralization The process by which acid or alkaline properties of a solution are altered by addition of certain reagents to bring the hydrogen ion and hydroxyl ion concentrations to an equal value; sometimes referred to as *p*H 7, the value of pure water.

Neutralization surface Surface impoundments that (1) are used to neutralize wastes considered hazardous solely because they exhibit the characteristic of corrosivity; (2) contain no other wastes; or (3) neutralize the corrosive wastes sufficiently rapidly so that no potential exists for migration of hazardous waste from the impoundment.

Neutralize To make harmless anything contaminated with a chemical agent. More generally, to destroy the effectiveness.

NFPA *National Fire Protection Association*. An international voluntary membership organization to promote improve fire protection and prevention and establish safeguards against loss of life and property by fire. Best known on the industrial scene for the maintenance of National Fire Codes, (i.e., 16 volumes of codes, standards, recommended practices, and manuals) and periodically updated by NFPA technical committees.

NIOSH *National Institute for Occupational Safety and Health* of the Public Health Service, United States Department of Health and Human Services (DHHS). Federal agency which, among other activities, tests and certifies respiratory protective devices and air sampling detector tubes, recommends occupational exposure limits for various substances and assists OSHA and MSHA in occupational safety and health investigations and research.

Nonflammable gas Any material or mixture, in a cylinder or tank, other than poison gas, or flammable gas having in the container an absolute pressure exceeding 40 psi at 70°F, or having an absolute pressure exceeding 104 psi at 130°F (Title 49 CFR and CGA).

Nonpoint sources (CWA usage) Ill-defined runoff that enters waterways. (More stringent future regulation is likely.)

NOS or n.o.s. *Not otherwise specified* (DOT usage).

NPDES *National Pollutant Discharge Elimination System* (water quality usage).

NPTN *National Pesticides Telecommunication Network.* A national pesticide poison control center restricted to use by health professionals. The network assists the health professional in diagnosing and managing pesticide poisoning. Services include product active ingredient identification, symptomatic review, toxicologic review, specific treatment recommendations, physician consultation, and referrals for laboratory analyses. These services are provided 24 hours a day.

NQT *Nonquenched and tempered.*

NRC (a) *National Response Center* (AC 800-424-8802).
(b) *Non-reusable container.* (See Title 49 CFR 173.28 and Title 49 CFR 178.8)
(c) *Nuclear Regulatory Commission* (10 CFR usage).

Nuclide Any individual nuclear species, such as ^{14}C (carbon-14), ^{32}P (phosphorus-32), ^{131}I (iodine-131), etc., irrespective of whether or not the nuclide has other isotopes. The term *isotope* is frequently misused for nuclide but the strict meaning of the former as originally defined by Soddy (1914) is of the same place, i.e., in the same position in the periodic table. Thus, one may say that the nuclide phosphorus-32 is an isotope of phosphorus, or even more specifically of, say, phosphorus-33. A radioactive nuclide is often referred to as a *radionuclide*

Nuisance The class of wrongs that arise from the unreasonable, unwarranted or unlawful use by a person of his own property, either real or personal, or from his own lawful personal conduct working an obstruction or injury to the right of another, or of the public and producing material annoyance, inconvenience, discomfort or hurt.

OBA *Oxygen breathing apparatus.*

OHMR *Office of Hazardous Materials Regulation*, formerly within DOT's Materials Transportation Bureau. Now known as OHMT.

OHMT *Office of Hazardous Materials Transportation* of the Research and Special Programs Administration of DOT.

Olfactory Relating to the sense of smell.

On site The same or geographically contiguous property that may be divided by public or private right-of-way, provided the entrance and exit between the properties is at a crossroads intersection, and that access is by crossing as opposed to going along the same right-of-way. Noncontiguous properties owned by the same person but connected by right-of-way that he controls and to which the public does not have access is also considered on-site property [Title 40 CFR 260.10 (a) (48)].

Oral toxicity Adverse effects resulting from taking a substance into the body through the mouth.

Organic peroxide Any organic compound containing the bivalent $-O-O-$ structure and that may be considered a derivative of hydrogen peroxide where one or more of the hydrogen atoms have been replaced by organic radicals.

ORM (A-E) (DOT usage) *Other Regulated Materials.* Several classes of ORM materials are recognized (i.e., ORM-A, ORM-B, ORM-C, ORM-D, and ORM-E).

OSC *On-scene coordinator* in emergency response actions.

OSHA *Occupational Safety and Health Administration* of the United States Department of Labor. Federal (or State) agency with safety and health regulatory and enforcement authorities for most United States industry and business.

Outside packaging A packaging plus its contents. (See Title 49 CFR 171.8.)

Overpack Except when reference to a packaging specified in Title 49 CFR Part 178, means an enclosure used by a single consignor to provide protection or convenience in handling of a package or to consolidate two or more packages. Overpack does not include a freight container.

Oxidizer A chemical other than a blasting agent or explsoive as defined in Title 29 CFR 1910.109 (a), that initiates or promotes combustion in other materials, thereby causing fire either of itself of through the release of oxygen or other gases.

Package According to the United Nations definition, a complete product of the packing operation, consisting of the packaging and its contents prepared for transport.

Packaging The assembly of one or more containers and any other components necessary to assure compliance with minimum packaging requirements; includes containers (other than freight containers or overpacks), and multiunit tank car tanks (Title 49 CFR 171.8), also restates the methods and materials used to protect items from deterioration or damage; this includes cleaning, drying, preserving, packaging, marking, and unitization.

Packing Assembly of items into a unit, intermediate, or exterior pack with necessary blocking, bracing, cushioning, weather-proofing, reinforcement and marking.

Pallets A low portable platform constructed of wood, metal, plastic, or fiberboard, built to specified dimensions, on which supplies are loaded, transported, or stored in units.

Part A Interim permit for TSDF of hazardous waste prior to 1981 (RCRA usage).

Part B Final permit for TSDF (RCRA usage).

Pathogen Any microorganism capable of causing disease.

PCB *Polychlorinated biphenyl* (see Title 40 CFR 761.3).

PCB-contaminated electrical equipment Any electrical equipment, including transformers that contains at least 50 ppm but less than 500 ppm PCB. (Title 40 CFR 761.3).

PCB item An item containing PCBs at a concentration of 50 ppm or greater (Title 40 CFR 761.3). (The concentration requirement may vary by State.)

PCB transformer Any transformer that contains 500 ppm PCB or greater (Title 40 CFR 761.3).

PCDF *Polychlorinated dibenzofurans:* A class of toxic chemical compounds occurring as a thermal degradation product of PCBs.

PCP (a) Abbreviation for *pentachlorophenol* (q.v.), a wood preservative used on military ammunition boxes and telephone poles.
(b) 1-(1-Phenylcylohexyl) piperidine or angel dust or HOG, an analgesic and anesthetic that may produce serious psychologic disturbances.

PEL *Permissible exposure limit.* An exposure limit established by OSHA regulatory authority. May be a time weighted average (TWA) limit or a maximum concentration exposure limit. (See also Skin.)

PEP *Preventive engineering practices.*

Pesticide Any liquid, solid, or gaseous material that demonstrates an oral LD_{50} of greater than 50 mg/kg but less than 5,000 mg/kg, or an inhalation LC_{50} of greater than 0.2 mg/L, but less than 20 mg/L, or a dermal LD_{50} of greater than 200 mg/kg but less than 20,000 mg/kg (Title 40 CFR 162).

PF *Protective factor*. Refers to the level of protection a respiratory protective device offers. The PF is the ratio of the contaminant concentration outside the respirator to that inside the respirator.

***p*H** A measure of hydrogen ion concentration $[H^+]$. A *p*H of 7.0 is neutrality; higher values indicate alkalinity and lower values indicate acidity.

Phase I (RCRA usage) The regulations issued in May 1980 include the the identification and listing of hazardous waste, standards for generators and transporters of hazardous waste, standards for owners and operators of facilities that treat, store, or dispose of hazardous waste; requirements for obtaining hazardous waste facility permits, and rules governing delegation of authority to the States.

Phase II (RCRA usage) Technical requirements for permitting hazardous waste facilities. Sets specific standards for particular types of facilities to ensure the safe treatment, storage, and disposal of hazardous waste on a permanent basis by methods that will protect human health and the environment. Phase II standards enable facilities to move from "interim status" to final facility permits.

Pneumoconiosis Producing dust: Dust which, when inhaled, deposited, and retained in the lungs, may produce signs, symptoms and findings of pulmonary disease.

Pneumonitis Inflammation of the lungs characterized by an outpouring of fluid in the lungs. Pneumonia is the same condition, but involves greater quantities of fluid.

Pretreatment standards (CWA usage) Specific industrial operation or pollutant removal requirements in order to discharge to a municipal sewer.

Point sources (CWA usage) Well defined places at which pollutants enter waterways.

Poison class A Poisonous gases or liquids of such a nature that a very small amount of the gas, or vapor of the liquid, mixed with air is dangerous to life (Title 49 CFR 173.326).

Poison class B Demonstrates an oral LD_{50} of up to and including 50 mg/kg, or in inhalation LC_{50} of up to and including 2 mg/L, or a general LD_{50} of up to and including 200 mg/kg; *or* Is either classed as a poison class B per Title 49 CFR 173.343, or qualifies as a category I pesticide per Title 40 CFR Part 162 excluding the corrosivity criteria.

Poison control centers A nationwide network of poison control centers has been set up with the aid of the United States Food and Drug Administration and Department of Health and Human Services. The centers, usually established in local hospitals, are now widely distributed and available by phone from most parts of the country. Staff members are specially trained in the treatment of poisoning cases.

Poison information center (pesticide) See NPTN.

Pollution Contamination of air, water, land, or other natural resources that will, or is likely to, create a public nuisance or render such air, water, land, or other natural resources harmful, detrimental, or injurious to public health, safety, or welfare, or to domestic, municipal, commercial, industrial, agricultural, recreational, or other legitimate beneficial uses, or to livestock, wild animals, birds, fish, or other life.

Polychlorinated biphenyl (PCB) Any of 209 compounds or isomers of the biphenyl molecule that have been chlorinated to various degrees (includes monochlorinated compounds). (See PCB.)

Polymers Large molecules formed by the combination of many smaller molecules.

Polymerization A chemical reaction, usually carried out with a catalyst, heat, or light, and often under high pressure. In this reaction a large number of relatively simple molecules combine to form a chainlike macromolecule. This reaction can occur with the release of heat. In a container, the heat associated with polymerization may cause the substance to expand and/or release gas and cause the container to rupture, sometimes violently. The polymerization reaction occurs spontaneously in nature; industrially, it is performed by subjecting unsaturated or otherwise reactive substances to conditions that will bring about their combination.

POTW *Publicly owned treatment works.*

ppb *Parts per billion.* A unit for measuring the concentration of a gas or vapor in air; parts (by volume) of the gas or vapor in a billion parts of air. Usually used to express measurements of extremely low concentrations of unusually toxic gases or vapors. Also used to indicate the concentration of a particular substance in a liquid or solid.

PPE *Personal protective equipment.*

ppm *Parts per million.* A unit for measuring the concentration of a gas or vapor in air; parts (by volume) of the gas or vapor in a million parts of air. Usually used to express measurements of extremely low concentrations of unusually toxic gases or vapors. Also used to indicate the concentration of a particular substance in a liquid or solid.

PPP *Preparedness and prevention plan* (RCRA Usage).

Premanufacture notification (PMN) A major control mechanism exercised under the toxic substances control act to allow EPA to assess the safety of new chemicals before manufacture.

Pretreatment standards (CWA usage) Specific industrial operation or pollutant removal requirements in order to discharge to a municipal sewer.

PSD (CAA usage) *Prevention of significant deterioration* (of air quality).

psi *Pounds per square inch.*

psia *Pounds per square inch absolute.*

psig *Pounds per square inch gauge.*

Proper shipping name The name of the hazardous material shown in Roman print (not italics) in Title 49 CFR 172.101 or 172.102 (when authorized).

Pulmonary Pertaining to the lungs.

Pyrophoric A chemical that will ignite spontaneously in air at a temperature of 130°F (54.4°C) or below.

rad A unit for the measurement of radioactivity. One rad is the amount of radiation that results in the absorption of 100 ergs of energy by 1 g of material.

Radioactive material A material that might or might not require the issuance of a license, according to 10 CFR, to persons who manufacture, produce, transfer, receive, acquire, own, possess, or use byproduct materials.

RAM *Radioactive material.*

RAM licensed exempt Any radioactive material, the radionuclide of which is not subject to the licensing requirement of Title 10 CFR.

RCRA *Resource Conservation and Recovery Act* (1976).

Recovery drum A nonprofessional reference to a drum used to overpack damaged or leaking hazardous materials. (See Disposal drum.)

Relative biological effectiveness (RBE) A measure of the relative effectiveness of absorbed doses of radiation.

rem A measure of radiation dose meaning roentgen equivalent man. The dose in rems is calculated by multiplying the dose in rads by the relative biological effectiveness of the radiation considered.

Reportable quantity (DOT and EPA usage) The quantity specified in column 2 of the Hazardous Materials Table in Title 49 CFR 172.101, for any material identified by the letter E in Column 1 (Title 49 CFR 171.8), or any material identified by EPA in Table 117.3, Reportable Quantities of Hazardous Substance, in Title 40 CFR 173. The letter E in Column 1 (Title 49 CFR 172.101) identifies this material as a potential hazardous substance.

Residue As related to Title 49 CFR 171.8, residue is the hazardous material remaining in a packaging after its contents have been emptied and before the packaging is refilled, or cleaned and purged of vapor to remove any potential hazard. Residue of a hazardous material, as applied to the contents of a tank car (other than DOT Specification 106 or 110 tank cars), is a quantity of material no greater than 3 percent of the car's marked volumetric capacity.

Respiratory system Consists of (in descending order) the nose, mouth, nasal passages, nasal pharynx, pharynx, larynx, trachea, bronchi, bronchioles, air sacs (alveoli) of the lungs, and muscles of respiration.

Risk assessment An investigation of the potential risk to human health or the environment posed by a specific action or substance. The assessment usually includes toxicity, concentration, form, mobility and potential for exposure of the substance.

Roentgen A measure of the charge produced as the rays pass through air.

Roentgen equivalent man or rem The product of the absorbed dose in rads multiplied by the RBE.

RQ See Reportable quantity.

RSPA *Research and Special Programs Administration* (of DOT).

SADT *Self-accelerating decomposition temperature test*. A test which establishes the lowest temperature at which a peroxide, in its largest commercial package, will undergo self-accelerating decomposition.

Salvage drum A drum with a removable metal head that is compatible with the lading used to transport damaged or leaking hazardous materials for repackaging or disposal. (See Title 49 CFR 173.3.) (Also referred to as *disposal* or *recovery drum*.)

Salivation An excessive discharge of saliva; ptyalism.

SCBA Self-contained breathing apparatus. (See Full protective clothing and Fully encapsulating suits).

SDWA *Safe Drinking Water Act* (1974).

Secondary materials Spent materials, sludges, byproducts, scrap metal and commercial chemical products recycled in ways that differ from their normal use.

Sensitizer A substance which on first exposure causes little or no reaction in man or test animals, but which on repeated exposure may cause a marked response not necessarily limited to the contact site. Skin sensitization is the most common form of sensitization in the industrial setting although respiratory sensitization to a few chemicals is also known to occur.

Significant new use rule (SNUR) (TOSCA usage) Stipulation (usually applied as a criterion for manufacture of a specific chemical) that EPA must be notified of significant new use.

SIP *State implementation plan.* CAA Section 110.

"Skin" A notation, sometimes used with PEL or TLV exposure data; indicates that the stated substance may be absorbed by the skin, mucous membranes, and eyes—either airborne or by direct contact—and that this additional exposure must be considered part of the total exposure to avoid exceeding the PEL or TLV for that substance.

Sludges High moisture content residues from treating air or waste water or other residues from pollution control operations.

Smoke An air suspension (aerosol) of particles, often originating from combustion or sublimation. Carbon or soot particles less than 0.1 μ (micron) in size result from the incomplete combustion of carbonaceous materials such as coal or oil. Smoke generally contains droplets as well as dry particles.

SOP *Standard operating procedures.*

SPCC plan (CWA usage) *Spill prevention, control and countermeasure plan.*

SPM *Spill prevention management.*

Spontaneously combustible (IMDG code) Solids or liquids possessing the common property of being liable spontaneously to heat and to ignite.

SRP *Spill response plan.*

SRT *Spill response team.*

STC *Single trip container.* (See Title 49 CFR 173.28 and Title 49 CFR 178.8.)

Storage When used in connection with hazardous waste, it means the containment of hazardous waste, either on a temporary basis or for a period of years, in such a manner as not to constitute disposal of such hazardous waste.

Storage facility Any facility used for the retention of HW prior to shipment or usage, except generator facilities (under Title 40 CFR), which is used to store wastes for less than 90 days, for subsequent transport.

Storage tank Any manufactured, nonportable, covered device used for containing pumpable hazardous wastes.

Strict liability The defendant may be liable even though he may have exercised reasonable care.

STEL *Short term exposure limit:* ACGIH terminology.

Surface impoundment Any natural depression or excavated and/or diked area built into or upon the land, which is fixed, uncovered, and lined with soil or a synthetic material, and is used for treating, storing, or disposing wastes. Examples include holding ponds and aeration ponds.

Synergism Cooperative action of substances whose total effect is greater than the sum of their separate effects.

TCDD *Tetrachlorodibenzodioxin.* A TCDD associated with the manufacturer of 2,4,5-T (Silvex) and occurring as a thermal degradation product of chlorinated benzenes.

Teratogen A substance or agent which can result in malformations of a fetus.

Threshold (OSHA usage) The level where the first effects occur; also the point at which a person just begins to notice a tone (sound) is becoming audible.

TI *Transport index.* Applicable to radioactive materials. [See Title 49 CFR 173.403 (bb).]

TLV *Threshold limit value.* An exposure level under which most people can work consistently for 8 hours a day (day after day) with no harmful effects. A table of these values and accompanying precautions is published annually by the American Conference of Governmental Industrial Hygienists (ACGIH).

Totally enclosed manner Any manner that will ensure no exposure of human beings or the environment to any concentration of PCBs.

Toxicity A relative property of a chemical agent and refers to a harmful effect on some biological mechanism and the condition under which this effect occurs.

Trade secret Any confidential formula, pattern, process, device, information or compilation of information (including chemical name or other unique chemical identifier) that issued in an employer's business, and that gives the empoyer an opportunity to obtain an advantage over competitors who do not know or use it.

TSCA or TOSCA *Toxic Substances Control Act* (1976).

TSDF *Treatment, storage, or disposal facility.*

TWA *Time weighted average exposure.* The airborne concentration of a material to which a person is exposed, averaged over the total exposure time—generally the total workday. (See also TLV.)

TWA-C *Time weighted average—ceiling limit.* The excursion limit placed on fast acting substances that limits all exposures below the applicable C limit. All time weighted average concentrations and peak exposures must be less than this limit.

UEL *Upper explosive limit.* The highest concentration of the material in air that can be detonated.

UFL *Upper flammable limit.* The highest concentration of the material in air that will support combustion.

UL *Underwriters Laboratories, Inc.*

UN number *United Nations identification number.* When UN precedes a four-digit number, it indicates this identification number is used internationally to identify a hazardous material.

Unitization Any combination of unit, intermediate or exterior packs of one or more line items of supply into a single load in such a manner that the load can be handled as a unit through the distribution system. Unitization (unitized loads, unit loads) encompasses consolidation in a container, placement on a pallet or load base, or securely binding together.

Unit pack The first tie, wrap, or container applied to a single item or a group of items which constitutes a complete or identifiable package. The unit pack should be overpacked for shipment unless the unit container is specifically designed to provide shipping protection.

UPS *United Parcel Service.*

UST *Underground Storage Tanks* (see LUST).

Vapor An air dispersion of molecules of a substance that is liquid or solid in its normal physical state, at standard temperature and pressure. Examples are water vapor and benzene vapor. Vapors of organic liquids are loosely called fumes; however, it is not technically appropriate to use the term *fume* for vapors of organic liquids.

Vapor density The ratio of the vapor weight of the commodity compared to that of air. Vapors will diffuse and mix with air due to natural air currents. In general, if the ratio is greater than 1, the vapors are heavier and may settle to the ground; if lower than 1, the vapors will rise.

Vapor pressure The pressure of the vapor in equilibrium with the liquid at the specified temperature. Higher values indicate higher volatility or evaporation rate.

X-radiation, X-rays Electromagnetic ionizing radiation originating outside the atomic nucleus. X-rays are indistinguishable from gamma rays of the same energy, the distinction being one of source.

Index

Access control points, 119
Acetaldehyde, 55
Acetic acid, 48-49, 55
Acid pointers, 49
Acids, 48-49, 54
   chemical reactions involving, 50
Acne, 101
Activated charcoal, 172
Active samplers, 170-77
Acute effects, of radiation, 76
Acute toxicity, 100
Additive effect, 98
Administrative controls, 113
Advection, 264-66
Agent Orange, 56
Air
   consumption of, with SCBA unit, 224
   percentage of oxygen in, 71
Airborne particulates, filter media for, 175-77
Airline respirators. *See* supplied air respirators
Air monitoring
   initial, 18
   periodic, 18
   purpose of, at hazardous waste sites, 155-56
Air-purifying respirators (APRs), 182-83
   advantages of, 186-87
   considerations in using, 189-90
   disadvantages of, 187-89
   inspection of, 230
   storage of, 231
Albumin, in urine, 39
Alcohol, 54
Aldehyde, 54
Aldosterone, 37-38
Alkali metals, 46-47
Alkaline phosphatase and 5'-nucleotidase in blood test, 37
Allergic contact dermatitis, 101
Alpha counter, 77
Alpha emitters, 73
Alpha particles, 73
Alpha radiation, 73-74
Ambient temperature extremes, effect of, on work duration, 226

American Conference of Governmental Industrial Hygienists, 108
Ames test, 100
Amines, 50
Amino, 55
Anaphylactic shock, first aid for, 307
Animals, and toxicity, 93-94
Animal studies, of toxicity, 96
Antagonism, 99
Aprons, 205
Aquifers, 259
   confined, 260-61
   perched, 261
   unconfined, 259-60
Area sampling, 170
Argon, 47
Aromatic compounds, 54
Aromatic hydrocarbons, health effects of, and medical monitoring, 27
Artesian well, 261
Asbestos, health effects of, and medical monitoring, 27
Asphyxiation, 102-3
Assigned protection factors, 195-97
Atmosphere-supplying respirators, 183-84
Atomic number, 43
Atomic particles, properties of, 43
Atopic disease, 30
Audiometric testing, 30
Autoignition temperature (AIT), 58, 59, 110

Backhoe spike, 149
Bacon bomb, 244
Bacteria, in urine, 40
Bag sampling, 174-75
Base pointers, 50
Bases, 50
   chemical reactions involving, 50
Benefits analysis, 345-46
Beta emitters, 74
Beta particles, 74
Beta radiation, 74-75
BF3 counter, 78
Bioaccumulation, and radioactive isotopes, 76
Biohazards, 93

377

Biological decay of contaminant, 265
Biological effects of radiation, 76
Biological hazards, 90-92
  animals, 93-94
  microorganisms, 92-93
  plants, 93-94
Biological response, to exposure to more than one chemical, 98-99
Biological toxic effects, 104
Blast and fragmentation suit, 209
Bleeding, first aid for, 306-7
Blood, in urine, 39
Bloodstream, effects of chemicals in, 103
Blood tests, interpretation of the significance of alkaline phosphatase and 5'-nucleotidase, 37
  BUN (Blood Urea Nitrogen), 36
  calcium, 35
  carbon dioxide, 38
  chloride, 38
  creatinine, 36
  phosphorus, 35-36
  potassium, 38
  serum bilirubin, 36-37
  serum electrolytes, 37-38
  serum glucose, 38
  sodium, 37-38
  total cholesterol and triglycerides, 36
  uric acid, 36
Body water loss, 227
Boiling point, 109
Boots, 204
Breakthrough time, 199
Brisance, 58
Bromine, 46
Bromo, 55
Bucket-type bailer, sampling monitor wells with, 273-75
Buddy system, 141-42
Bulging/swelling containers, 148
Bulking of waste, 151-52
Bulk materials sampling, 241
BUN (blood urea nitrogen), in blood test, 36
Bung wrench, 149
Burns, first aid for, 308
Butanol, 55
Butyl, 54
n-Butyl amine, 56
Butyraldehyde, 55
Butyric acid, 55

Cadmium, 47
Calcium, in blood test, 35
Calibration, of atmospheric sampling systems, 178
Carbamate (SEVIN), 56
Carbon dioxide
  as fire suppressant, 297
  in blood test, 38
Carbon tetrachloride, as fire suppressant, 297
Carcinogenesis, 104
  tests for, 99-100
Carcinogens, 106-7
Cartridges, color code for gas mask canisters and, 189
CAS (chemical abstract service) number, 112
Casts, in urine, 40
Catalyst, 58
Caustics, 50
Cellulose, 175
Center for Disease Control, 93
Centrifugal separation, 175
Chain-of-custody record, 253, 255
Chemical fires, methods of extinguishing, 297
Chemical hazards, 42-56
  signs and symptoms of indicating potential medical emergencies, 33
Chemical protective clothing (CPC)
  attacks on, 199
  availability of information on performance characteristics of, 199-201
  basic principles of selection, 202
  chemical resistance, 199
  inspecting, 229
  materials and technologies, 202-3
  problems with information available on, 201-2
  responsibility of the employer, 203
  selection of, 198
  storage of, 231
  types of, 203-7
Chemical reactions, 58
  controlling, 65-66
  involving acids and bases, 50
Chemicals, water reactive, 47
Chemical toxicants, health effects of, and medical monitoring, 27-29
Chemistry, 42
Chemrel suit by Chemron, Inc., 203
Chloracne, 101

Chlorate, 51
Chlordane, 56
Chloride, in blood test, 38
Chloride ion, 44, 45
Chlorinated hydrocarbon pesticides (DDT, Chlordane, Mirex), 56, 103
Chlorine, 44, 46
Chloro, 55
Cholesterol, in blood test, 36
Chromium, 47
Chronic effects, of radiation, 76
Chronic toxicity, 100
Cirrhosis, 100
Civil defense authorities, 81
Cleaning, of work area, 141
Closed-circuit SCBAs, 191
Clothing. See also Chemical protective clothing; Personal protective clothing
  choice of, as unsafe condition, 135–36
  firefighters', 208
Cobalt, 47
Cold, 85–86
  contributing factors in, 86–87
  monitoring of, 89
  normal mechanisms in, 86
  prevention of problems in, 88–89
  problems related to, 87–88
  treatment of problems in, 89–90
COLIWASA, 243
Color, in urine, 39
Color code, for cartridges and gas mask canisters, 189
Combustion, 59
Combustible, 58
Compatibility, categorization of hazardous wastes by, 66–67
Compatibility staging, 67, 69, 150–51
Compatibility testing, 151–52
  detection methods for, 69
Compounds, 44
Comprehensive Environmental Response, Compensation and Liability Act (1980), 1, 2, 311
Compressed gas cylinder, 152
Confined aquifers, 260–61
Confined spaces, 139–40
  entry procedures for, in site safety plan, 336
Contact dermatitis, 101
Containerized liquids, 242–44

Containers, 19. See also Drums
  handling of
    minimization of danger, 145
    monitoring, 146
    occasions for, 145
    planning for, 147
    preliminary classification, 147
    reasons for concern, 145
    subsurface investigation, 146
    visual inspection of, 146
  labeling of, for hazardous materials, 316–17
  moving of, 147–48
  opening of, 148–50
  staging of, 150–51
Contaminant plume, 262–63
Contamination
  of groundwater, 261–62
    biological decay, 265
    factors affecting contaminant migration, 264–66
    movement of contaminants, 262, 264
    predicting contaminant migration, 266
    sources of, 262, 263
Contamination reduction corridor (CRC), 119
Contamination reduction zone (CRZ), 119
Continuous-flow respirators, 197
Coolant supply, 226
Cooling garment, 210
Corrosion, 101
Covalent bonding, 45
Creatinine, in blood test, 36
Crystals, in urine, 40
Cushing's syndrome, 38
Cyclones, 175

DBCP (dibromochloropropane), 104
DDT, 56, 103
Decontamination, 19–20, 322
  definition of, 322
  factors in plan for, 322–24
  for ill or injured worker, 324
  measuring effectiveness of, 332–33
  methods of, 324–25
  of protective clothing, 325–30
  in site safety plan, 336
  protecting personnel during, 332
  of sampling equipment, 245–46
  of tools, 331
  of vehicles and heavy equipment, 331–32
Decontamination line, 330

Decontamination personnel, protecting, 332
Deep frostbite, 88
Degradation, 199
 of ensemble components, 230-31
Detoxification, 103
Dichromate, 51
Diffusion samplers, 177
Diking, 126-28
 of spills, 129, 303
Dimethyl ketone, 56
Direct-reading air monitoring instruments, 154-55
 considerations in selecting, 156
  inherent safety, 157
  portability, 156
  reliable, useful results, 156
  selectivity/sensitivity, 157
 for oxygen deficient/combustible atmospheres
  combustible gas indicator, 159-61
  oxygen meter, 158-59
 purpose of, at hazardous waste sites, 155-56
  conditions to monitor, 156
  periodic monitoring, 156
  preliminary site survey, 155-56
 for radioactive atmospheres, 157-58
 for toxic atmospheres, 161
  flame ionization detector, 164-67
  photoionization detector, 162-64
Disposable overgarments, 204, 207
Disposable shoe or boot covers, 206
Dose rate effects, of radiation, 76
Dosimeter Corp. model 3700, 157
DOT. *See* Transportation, U.S. Department of (DOT)
Drager bellows pump, 169
Drill rig, safety with, 136-38
Drum deheader, 149-50
Drums. *See also* Containers
 handling, 19, 142-43
 moving buried, 148
 moving leaking, open, and deteriorated, 148
 opening, 143-44

Ear muffs, 82, 211
Ear plugs, 82, 211
Electrical accidents, prevention of, 84-85
Electrical hazards, categories of, 83
Electricity, 82-83
 nature of, 83
 resistance to, 83
Electromagnetic spectrum, 74
Electrons, 43
Electron sharing, 45
Element, 43
Elevated tanks, 152
Embryo damage, tests for, 100
Emergency procedures, 276
 fire extinguishing, suppressants and protection, 292-99
 legal requirements for
  in community, 276-89
  site emergencies, 290
 medical response and first aid, 305-8
 spills/spill response, 299-305
 training in, 20-21
 understanding and responding to, 290-92
Emergency response plan, 20-21
 in site safety plan, 336
Employee
 informing, of site safety plan provisions, 337
 responsibilities of, under OSHAct, 9
 rights of, under the OSHAct, 8-9
Engineering analysis, 128
Engineering controls, 82, 113-14
 cave-in hazards, 124
 construction of systems, 129
 diking, 126-28
  assessing stability of existing, 128
 and personal protective equipment, 17-18
 safety precautions, 123-24
 site characterization,
  databases, 116-17
  hazard assessment, 118-19
  historical research, 115-16
  on-site survey, 118-19
  perimeter reconnaissance, 117-18
  site map, 117
  zoning, 119-21
 spill containment, 129
 trenching, 122-23
Environmental samples, 234-35, 248
 marking/labeling, 249, 251-53
 packaging of, 248, 251
 procedures for handling, 251
 shipping papers for, 249
 transportation of, 249
 versus hazardous materials samples, 248

INDEX 381

Environmental Protection Agency (EPA), 1
 categorization of hazardous wastes by, 311, 312
 Chemical Emergency Preparedness Program (CEPP) of, 276
 classification of flammables by, 60
 definition of hazardous materials by, 311
 and inventory of toxic chemical emissions, 279
 review of emergency system by, 279
 and risk assessment, 42
 role of, in site safety plan, 334
Epidemiological studies, of toxicity, 96
Equipment. *See also* Personal protective equipment (PPE); Tools
 area safety, 141
 heavy, 331-32
 material handling, 144
 for removing groundwater, 269
 respiratory protective, 184, 197-98
 sampling, 238-46
Ester, 55
Ethanol, 55
Ether, 55
Ethyl, 54
Ethyl amine, 56
Etiologic agents, packaging and labeling of, 92
Evaporation, 258
Evaporative heat loss, 86
Excavation procedures, in site safety plan, 336
Exclusion zone, 119
Excretion, 103
Explosion, 58
 control of, 60
Explosive waste, moving containers with, 148
Explosive range, 59
Explosives
 classifications for, 63-65
 definition of, 62-63
Extended bottle sampler, 242
Extinguishing media, 110
Extremely Hazardous Substances (EHS), 278
Eye exposure, 101

Facepieces, use of, with respiratory protective equipment, 182
Face shield, 211
Federal Emergency Management Agency (FEMA), 279
Fiber filters, 175

Film, 77-78
Filter media, for airborne particulates, 175-77
Fire, 58
 basic considerations in fighting, 292
 control of, 60
 methods of fighting, 292, 297
 phases in fighting, 293
 prevention of, 136, 298-99
 responding to, 136
 type of, 60, 61
Fire extinguisher, types of, 296
Firefighters' protective clothing, 208
First aid, 305-8
Flame/fire retardant coveralls, 209
Flame ionization detector, 164-67
Flammability, 58, 141
Flammable chemicals, properties of common, 293-95
Flammable liquids
 DOT classification of, 62
 EPA classification of, 60
 NFPA classification of, 62-63
 transportation of, 313
Flammable liquids/solids, 251
 marking/labeling, 251-53
 packaging, 251
 shipping papers for, 253
 transportation of, 253-54
Flammable or explosive limits (LEL & UEL), 58
Flashpoint, 58, 59, 110
Flotation gear, 210
Fluorine, 46
Fluoro, 55
Formaldehyde, 55
Formic acid, 55
Foxboro organic vapor analyzer (OVA), 165
 colormetric tubes, 167-69
 gas chromatography mode, 165-67
 survey mode, 165
Fractures, first aid for, 308
Fritted bubblers, 173-74
Frostbite, 87-88
Frost nip, 87
Full facepieces, 182
Fully encapsulating suits, inspecting, 229

Gamma radiation, 73
Gas chromatogram, 165, 166
Gases, active samplers for collecting, 171-75

Gas mask canisters, color code for cartridges and, 189
Gas pressure displacement system, purging with, 270-71
Geiger counter, 157-58
Geiger-Muller counter, 77
Geologic units, properties of, 258-59
Geotechnical investigation, 128
Glass, 175
Glass-bead column, 174
Glass tubes (drum thieves), 240, 242-43
Gloves, 204, 206
  inspecting, 229
Glucose, in urine, 39
Goggles, 211
Grain thief, 241
Gravity corer, 240-41
Groundwater, 258
Groundwater contamination
  biological decay, 265
  factors affecting contaminant migration, 264-66
  movement of contaminants, 262, 264
  predicting contaminant migration, 266
  sources of, 262, 263
Groundwater hydrology, 257
Groundwater monitoring, 256
Groundwater sampling procedures, 267
  collecting samples, 270
  equipment used for removing groundwater, 269
  purging monitoring wells
    prior to sampling, 268-69
    with gas pressure displacement system, 270-71
    with peristaltic pump, 271-72
  sampling monitor wells with bucket-type bailer, 273-75
  peristaltic pump, 272-73
Groundwater systems, 257, 258
  aquifers, 259
    confined, 260-61
    perched, 261
    unconfined, 259-60
  effects of man's activities on hydrologic systems, 261-62
  factors affecting, 258
  groundwater flow, 258
  hydrologic cycle, 257-58
  water beneath the earth's surface, 258

Guide for Infectious Waste Management (EPA), 93

Half-life, and radioactive isotopes, 75
Half masks, 182
Halogenated aliphatic hydrocarbons, health effects of, and medical monitoring, 27
Halogen group, 46
Halons, as fire suppressant, 297
Hand auger, 239
Hand corer, 240
Hazard assessment, 118-19
Hazardous awareness review, 347-49
Hazardous chemicals
  as defined by OSHA, 310
  definition of, 310
Hazardous materials, definition of, 310-11
Hazardous material samples, 249-51
  versus environmental samples, 248
  shipping papers, 253
  transportation, 253-54
Hazardous Materials Emergency Planning Guide, 278
Hazardous materials response (HAZMAT) teams, 16
Hazardous Materials Table, 313, 314-15
Hazardous Materials Transportation Act (1975), 311
Hazardous substances, definition of, 311
Hazardous wastes
  categorization of, by compatibility, 66-67
  classifications of, 2
  definition of, 311
  samples of, 235
  transportation of. *See* Transportation of hazardous wastes
Hazardous Waste Manifest, 2
Hazardous waste site, 41
  OSHA safety standards, 9-24
  potential radiation sources at, 79-80
  problems with drilling operations at, 138-39
  purpose of air monitoring at, 155-56
  trucks on, 320
Hazard recognition, 41
  chemical hazards, 42-56
  risk assessment, 41-42
Headphones, 212
Heart rate, monitoring of for heat stress, 227
Heat, 85-86
  contributing factors in, 86-87

monitoring of, 89
normal mechanisms in, 86
prevention of problems in, 88–89
problems related to, 87
treatment of problems in, 89–90
Heat cramps, 87
Heat exhaustion, 87
    signs and symptoms of indicating potential medical emergencies, 33
Heat injury prevention, 228
Heat rash, 87
Heat stress, 226
    monitoring for the effects of, 226–27
        body water loss, 227
        heart rate, 227
        oral temperature, 227
        personal factors affecting respirator use, 228
Heat stroke, 87
    signs and symptoms of indicating potential medical emergencies, 33
Heavy equipment
    decontamination of, 331–32
    safety with, 136–38
Heavy metals, 47
    health effects of, and medical monitoring, 27–28
    screening for, in urine, 40
Helium, 43, 44, 47
Herbicides, health effects of, and medical monitoring, 28
Hoisting device, 141
Hood, 211
Hotline, 119
Hot work, 142
Hot zone, 119
Hybrid SCBA/SAR combination systems, 191
Hydraulically operated drum piercer, 149
Hydraulic conductivity, 259
Hydrides, 47
Hydrocarbons, 52
Hydrochloric acid, 49
Hydrodynamic dispersion, 264
Hydrogen, 43, 44
Hydrogen ions, 48
Hydrologic systems, effects of man's activities on, 261–62
Hydroxy, 54
Hyperglycemia, causes of, 38
Hypoglycemia, causes of, 38

Immediately dangerous to life and health (IDLH) atmosphere, 18
Impinger, 173
Incipient frostbite, 87
Infiltration, 257
Informational programs, 19
Instantaneous grab-type samples, 170
Integrated samples, 170
Involuntary muscle contractions, 86
Iodine, 46
Iodo, 55
Ionic chemical reactions, 44
Ionization chamber, 77
Ionization potential, 162
Ionizing radiation, 73
Irritant smoke tests, 193, 195
Isolation, 141

Kemmerer bottle, 242
Kemmerer sample, 244
Ketones, 54
    in urine, 39
Knife, 212
Krypton, 47

Labeling
    of containers for hazardous materials, 316–17
    of environmental samples, 249, 251–53
    of etiologic agents, 92
    of flammable liquids/solids, 251–53
    of poison samples, 254
Lab packs, 144–45, 148
Ladders, 141
Landfilled materials, bacterial decomposition of some types of, 93
Latency period, 106
LD_{50} concept, in toxicity, 98
Leggings, 205
Liquid absorbers, 172
Liquids sampling, 241–42
Lithium, 46
Lithium chloride, 46
Local Emergency Planning Commission (LEPC), 279
Local Emergency Planning Committees (LEPC), 277–78
Lockout, of work area, 141
Lower explosive limit (LEL), 59, 110
Lung exposure, 102–3

384 INDEX

Manifest system, 317-18, 319
Manually operated opening devices, 149
Material handling equipment, 144
Material Safety Data Sheets (MSDS), 105, 108, 279, 310
　information on, 108-12
Matter, definition of, 42-43
Medical emergencies
　primary survey, 305-7
　secondary survey, 307-8
　signs and symptoms of chemical exposure and heat stress that indicate potential, 33
Medical examination
　examination followup/consultation, 32
　periodic, 32-34
　pre-employment, 29
　　baseline information, 30
　　family history, 30
　　laboratory tests, 31-32
　　occupational history, 30
　　past medical history, 30
　　personal and social data, 30
　　physical examination, 30-31
　　systems review, 30
Medical monitoring, common chemical toxicants found at hazardous waste sites and their health effects, 27-29
Medical records, 34-35
Medical surveillance, 16-17, 25
　evaluation and update of program, 35
　examination followup/consultation, 32
　general characteristics of program, 26
　medical records, 34-35
　objectives of, 25
　periodic medical examination, 32-34
　pre-employment medical examination, 29-32
　reasons for conducting, 26
　in site safety plan, 335
　termination examination, 34
Melting point, 109
Membrane filters, 175, 177
Mercury, 47
Mesothelioma, 102
Metabolic poisons, 104
Metabolism, 103
Methane, 46
　relative response of OVA calibrated, 166
Methane gas, 93

Methanol, 55
Methyl, 54
Methyl amine, 56
Methyl ethyl ketone, 56
Microorganisms, 92-93
Microscopic examination, in urine, 40
Mineral acids, 49
Mirex, 56
Molecule, 45
Monitoring requirements, in site safety plan, 336
Monitoring wells
　purging, prior to sampling, 268-69
　sampling with
　　bucket-type bailer, 273-75
　　peristaltic pump, 272-73
Mountains in Minutes, 303
MSA 260 CGI, calibration of, to pentane, 160
Multistix reagent strips, for urine testing, 40
Muriatic acid, 49
Mutagenesis, tests for, 99-100

National Fire Protection Association (NFPA)
　creation of minimum standards for electrical devices, 157
　classification of flammable liquids by, 62-63
　hazard identification system of, 63-65
　on personal protective equipment, 218
National Institute for Occupational Safety and Health, 7, 105
National pollution discharge elimination system (NPDES) wastewater permit, 116
National Response Center (NRC), 279
　reporting transportation accidents to, 320
National Response Team, 278
NAWDEX, 117
Negative-pressure respirators, 196-97
Negative-pressure tests, 192
Neon, 47
Neurotoxicity, 104
Neutrons, 43
New technologies programs, 22-24
　in site safety plan, 336
Nickel, 47
NIOSH manual of analytical methods, 171
Nitrile, 55
Nitro, 55
Noble gases, 47

INDEX

Noise, 81
  effects of, 81
  measurement of, 81–82
  protection from, 82
Non-encapsulating chemical protective suit, 204, 205
Non-ionizing radiation, 75
Nuclear Regulatory Commission (NRC), 80

Observer system, 141–42
Occupational Safety and Health Act (OSHAct) of 1970, 6. *See also* Occupational Safety and Health Administration (OSHA); 29 CFR Part 1910.120
  background and intent of, 6–7
  organizations created by, 7
  rights and responsibilities of employees under, 8–9
  standards, 7–8
Occupational Safety and Health Administration (OSHA), 2, 7. *See also* Occupational Safety and Health Act (OSHAct)
  definition of hazardous chemicals by, 310
  respiratory protection program under, 181
  role of, in site safety plan, 334
  safety standards for hazardous waste sites, 9–24. *See also* 29 CFR Part 1910.120
Odorous vapor tests, 193, 195
Off-site characterization, 114
Ongoing monitoring and hazard assessment program, 114
On-site survey, 114, 118
Open-circuit SCBAs, 191
Opening devices
  manually operated, 149
  remotely controlled, 149
Oral exposure, 102
Oral temperature, 227
Organic chemicals, 45, 51–54
  production of, 53
Organic peroxides, 56
Organochlorine insecticides, health effects of, and medical monitoring, 28
Organophosphate insecticides, 56, 104
  and medical monitoring, 28
Oxidizing agents, 58
  reactions involving strong, 51
  relative strengths of, 52
Oxygen
  importance of, 70–71
  measuring concentration of, 72
  percentage of in air, 71
Oxygen content, 141
Oxygen deficiency, 69–70
  combustible atmospheric use of oxygen meter to evaluate, 158–59
  combustible gas indicator to evaluate, 159–61
  factors leading to, 71
  importance of oxygen, 70–71
  measuring oxygen concentration, 72
  percentage of oxygen in air, 71
Oxygen meter, 158–59

Particulates, active sample for collecting, 175–77
Passive samplers, 170, 177–78
Patch products, to contain spills, 302
PCBs, medical monitoring of, 28–29
Penetration, 199, 225–26
Pentane, calibration of MSA 260 CGI to, 160
Perched aquifers, 261
Perchlorate, 51
Perimeter reconnaissance, 117–18
Periodate, 51
Periodic inspection, 128
Periodic monitoring, of air, 156
Periodic Table of the Elements, 45
Peristaltic pump, 242, 243
  purging with, 271–72
  sampling monitor wells with, 272–73
Permanganate, 51
Permeability, 259
Permeation, 199, 226
Permeation rate, 199
Permeation samplers, 178
Permissible exposure limits (PELs), 42, 108, 114
Peroxides, 51
Personal dosimeter, 212
Personal factors affecting respirator use, 228
Personal locator beacon, 212
Personal protective clothing (PPC). *See also* Chemical protective clothing; Personal protective equipment (PPE)
  decontamination of, 325–30
  level A protection, 207, 215–16
  level B protection, 213, 217
  level C protection, 213–14, 221
  level D protection, 214, 222

386 INDEX

Personal protective clothing (PPC) (*Continued*)
  modified levels of protection, 214
  and thermoregulation, 86–87
  for unique hazards, 207, 208–12
Personal protective equipment (PPE), 207, 244. *See also* Personal protective clothing
  for confined spaces, 141
  doffing, 231
  donning, 228–29
  impact of heat stress on ability to use, 226–28
  inspection of, 229–30
  in-use monitoring of, 230–31
  level A protection, 207, 215–16
  level B protection, 213, 217
  level C protection, 213–14, 221
  level D protection, 214, 222
  maintenance of, 232
  modified levels of protection, 214
  preclusion of use of, by medical conditions, 26
  respirators, 180–98
  reuse of, 231–32
  selection and use of, 17
  in site evaluation, 13
  in site safety plan, 335
  specific requirements, 18
  storage of, 231
  training in use of, 223–24
  work mission duration, 224–26
    air supply consumption, 224
    ambient temperature extremes, 226
    permeation and penetration of protective clothing or equipment, 225–26
  written program for, 219, 223
Personal sampling, 170
Personal sampling instruments, 169
  active samplers, 170–71
    sampling pumps, 171
  calibration, 178
  choosing sampling methods, 169–70
  passive samplers, 177
    diffusion samplers, 177
    permeation samplers, 178
  personal sampling plan, 178–79
  sample collection devices, 171
    for gases/vapors, 171–75
    for particulates, 175–76
Personal sampling plan, 178–79

Personnel monitoring, 18–19, 79
Pesticides, 56
pH, 109
  in urine, 39
Phenobarbitol, 99
Phenyl, 55
Phosgene, and fire prevention, 298
Phosphorus, in blood test, 35–36
Photoionization detector, 162–64
pH scale, 48
Physical decontamination, 324–25
Physical properties, 45
Pick axe/shovel/scoop, 239
Plants, and toxicity, 93–94
Plug products, to contain spills, 302
Plume shapes, 264
Pneumatically operated bung remover, 149
Pocket ionization chamber and dosimeter, 78
Poison ivy, 101
Poison samples, 254
  labeling of, 254
  packaging of, 254
  sample identification of, 255
  shipping papers for, 255
  transportation of, 255
Polychlorinated biphenyls (PCBs), and medical monitoring of, 28–29
Polymerization, 58
Polymers, 54
Ponar grab, 241
Pond sampler, 240, 242
Ponds/lagoons, 153
Porosity, 259
Positive-pressure respirators, 197
Positive-pressure tests, 192–93
Positive social reinforcement, 133
Post hole digger, 239
Posting, 142
Potassium, 46
  in blood test, 38
Potassium bicarbonate, as fire suppressant, 297
Potassium chloride, 46
Potentiation, 99
Precipitation, 257
Pre-employment medical examination, 29–32
Preliminary site survey air monitoring for, 155–56
Presbycusis, 82
Pressure-demand respirators, 197

INDEX 387

n-Propyl amine, 56
Propanol, 55
Propionaldehyde, 55
Propionic acid, 55
Propyl, 54
Protective accessories, 207
Protein, in urine, 39
Protons, 43
Proximity garment (approach suit), 208
PVC tubing, 243

Qualitative fit tests, 192–93, 195
Quantitative fit tests, 195

Rabbit ear test, 100
Radiant heat loss, 86
Radiation, 72–73
  alpha, 73–74
  beta, 74–75
  biological effects of, 76
  container, 79
  gamma, 73
  ground, 79
  ionizing, 73
  measuring, 79
Radiation-contamination protective suit, 209
Radiation detecting devices, 77–78
Radiation exposure, federal guidelines for, 157
Radiation sources
  ionizing, 79–80
  non-ionizing, 80
Radioactive atmospheres, use of geiger counter to measure, 157–58
Radioactive isotopes, 73
  bioaccumulation, 76
  half-life, 75
  specific activity, 75–76
Radioactive waste, 144
  moving containers with, 147–48
Radioactivity, cautions regarding, 76
Reactive intermediates, formation of, 103
Reactivity, 58
Reconnaissance investigation, 128
Reducing agent, 58
Regulatory agencies, 80–81
Relative response, 160
Remotely controlled opening devices, 149
Rescue procedures, 142

Resistance, to electrical current, 83
Resource Conservation and Recovery Act (RCRA) (1976), 1–2, 311
Respirators
  air-purifying, 182–83, 186–90
  atmosphere-supplying, 183–84
  classification of respiratory protective equipment, 181
  facepiece type, 182
  hybrid SCBA/SAR combination systems, 191
  importance of respirator fit
    assigned protection factors, 195–97
    fit and fit testing, 191
    positive versus negative-pressure modes of, 196
  inspecting, 229
  need for respiratory protection, 180–81
  operation of, 196
  OSHA requirements for, 181
  respiratory protective equipment, 197–98
  selection of respiratory protective equipment, 184
  self-contained breathing apparatus, 190–91
  supplied-air, 183, 190, 230, 231
Respiratory arrest, first aid for, 306
Respiratory protection
  need for, 180–81
  OSHA program for, 181
Respiratory protective equipment, effectiveness of, 197–98
Responder by Life Guard, Inc., 203
Risk assessment, of hazardous wastes, 41–42
Runoff, 257

Safety boots, 206
Safety glasses, 211
Safety harness and lifeline, 141
Safety helmet (hard hat), 210
Safe work practices
  area safety equipment, 141
  atmospheric testing, 141
  clothing, 135–36
  confined spaces, 139–40
  drums and containers, 142–43
  entry permit, 140–41
  fire prevention and response, 136
  fuel, 136
  heavy equipment and drill rig safety, 136–38

388 INDEX

Safe work practices (*Continued*)
  hot work, 142
  observer (buddy) system, 141–42
  personal protective equipment, 141
  posting, 142
  preparation of work area, 141
  problems with drilling operations at hazardous waste sites, 138–39
  rescue procedures, 142
  safe entry procedures, 140
  site, 134–35
  tool use, 136, 142
  unsafe acts, 131–34
  unsafe conditions, 134–36
Salt, as additive to water in suppressing fire, 295
Sampling
  purpose of, 234–35
  reasons for dangers in, 235
  use of, 235
Sampling plan
  background information, 236
  bulk materials sampling, 241
  container compatibility, 237
  containerized liquids, 242–44
  decontamination of sampling equipment, 245–46
  depths between 3 and 16 feet, 239
  development of, 235–46
  greater depths, 239
  implementation of, 246
  liquids sampling, 241–42
  personal protective equipment, 244
  recordkeeping, 246
  resistance to breakage, 237
  sample containers, 237
  sample integrity, 244
  sample location, 236
  samples per sampling point, 236
  selection of sampling equipment, 238–44
  shallow depths, 239
  sludge/sediment sampling, 239–41
  soil sampling, 238–39
  standard operating procedure (SOP), 244
  volume, 237
  volume per sample, 237
Sampling procedures, 145
Sampling pumps, 171
Sampling trier, 239, 241
Sanitation at temporary workplaces, 21–22
Saturated zone, 258

Scintillation counter, 77
Scoop, 239, 240, 241
Sediments, 239
Self-contained breathing apparatus (SCBAs), 18, 114, 183, 190–91
  advantages of, 190–91
  and air supply consumption, 224
  considerations in using, 191
  disadvantages of, 191
  inspecting, 230
  in site evaluation, 13
  storage of, 231
Sequential multiple analyzer, 35
Serum bilirubin, in blood test, 36–37
Serum electrolytes, in blood test, 37–38
Serum glucose, in blood test, 38
Setaflash (SETA), 110
Shock, first aid for, 307
Shock-sensitive waste, 144
  moving containers with, 148
Shoring, 124
Side-scan radar mosaic imagery (SLAR), 117
Silica gel, 172
Silver, 47
Site characterization and analysis, 13, 114–19
Site control, 14
Site control program, 14
Site entry, checklist for safe, 349–51
Site illumination, 21
Site map, 117
Site monitoring, 18–19
Site safety plan, 11–13, 334
  anatomy of plan description, 334–36
  benefits analysis, 345–46
  checklist for safe site entry, 349–51
  hazardous awareness review, 347–49
  informing employees of provisions of, 337
  methods of interpreting, 344–45
  putting into action, 336–37
  role of OSHA and EPA, 334
  sample, 337–44
Site water supplies, at temporary workplace, 22
Skin cancers, 101
Skin exposure, 101
Sleeping quarters, at temporary workplace, 22
Sleeve protectors, 205
Sleeves, 206
Sloping, 124
Sludge, 239
Sludge/sediment sampling, 239–41

INDEX 389

Small peristaltic pump, 242
Snakebites, 94
Sodium, 44, 46
  in blood test, 37–38
Sodium bicarbonate, as fire suppressant, 297
Sodium chloride, as fire suppressant, 297
Sodium ion, 44, 45
Soil sampling, 238–39
Solid sorbents, 171–72
Solubility, 59
  in water, 109
Sorbent materials, to contain spills, 303
Sorption, 264
Specific activity, and radioactive isotopes, 75–76
Specific gravity, 109
  in urine, 39, 40
Spider bites, 93
Spills/spill response, 299
  anatomy of, 299–301
  containment of, 301–5
  contaminant diversion in waterways, 303–5
  diking, 129, 303
  patch and plug products to contain, 302
  in site safety plan, 336
  sorbent materials to contain, 303
Spiral and helical absorbers, 174
Splash hood, 211
Split spoon, 239
Stability analysis, 128
Standard operating procedure (SOP), for sampling, 244
State Emergency Response Commission (SERC), 279
Storage, 103
Storage tanks, contamination of groundwater due to leaking, 262
Strong acid, 49
Submerge sample container, 241
Submerge stainless steel beaker or scoop, 241
Submersible pump, 244
Sulfuric acid, 49
Superficial frostbite, 87
Superfund Act, 1, 2, 311
Superfund Amendment and Reauthorization Act (SARA) of 1986, 1, 2
  Title III, Emergency Planning and Right to Know, 276–77
  Title III, Sections 301–303, Emergency Planning, 277–78
  Title III, Section 304—Emergency Notification, 278–79
  Title III, Section 305, 279
  Title III, Section 322, 279–89
  Title III, Sections 311.312—Community Right-to-Know Reporting Requirements, 279
  Title III, Section 313, 279
Superfund sites, monitoring of, 72
Supplied-air respirators (SARs), 183, 190
  advantages of, 190
  considerations in using, 190
  disadvantages of, 190
  inspecting, 230
  storage of, 231
Support zone, 119
Suppressants, for fires, 295–98
Surface water, 258
Surface water hydrology, 257
Surfactants, as additive to water in suppressing fire, 295
Sweetener tests, 193, 195
Symbol, chemical, 43
Synergistic effect, 98
Systemic hypothermia, 88

Tag closed cup (PMCC), 110
Tanks/vaults, 152
  entry procedures for, 145
Target tissue, and toxic chemicals, 103–4
Temporary workplaces, sanitation at, 21–22
Teratogenesis, 104
  tests for, 100
Termination examination, 34
Thermoregulatory mechanisms, 86
Thickening agents, as additive to water in suppressing fire, 295
Threshold limit value (TLV), 105–6, 108
  ceiling, 106
  short term exposure limit, 106
  time weighted average, 106
Threshold planning quality (TPQ), 278
Tick, 93
Toilet facilities, at temporary workplace, 22
Tools, 142. *See also* Equipment
  decontamination of, 331
  and safe work practices, 136, 142
Total cholesterol and triglycerides, in blood test, 36
Totally encapsulating chemical protective suit (TECP), 18, 204, 205

Toxicants, limiting exposure to, 104–5
Toxic atmospheres
  monitoring of, 161
    with colorimetric tubes, 167–69
    with flame ionization detector, 164–67
    with photoionization detector, 162–64
Toxic chemicals, and target tissues, 103–4
Toxic effects, tests for, 99–100
Toxicity, 141
  acute versus chronic, 100
  definition of, 95
  measurement of, 96–97
Toxicology, 95
  biological response, 98–99
    additive effect, 98
    antagonism, 99
    potentiation, 99
    synergistic effect, 98
  factors affecting response of humans and laboratory animals, 96–97
  forms of toxic substances, 97–98
  LD_{50} concept, 98
  measurement of toxicity, 96–97
  routes of entry, 100–101
    eye exposure, 101
    lung exposure, 102–3
    oral exposure, 102
    skin exposure, 101
  threshold limit values, 105–6
Toxic substances
  effects of inhalation of, 102
  forms of, 97–98
Training, 14
  emergency response, 20–21
  general requirements, 14–15
  in personnel protective equipment use, 223–24
  scope of, 15–16
Transpiration, 258
Transportation of hazardous wastes, 310–11, 315–16
  accidents in, 320
  cooperation with the TSD facility, 320–21
  government regulations for, 311–16
  labeling of containers, 316–17
  manifest system, 317–18, 319
  placarding vehicles, 317
  trucks on site, 318, 320
Transportation, U.S. Department of (DOT)
  classification
    of flammable liquids, 62
    of hazard classes, 250
  definition of hazardous materials by, 311
  regulations for hazardous materials, 311–16
Transportation Safety Act (1974), 311
Treatment, storage, and disposal (TSD) facility, 311
  cooperation with, 320–21
Trenching, 122
  in site safety plan, 336
Triglycerides, in blood test, 36
Trucks
  on hazardous waste site, 318, 320
  vacuum, 152
29 CFR PART 1910.120, 10
  air monitoring, 18
  decontamination, 19–20
  emergency response, 20–21, 290
  engineering controls, 17
  general requirements of, 11
  handling drums and containers, 19, 143
  informational programs, 19
  intent and applicability, 10–11
  medical surveillance, 16–17
  new technology programs, 22–24
  personal protective equipment, 17–18
  respiratory protection program under, 181
  personnel monitoring, 18–19
  sanitation at temporary workplaces, 21
  site characterization and analysis, 13
  site control, 14
  site illumination, 21
  site monitoring, 18–19
  on site safety plan, 11–13, 334–36
  training, 14–16
  work practices, 17
Two-way radio, 212
Tyvek, 203

Unconfined aquifers, 259–60
UN hazard classes, 313
U.S. Army Corps of Engineers, role of, in site safety plan, 334
Unknown sample, shipping of, 250–51
Unsafe acts, 131–34
Unsaturated zone, 258
Upper Explosive Limit (UEL), 110
Uric acid, in blood test, 36
Urine tests
  bacteria, 40
  blood, 39
  casts, 40
  color, 39
  crystals, 40

glucose, 39
heavy metal screening, 40
ketones, 39
microscopic examination, 40
multistix reagent strips, 40
pH, 39
protein (albumin), 39
routine examination of, 40
specific gravity, 39, 40

Vacuum trucks, 152
Vapor buildup, 242
Vapor density, 109
Vapor pressure, 59
Vapors, active samplers for collecting, 171-75
Vasoconstriction, 86
Vehicles. *See also* Trucks
  decontamination of, 331-32
  placarding, 317
Venomous insects, 93
Ventilation of work area, 141
Viemeyer sampler, 239
Volatile percent, 59

Washing facilities, at temporary workplace, 22
Waste. *See also* Hazardous wastes
  bulking of, 151-52
  explosive, 148
  radioactive, 144, 147, 148
  shock-sensitive, 144, 148
Waste disposal, contamination of groundwater due to improper, 262
Waste site monitoring wells, specific considerations for, 266-67
Water
  as fire suppressant, 295-96
  supply of, at temporary workplace, 22
Water reactive chemicals, 47
Water table, 258
Waterways, containment of spills in, 303-5
Weak acids, 48
Weighted bottle, 242
Wet bulb globe temperature (WBGT) index, 89
Wet globe thermometer (WGT), 89
Wind chill factor, 91
Work area preparation, 141
Workers, decontamination for ill or injured, 324
Work mission duration, and use of personal protective equipment, 224-26
Workplace
  confined spaces in, 139-40
  preparation of safe, 141
  sanitation of temporary, 21-22
Wounds, first aid for, 308

Zoning, 119-21